MOLECULAR MODELS FOR FLUIDS

This book presents the development of modern molecular models for fluids from the interdisciplinary fundamentals of classical and statistical mechanics, of electrodynamics, and of quantum mechanics. The concepts and working equations of the various fields are briefly derived and illustrated in the context of understanding the properties of molecular systems. Special emphasis is devoted to the quantum mechanical basis, because this is used throughout in the calculation of the molecular energy of a system. The fundamentals are then used to derive models for various types of applications, ranging from ideal gas properties to excess functions and, finally, equations of state. The book is application-oriented and stresses those elements that are essential for practical model development. It is suited for graduate courses in chemical and mechanical engineering, physics, and chemistry, but may also, by proper selection, be found useful on the undergraduate level.

Professor Lucas has been the Chair of Thermodynamics at RWTH Aachen University since 2000. He previously held positions at the University of Stuttgart, at the University of Duisburg, and as the Scientific Director for the Institute of Energy and Environmental Technology, Duisburg Rheinhausen. He is a member of the Berlin Brandenburg Academy of Sciences and Humanities and acatech, German National Academy of Engineering, and is a consultant to various industrial companies in the fields of energy and chemical engineering. His research interests include the molecular modeling of fluids, energy systems analysis, and phase and reaction equilibria in fluids. He is the author or coauthor of more than 150 technical articles in scientific journals and is the author of a popular thermodynamics textbook.

MOLECULAR MODELS FOR FLUIDS

KLAUS LUCAS

RWTH Aachen University

CAMBRIDGE UNIVERSITY PRESS
Cambridge, New York, Melbourne, Madrid, Cape Town,
Singapore, São Paulo, Delhi, Tokyo, Mexico City

Cambridge University Press
32 Avenue of the Americas, New York, NY 10013-2473, USA

www.cambridge.org
Information on this title: www.cambridge.org/9781107402515

First published 2007
First paperback edition 2011

A catalog record for this publication is available from the British Library

Library of Congress Cataloging in Publication data

Lucas, Klaus.
Molecular models for fluids / Klaus Lucas.
    p.   cm.
Includes bibliographical references and index.
ISBN-13   978-0-521-85240-1 (hardback)
ISBN-10   0-521-85240-4 (hardback)
1. Fluid dynamics – Mathematical models.   1. Title.
TA357.L78   2006
620.1′ 064015118 – dc22   2006018286

ISBN 978-0-521-85240-1 Hardback
ISBN 978-1-107-40251-5 Paperback

To Gabi, Hanno, and Elena

Contents

Nomenclature

a)   **Roman Letters**

A	free energy, Helmholtz potential, surface area
a_i	activity of component i
b	molecular size parameter
B	grand canonical potential, second virial coefficient
C	third virial coefficient
c	speed of light
C_p	isobaric heat capacity
C_v	isochoric heat capacity
d	hard sphere diameter
e	elementary charge
E	energy, electrical field strength
F	force
f	fugacity, Mayer function
G	Gibbs free energy and nonrandomness factor
g	degeneracy
$g(r)$	pair correlation function
H	enthalpy, Hamilton function
$\hat{H}$	Hamilton operator, Hamiltonian
H_i	Henry coefficient of component i
h	Planck's constant
I	moment of inertia, ionization potential
K	chemical equilibrium constant
k	Boltzmann constant
l	rotational quantum number
M	molar mass
m	mass
N	number of moles, molecules, surface contacts

N_A	Avogadro number
n	number density, number of segments or groups, translational quantum number
P	probability
p	pressure, momentum
Q	canonical partition function, charge
Q^C	configurational partition function
q	molecular partition function, nondimensional external surface, charge
R	gas constant
r	distance, volume segment of a molecule
S	entropy
T	thermodynamic temperature
U	internal energy, potential energy
V	volume
$\hat{V}$	interaction operator
v	molar volume, vibrational quantum number
x	mole fraction, coordinate
y	mole fraction, coordinate, packing density
Z	compressibility factor, charge number, state quantity
z	coordinate, coordination number

b)   Greek Letters

α	dipole polarizability, nonsphericity parameter
β	$1/kT$
γ_i	activity coefficient of component i
ε	energy potential parameter, permittivity
η	packing density
θ	quadrupole moment, characteristic temperature, angle, surface fraction
λ	wave length, eigenvalue, perturbation parameter
μ	chemical potential, dipole moment, Joule–Thomson coefficient
ν	stoichiometric coefficient, frequency
Ξ	grand canonical partition function
Π	polarizability tensor
ϱ	charge density distribution, density
σ	distance potential parameter, surface charge density
Φ	Slater determinant
ϕ	pair potential energy, angle, orbital, fugacity coefficient, volume fraction
ψ	wavefunction, electrostatic potential
Ω	octopole moment, molecular arrangement
ω	orientational angle, acentric factor, exchange energy

c) Indices, Superscripts, Subscripts

a,b	interaction sites, charges
att	attractive
C	configurational
c	critical point
corr	corrections
d	hard sphere
E	excess
el	electronic
G	gas
h	hard body
i, j	component i, j in the mixture, quantum states
ig	ideal gas
ids	ideal dilute solution
ir	internal rotation
is	ideal solution
L	liquid
M	mixing
ntr	nontranslatory
P	perturbation contribution
r	rotation, reduced
ref	reference state
rep	repulsive
res	residual
s	saturated
t	triple point
tr	translatory
v	vibrational
0	reference
0i	pure component
$''$	saturated vapor
$'$	saturated liquid
$*$	conjugate complex, dimensionless
$\hat{}$	operator
α, β, κ	segments, groups, components

Preface

Many important industrial applications, as well as insight into the phenomena of nature, rely crucially on knowledge about fluid phase behavior. In space and other high-temperature industries, as well as in combustion processes, the properties of gases manifesting various types of reaction, including dissociation and ionization, are required. In chemical and environmental science and technology, phase and reaction equilibria of multicomponent mixtures form the basis of understanding the phenomena and designing synthesis, separation, and purification processes. Biotechnological downstream processing relies on the distribution properties of biomolecules in different phases of aqueous and organic solutions. Even in standard mechanical engineering equipment technology, such as refrigerator design, lack of data for new environmentally friendly refrigerants has proved to be a severe obstacle to technological progress. In all these cases, and many others, fluid phase properties form the basis of modern technological processes and detailed and quantitative knowledge of their properties is the premise of innovation. Experimental studies alone, although indispensible in the field of fluid system science, cannot serve these needs. The project of studying the fluid phase behavior of a multicomponent system experimentally is hopeless in view of the large number of data that would be needed. Instead, molecular models, which can be evaluated on a computer and make use of the limited data available to predict the fluid phase behavior in the full range of interest, are needed. Due to the broad availability of high-speed computers, such models can be quite ambitious, including use of quantum-chemical and molecular simulation computer codes.

Designing molecular models for fluid systems is an interdisciplinary field having its roots in classical mechanics, quantum chemistry, statistical physics, and electrodynamics. All models derived in this book originate from this basis. Their application to the computation of fluid phase behavior is executed with the help of the network of classical thermodynamic equations.

Chapter 1 is designed as an introduction to the subject. Some examples of macroscopic fluid behavior to be addressed in later chapters are discussed.

The variety of important phenomena and also the inadequacy of a purely macroscopic-experimental approach are clearly evident. Also, a first qualitative look is taken at molecular models and their relation to macroscopic properties.

Chapter 2 is devoted to the foundations of the field. The first section reviews the network of classical thermodynamic equations needed to compute the various aspects of fluid phase behavior. It is followed by a brief exposition of statistical mechanics, which provides the formal link between the macroscopic properties and a molecular model. The following section introduces the concepts of classical mechanics and applies them to established models of single molecule energy modes. The mechanical view of a molecular system is extended in the fourth section on classical electrostatics to include electrostatic interactions between molecular charges as a basis for the modeling of intermolecular forces. A section on quantum mechanics provides the necessary corrections and extensions to the classical mechanical and electrostatic results that make them applicable to molecular systems. In particular, it summarizes the quantum-chemical procedures now available for obtaining information about the geometrical structures of molecules as well as about charge distributions and associated molecular properties. A final section completes the foundations by developing the numerical path from the microscopic to the macroscopic world via computer simulation.

Chapter 3 considers as a first class of important applications the properties of the ideal gas. Besides deriving the ideal gas equation of state and general equations for the thermodynamic functions, it gives examples of predicted properties such as heat capacities and equilibrium constants on the basis of molecular properties derived from spectra or quantum-chemical computer codes.

Chapter 4 is concerned with models for the excess functions of liquid mixtures. The molecules of a liquid are assumed to interact via contacts over their surfaces. A statistical mechanical model is derived under well-defined approximations that can be used to predict the fluid phase behavior of liquid mixtures from information of the contact energies and its relation to some frequently used semiempirical models is discussed. The contact energies are either obtained from fitting the model to a large data base or, more recently, from quantum-chemical calculations.

More general applications over a large density range are considered in Chapter 5 on the basis of equation of state models. The intermolecular interactions must now be formulated as distance- and orientation-dependent functions. For low to moderate densities, the equation of state can be generally formulated in terms of a density expansion, with the expansion coefficients being expressed as integrals over the intermolecular potential functions. In the full density range, conformal potential models are derived, which can be corrected for specific intermolecular interactions by a perturbation approach. Rather sophisticated models with good predictive capacity can be formulated.

The text is designed for teaching first-year graduate courses within the curricula of mechanical and chemical engineering, physics, and chemistry. It aims at introducing the reader to the interdisciplinary scientific basis of the field, while at the same time making him or her acquainted with those approaches that have proven to be generally successful and are expected to be of lasting value. No effort is made to review the enormous plenitude of semiempirical models that have been published. Instead a unifying perspective is taken in deriving exemplary models systematically on the basis of the universally valid foundations. Actual calculations of molecular and macroscopic properties are today frequently done with the help of sophisticated computer codes. Although it does not discuss all details of the refined models included in such codes, the material in this book is meant to provide a basic understanding of them and lay the scientific basis on which future progress can be made.

During the preparation of the manuscript, of which some parts are based on my earlier book *Applied Statistical Thermodynamics* (Springer-Verlag, 1991), I have benefited from the contributions and the advice of some of my co-workers and colleagues. I wish to particularly acknowledge extensive discussions with K. Leonhard during the whole period of preparation of the book, as well as advice on important details by R. Bronneberg, R. Heggen, V. N. Nguyen, M. Singh, and J. Veverka. My colleagues D. Andrae, J. Groß, A. Klamt, G. Sadowski, and J. Vrabec critically read some parts of the manuscript and made helpful suggestions. The figures were prepared by T. Ameis and final production was executed by E. Frach and M. Lipková. I am very grateful to all of them.

Klaus Lucas
Aachen, Spring 2006

1 Introduction

Our society relies on the use of energy and matter in a plenitude of different forms. They are produced from natural resources by technical processes of energy and matter conversion that have to be designed in an economically and ecologically optimum way. In these processes it is the fluid state of matter that dominates the relevant phenomena. In particular, the properties of fluid systems in equilibrium enter into the fundamental process equations and control the feasibility of the various process steps. Models for fluids in equilibrium are thus a prerequisite for any scientific process analysis. Although fluid models can be constructed entirely within the framework of a macroscopic theory on the basis of experimental data, it is clear that this approach is limited to those few systems for which enough data can be obtained. Typical examples are the working fluids of the standard power generation and refrigeration processes. The vast majority of technically relevant processes are, however, concerned with complex fluid systems that cannot be analyzed experimentally in sufficient detail with a reasonable effort. In such cases one must turn to the microscopic basis of matter and design a theory based on the molecular properties of a fluid that requires only few data or is even fully predictive. In this introductory chapter we present an overview of the challenges of this approach by presenting a review of macroscopic fluid phase behavior in equilibrium, along with the problems associated with obtaining the necessary information from data. We also give a first introduction to the primary concepts of the microscopic world, including a brief glance at the properties of real molecules and the philosophy behind formulating molecular models.

1.1 The Macroscopic World

Energy and matter conversion processes are part of the macroscopic world. They are carried out without taking notice of the underlying microscopic phenomena, much like breathing air and handling materials in every day life. Within the limits of the macroscopic world the properties of the associated fluid

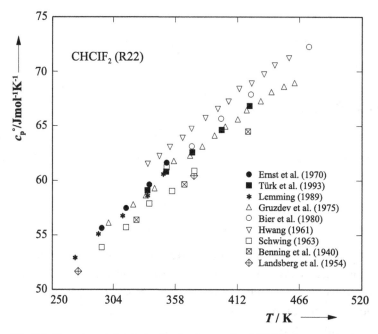

Fig. 1.1. Experimental data for the heat capacity of R22 (references to data given in [1]).

systems must be obtained from experiment. Fluid phase behavior is very plentiful, when studied over a wide range of temperature, pressure, and composition, and there are many fluids of technical interest. A few examples may illustrate this wealth of phenomena and the limitations of the macroscopic approach in providing the property information necessary to design technical processes.

An important property related to the energy conversion of a process is the heat capacity c_p. Fig. 1.1 shows the temperature dependence of the zero-pressure gaseous heat capacity of refrigerant R22 [1]. For calculations of refrigeration processes this property should be known to an accuracy of about 1%. The available experimental data reveal a spectrum of values differing by up to 10%. Little can be said about the reliability of the individual data sources on a purely macroscopic basis because experiments as a rule are not described in enough detail to assess their potential inaccuracies. Frequently, more recent data are given more credit than older data, and data from well-known laboratories are preferred over those from no-name sources, but this may be misleading, as was also found for R22. So, within the macroscopic approach to fluid properties and without an independent theoretical basis for evaluation of accuracy, we have to rely on the measured data as they are. We shall see, however, in Chapter 3 that molecular theory allows an unambiguous decision as to which set of data is correct and even a prediction when no data are available.

Proceeding to the large area of combustion processes, we need the heat capacities at much higher temperatures, but in particular we need properties related

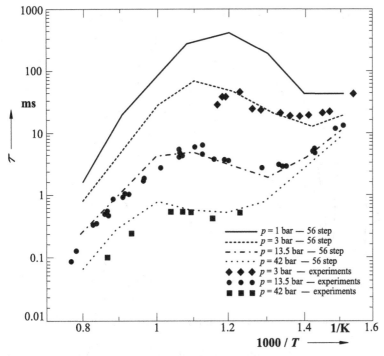

Fig. 1.2. Comparison of experimental ignition delay times with predicted values for heptane [2].

to the reactivity of the gaseous mixtures. An example is the ignition of heptane with stoichiometric air. In Fig. 1.2 we compare experimental ignition delay times for various pressures with model calculations on the basis of a 56-step skeletal reaction mechanism assumed to describe this combustion process [2]. The ignition delay time τ, i.e., the time that elapses until autoignition starts in a premixed fuel–air mixture, is plotted in milliseconds against the temperature of the mixture. Such information is relevant, e.g., in diesel engines. It can be seen that the ignition delay time first increases with decreasing temperature, but then shows the opposite behavior in an intermediate temperature range. To model such a combustion process, thermodynamic enthalpies of formation and standard entropies, as well as heat capacities, are required in the context of the law of mass action connecting forward and backward kinetic coefficients. The data are needed at high temperatures and for all relevant combustion intermediates, frequently radicals. Such data, although not readily obtained from experiment, can be calculated on the basis of the molecular models developed in Chapter 3.

Another example of high-temperature fluid phase behavior is provided in Fig. 1.3, which shows the composition of air at low density and at very high temperatures, including dissociation and ionization effects. The composition is given in terms of the number of particles per atom of air. The usual weight fraction of each constituent i is calculated from this number by multiplication

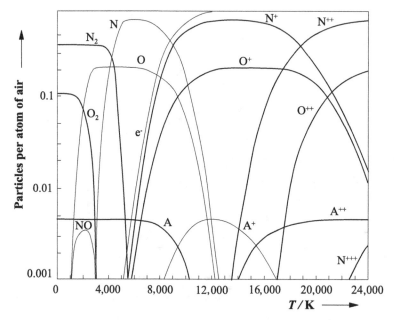

Fig. 1.3. Equilibrium composition of air at low density ($\varrho/\varrho_0 = 10^{-6}$; $\varrho_0 = 1.2927$ kg/m^3) [3].

by $1.991 M_i/M$, where M_i is the molar mass of i and M the molar mass of air. The availability of such data is relevant in aerospace engineering and other high-temperature applications. Because measurements are prohibitively costly, if possible at all, a molecular model is needed to provide insight into the physicochemical behavior of such systems as a basis of their prediction; cf. Chapter 3.

In chemical engineering applications we are frequently interested in fluid phase behavior at normal temperatures and pressures, controlled by mixing effects in liquids. An example is the heat effect that arises when two pure liquids are mixed. This effect is described in terms of the so-called excess enthalpy h^E. Fig. 1.4 presents experimental data on the excess enthalpy for the system benzene–hexane at 50°C, along with some predictions from a macroscopic model based on data. The technical interpretation of these data is that on mixing equal amounts of pure benzene and pure hexane, both liquid at 50°C, one has to withdraw a heat of about 800 J/mol if the mixture is to remain at the temperature of 50°C. It turns out, and will be formally derived in Section 2.1, that this aspect of fluid system behavior is related to the temperature dependence of the vapor–liquid equilibrium of the system. So one would expect to be able to predict h^E from data on the vapor–liquid equilibrium as a function of temperature. Fig. 1.4 shows that in practice this is not so with sufficient accuracy. Errors on the order of 25% are found when vapor–liquid equilibrium data on the system benzene–hexane are used to obtain the excess enthalpy of this system. Unavoidable uncertainties in the data and an inaccurate mathematical representation of them combine to produce these relatively large errors. Because such inaccuracy is not acceptable in process design, a direct experimental investigation appears necessary. Such experiments are costly, in particular in

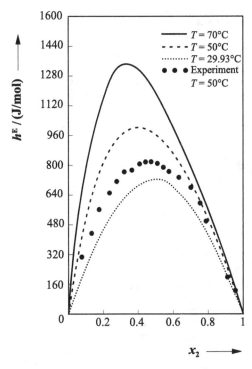

Fig. 1.4. Prediction of excess enthalpy h^E from data on the vapor–liquid equilibrium of the system benzene–hexane in the temperature range from 29.93 to 70°C. Data from [4–6].

multicomponent mixtures. It will be shown in Chapters 4 and 5, however, that an adequate molecular model, even in the absence of data, can give satisfactory results for the excess enthalpy, far more reliable than those obtained by formal management of the vapor–liquid equilibrium data.

Among the most important aspects of fluid phase behavior in chemical engineering applications is vapor–liquid equilibrium. It is the basis of many separation processes in the chemical industry. Fig. 1.5 shows data for the vapor–liquid equilibrium of three systems containing 1-chlorobutane at $T = 293.15$ K. Although the second component is quite similar in all three systems, we find rather different phase behavior. The system with 1,4-cyclohexadiene is quite nonspecific and can be represented by an almost ideal interpolation between the pure fluid vapor pressures. A change to cyclohexene induces a strong positive deviation from ideality and a further change to cyclohexane even leads to a strong azeotrope. Because simple estimation from the available pure fluid data thus appears hopeless in the general case, experimental data must be taken. This is costly and so there is definitely a need for prediction methods for such unexpected phase behavior. Such methods can indeed be designed on the basis of molecular theory, as seen in Chapters 4 and 5.

Although the estimation of vapor–liquid equilibria in mixtures from pure component data on a simple mean-value basis frequently gives at least a rough estimate, an even stronger challenge is the prediction of the gas solubility in a solvent. There is no simple correlation between gas solubility and pure

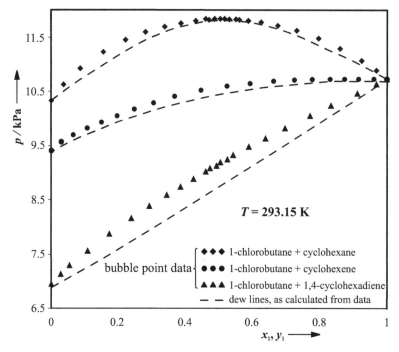

Fig. 1.5. Vapor–liquid equilibrium of three systems containing 1-chlorobutane. Data from [7].

component properties of either the solvent or the solute, while at the same time the solubility of different gases in the same solvent may vary over orders of magnitude. Fig. 1.6 illustrates this for the solubility of some simple gases in water over a range of temperatures, as described by the Henry coefficients [8]; cf. Section 2.1. The mole fraction of the gas dissolved in the liquid is roughly inversely proportional to the Henry coefficient. So pure oxygen will be roughly twice as solvable in water at room temperature as pure nitrogen, and pure carbon dioxide about 50 times as solvable. Clearly, such large differences have important consequences for the technical processes associated with gas solubility and reliable models to predict it are highly desirable. Such models can be formulated on the basis of molecular theory, as shown in Chapters 4 and 5. The problem of predicting the solution behavior of gases in liquids becomes even more challenging in electrolyte solutions. Whereas solubilities in simple systems can frequently be modeled on the basis of ideal dilute solution behavior, thus leading to an inverse linear relationship between solute concentration in the gas and in the liquid, controlled by the Henry coefficient, this is usually not so when electrolytes are involved. Nonidealities thus become noticeable even at rather small concentrations of the solute in the liquid. Experimental determination of Henry coefficients from solubilities in the ideally dilute region is then no longer sufficient, but rather data over the full range of partial pressures of the solute in the gas phase must be taken. For any new component

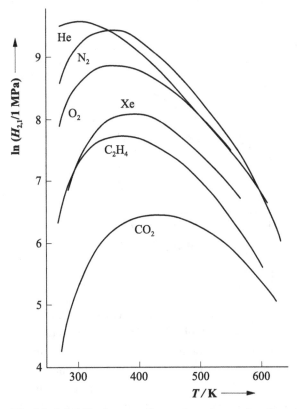

Fig. 1.6. Solubility for several gases in water as a function of temperature [8].

in the system a corresponding new series of measurements is required. Fig. 1.7 demonstrates this type of solution behavior for sulfur dioxide in the system $H_2SO_4 + H_2O$ [9]. Predictions from simple macroscopic concepts are bound to fail, but become possible on the basis of molecular models, as briefly touched on in Section 2.4.

Finally, another type of phase equilibrium that is important in technical applications is liquid–liquid equilibrium, i.e., the equilibrium between two liquid phases. Fig. 1.8 shows as an example this phase behavior for two systems containing ethanenitrile. Although in both systems the second component is rather similar, i.e., two isomers of octane, we find significant differences in the quantitative miscibility gaps. So, whereas the two liquids enthanenitrile and 2,4,4-trimethyl-1-pentene are completely miscible at $T = 320$ K, there is a heterogeneous two-liquid region at this temperature between about $x_1 = 0.25$ and $x_1 = 0.85$ when the second component is 2,4,4-trimethyl-2-pentene. Evidently, the detailed structures of the molecules have a significant influence on fluid systems' phase behavior. Many rather different types of liquid–liquid equilibria are known experimentally. The prediction of such phase behavior is a strong challenge for any fluid model. We shall readdress this problem within the context of molecular theory in Chapters 4 and 5.

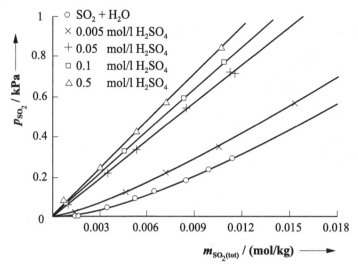

Fig. 1.7. Solubility of SO_2 in a H_2O-H_2SO_4 mixture at 298 K [9].

Even on the basis of the few examples discussed above we realize that fluid phase behavior manifests itself in a plenitude of different phenomena with applications in various different technologies. With an increasing number of components the phenomena become more complex. To design and operate technical processes we need quantitative knowledge of the relevant fluid phase properties. Experimental studies can only be done for a few systems

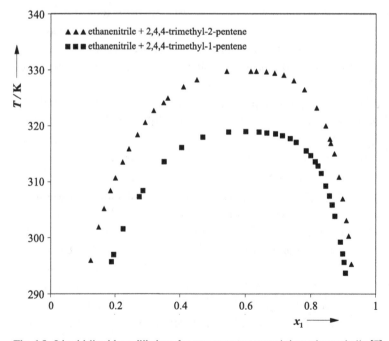

Fig. 1.8. Liquid-liquid equilibrium for two systems containing ethanenitrile [7].

of particular relevance. Generally, information on fluid phase behavior must therefore be obtained from a theory based on the molecular properties of the fluids under consideration.

1.2 The Microscopic World

Fluid systems, like all types of matter, consist of atoms and molecules. The existence of molecules had been inferred long before it could be proven experimentally. Today there are a variety of experimental techniques to make molecules visible. From such data we learn that molecules are generally rather complicated charge clouds and geometrical structures. Further, experiment tells us that a fluid owes its particular behavior not only to the properties of the single molecules but in most cases much more to the energetic interactions between them. Related to the interactions between the molecules and the resulting molecular motions we can differentiate between three states of matter, the gaseous, the solid, and the liquid. The state of a gas, i.e., at low density, is characterized by molecules moving about freely, hitting each other and the walls of the container only after having traveled relatively long distances without being disturbed by the presence of other molecules in the system. In a solid, on the other hand, molecules are packed close together and their motion is restricted to small vibrations about space-fixed positions. Thus a solid is a highly structured molecular system whereas a gas is essentially without order, i.e., chaotic. The molecular motion in liquids lies in between that of gases and that of solids. Neutron scattering experiments on liquids reveal that there is order at short distances from a selected molecule which, however, dies off rapidly with the distance increasing to more than a few molecular diameters.

The overwhelming plenitude of the macroscopic world consists of no more than about 100 elements. The applications addressed in this book even refer to molecules containing only a few of them, notably hydrogen, carbon, oxygen, nitrogen, fluorine, chlorine, and sulfur. Still, as will be seen, the majority of practically interesting fluids are covered by this small basis of atoms. This is a first indication of the potential of going to the atomic level when prediction methods for fluid phase behavior are to be set up.

Molecules are specific groupings of atoms. Rules that have their origin in the properties of the electrons underlie their formation. A covalent bond between two atoms results when they share a common pair of electrons. The two electrons thus serve as a kind of electrostatic glue. The number of bonds that one atom can form with other atoms depends on the number of electrons it can share with its neighbors, and is, for example, one for hydrogen, four for carbon, two for oxygen and three for nitrogen. Besides single bonds there are also double and triple bonds. The bond structure of a molecule can be visualized as shown in Fig. 1.9, where the molecule fumaric acid ($C_4H_4O_4$) is

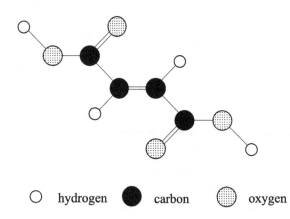

Fig. 1.9. Molecular structure of fumaric acid.

○ hydrogen ● carbon ⊙ oxygen

shown.[1] There are single bonds between a hydrogen atom and an oxygen atom or a carbon atom, as visualized by a simple bar. Double bonds are shown by double bars, and here appear at several locations between two carbon atoms and a carbon and an oxygen atom. Counting the bars reproduces the numbers of bonds associated with a particular atom. Whereas single bonds permit free internal rotation and folding, double and triple bonds are rather stiff and prevent internal rotation. Those electron pairs that do not take part in covalent bonds are called lone electron pairs and have an influence on the charge distribution of the molecule.

The arrangement of atoms or groups of atoms in a molecule, i.e., the molecular structure, is described by the terms configuration and conformation. The two notions are not identical. The configuration of a molecule represents the positions of the atoms as a well-defined geometrical structure. It is peculiar to a molecule. Molecules with identical sum formulae may display different configurations of their atoms and then are referred to as isomers. So octane (C_8H_{18}) can exist as *n*-octane and one of several isooctanes, with two isomer configurations as shown in Fig. 1.10. The atoms are bonded together differently in the two isomers, with the outside of the chain location of the carbon atom making isooctane stiffer than *n*-octane. The two molecules have different physicochemical properties, and there are various other isomers of octane depending on the particular arrangements of the atoms. Further, there may be different arrangements of molecular groups associated with discrete torsional angles around a stiff double bond, such as fumaric acid of Fig. 1.9 in comparison to maleic acid of Fig. 1.11, which both have the same sum formula. These are again two different molecules referred to as isomers, where fumaric acid represents the so-called *trans*-configuration and maleic acid the so-called *cis*-configuration. Again, the two isomers show quite different physicochemical

---

[1] The radius of a carbon atom is about 15 nm. Representing it by a circle of order of magnitude 1 cm in diameter thus means an increase by a factor of about 10^7.

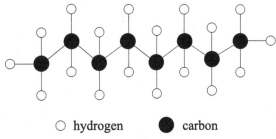

○ hydrogen   ● carbon

n-octane

Fig. 1.10. Molecular structures of *n*-octane and *i*-octane (4-methylheptane).

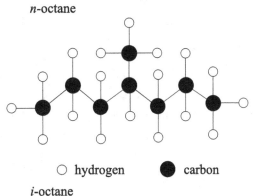

○ hydrogen   ● carbon

i-octane

properties. Finally, there are types of isomers referred to as optical isomers or enantiomers, important in biotechnological applications, where the two different configurations have almost identical physicochemical, but rather different biological properties. These phenomena make it clear that a molecular model for a fluid exclusively based on the properties of single atoms, although most attractive, cannot be expected to describe its macroscopic behavior properly. The structure of the molecule plays an important part.

The conformation of a molecule, on the other hand, describes the spatial arrangement of groups about one or more freely rotating bonds. The number of possible conformations of a complicated molecule can be large. Usually, various conformations coexist in a pure fluid, giving it the character of a mixture.

Fig. 1.11. Molecular structure of maleic acid.

○ hydrogen   ● carbon   ◌ oxygen

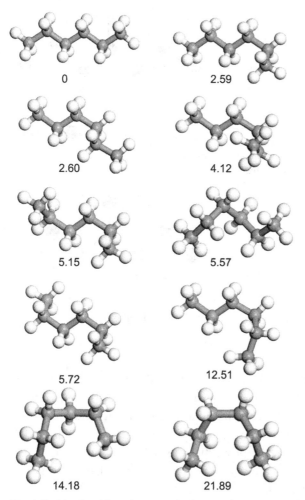

Fig. 1.12. The first 10 conformers of *n*-hexane (relative energies in kJ/mol).

Whereas the configurations are fixed by covalent bonding, the conformations are just stabilized by weak interactions between various parts of a molecule. The various conformations manifest themselves by local minima in the energy surface of a molecule. They are different manifestations of the same molecule. Since the energy hypersurface of a molecule can be calculated by the methods of quantum mechanics, cf. Section 2.5, the various conformations are accessible theoretically. Fig. 1.12 displays, as an illustration, 10 different conformers of the molecule *n*-hexane ordered in terms of increasing energy relative to that of the stretched conformation [10]. It can clearly be seen that these conformations are formed by various torsions around the single carbon–carbon bonds.

The structures of the molecules have an important influence on the energy of a molecular system, which in turn will be seen to be the crucial property determining the molecular model of a fluid. The energy of a molecular system is determined by the motions of the molecules in their mutual force fields.

Molecules are three-dimensional geometrical structures that consist of atoms, few or many, bonded together. As known from mechanics, they thus execute translational and rotational motions, and parts of them may rotate with respect to other parts and vibrate along intramolecular bonds. The associated energies, summarized as kinetic energy, contribute to the molecular energy of the system. Further, and most important, real molecules generally exert forces upon each other, which result from interactions of their charge distributions. This results in a potential energy of the molecules due to their positions in their intermolecular force fields, which is comparable to the potential energy of a mass point in the gravitational field of the earth. In contrast to the single-molecule contributions, this potential energy is not a property of a molecule but rather a property of the collection of molecules representing the system. The molecular energy of a fluid is thus that of all kinetic and potential energy contributions associated with the chaotic, undirected motion of the molecules. We formally introduce this by setting

$$E = E^{\text{kin}} + E^{\text{pot}}. \tag{1.1}$$

The kinetic part of the energy does not depend on the configuration coordinates of the molecular environment; i.e., it can be expressed exclusively in terms of the velocities or momenta of the molecules, such as those associated with the modes of translation, rotation, and vibration of single molecules. In contrast, the potential part takes those energy contributions into account that result from the presence of other molecules. Generally, all energy modes have a potential part in addition to the kinetic contribution.

Molecules can roughly be classified as simple and complicated, with important consequences for adequate molecular modeling of the potential energy. We first consider simple molecules such as the ones shown in Fig. 1.13 [11]. Fluids composed of such molecules tend to be technically interesting over a large density range. Applications are typically encountered in the gas industry. In a statistically averaged sense a well-defined and simple geometrical structure can be associated with each of them, which remains stable in any molecular environment. Although even such small molecules are capable of internal motions, such as vibrations and internal rotations, these are rapid and of small scale and so do not depend appreciably on the molecular environment. These energy contributions thus do not have a potential part and are rather taken care of fully by the kinetic term of (1.1); i.e., they are single-molecule properties and thus depend only on temperature, not on density. In contrast to this, the translational and the rotational motions of the molecule as a whole are clearly influenced by the molecular environment. They thus have a potential contribution in addition to the kinetic part and depend on density in addition to the dependence on temperature. Also, the molecules, due to their smallness, have such a simple geometry that it is easy to identify a molecular center, to

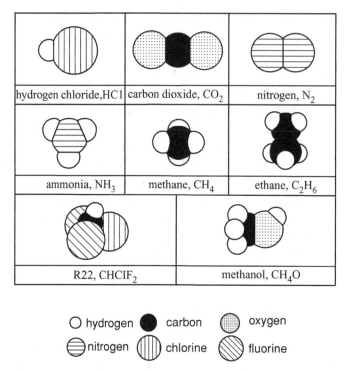

Fig. 1.13. Geometry of some simple molecules.

which their charge distribution can be referred. This is then a suitable basis for describing the intermolecular interactions between them.

This is not so for complicated molecules. We consider some examples of such molecules in Fig. 1.14 [11], which may stand for the whole plenitude of organic molecules relevant in chemical technology. Due to their largeness most of these molecules do not really have a molecular center. Rather, they appear as multicenter geometrical structures. This has consequences for the adequate representation of their charge distributions and their intermolecular potential energy, which will have to be based on the external surface rather than on the molecular center. Further, although well-defined shapes are associated with them in Fig. 1.14, these may transform into a number of conformations depending on the molecular environment; cf. also Fig. 1.12. So, generally, large and complicated molecules are flexible and frequently cannot be realistically modeled by rigid structures with respect to their intermolecular interactions. An example shown in Fig. 1.14 is octane, which can exist in a stretched geometry (a) but also easily rolls up into a globular form (b) due to rotations around the single carbon bonds. Many more intermediate conformations are relevant for this molecule. As a further consequence the internal motions, which are now slow and of large scale, will depend on the molecular environment, i.e., on density in addition to temperature, with severe consequences for designing a molecular model for a fluid composed of such molecules.

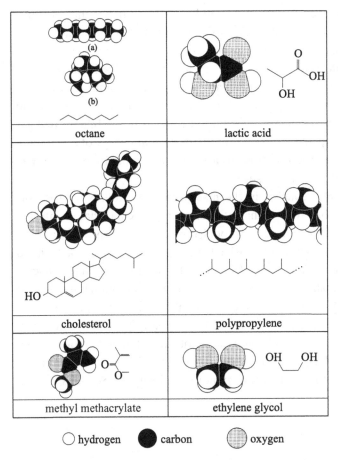

Fig. 1.14. Geometry of some complex molecules.

1.3 Molecular Models

In view of the complicated nature of molecular motions, it is clear that calculating the energy of a molecular system will be a difficult task. To make this feasible the complexity of the real molecular world must be simplified. So, instead of treating molecular reality, we set up molecular models. Such a model does not represent the true molecular behavior of a system. It is rather an artifact that is simple enough to be treated mathematically, while at the same time reproducing essential experimental facts. As a consequence, molecular models cannot be classified as being true or false. Rather, they must be considered useful or not useful in relation to the specific experimental situation to be considered and it is important to appreciate the constraints under which a particular model may be applied.

Much can be learnt about the general philosophy behind molecular models for fluids from a highly simplified version, the billiard ball gas. In the billiard ball model we consider the molecules to be very small billiard balls [12]. The only single-molecule energy mode modeled by a billiard ball is that of translation. Also, billiard balls do not exert any attractive forces upon each other. We further assume that they have such small volumes that there are also essentially no repulsive interactions between them, apart from occasional collisions with each other and with the container walls, which keep up uniform values of temperature and pressure. So we eliminate the potential energy in (1.1) and highly simplify the kinetic energy contribution. Evidently, the billiard ball gas cannot generally be expected to represent the molecular energy of a real fluid. However, as will be seen below, it is a reasonable approximation for a noble gas at room temperature and pressure.

The first fundamental concept in the macroscopic as well as microscopic theory of fluid phase behavior is that of energy, in particular that of internal energy. On the basis of the billiard ball model, this concept is easy to visualize. It is just the total macroscopic energy arising from the kinetic energies associated with the motions of the billiard balls; cf. Fig. 1.15. Even if the gas is at rest macroscopically the billiard balls are in permanent motion. This motion is incoherent or chaotic and thus does not have the effect of macroscopic motion of the gas. The macroscopic internal energy U of the system is clearly identified as the mean kinetic energy $< E^{kin} >$ associated with the chaotic translational motions of the billiard balls. On a molecular level the kinetic energy of a particle is preferably formulated in terms of its momentum $\boldsymbol{p}$, not its velocity; cf. Sections 2.2 and 2.3. We thus find

$$U = \frac{1}{2}\frac{N}{m}\langle \boldsymbol{p}^2 \rangle = N\langle E^{kin} \rangle, \tag{1.2}$$

where N is the number of billiard balls, m their mass, and $\langle \boldsymbol{p}^2 \rangle$ their mean squared momentum. Also, per general definition, the thermodynamic temperature T is related to the mean kinetic energy of the translating molecules. So, for the billiard balls, we have

$$T = \frac{2}{3}\frac{\langle E^{kin} \rangle}{k}, \tag{1.3}$$

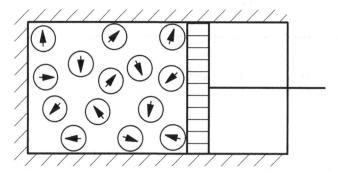

Fig. 1.15. The internal energy of a billiard ball gas.

where k is the Boltzmann constant as specified in App. 1. Combining (1.3) with (1.2) gives

$$U = \frac{3}{2}NkT \qquad (1.4)$$

as the internal energy of the billiard ball gas. Finally, we can easily relate the pressure p, as computed from the change of momentum of the billiard balls hitting the container walls, to the average of squared molecular momentum, $\langle \boldsymbol{p}^2 \rangle$, as

$$p = \frac{1}{3}\frac{N}{V}\frac{\langle \boldsymbol{p}^2 \rangle}{m}, \qquad (1.5)$$

which then, by using (1.3), yields

$$pV = NkT \qquad (1.6)$$

as the equation of state of the billiard ball gas. We note that the above results for the equation of state and the internal energy of a billiard ball system are fully adequate to model the properties of a noble gas at room conditions. We further note that they originate from simple kinetic formulations of the molecular concepts associated with the internal energy, the temperature, and the pressure.

The second fundamental concept in the macroscopic as well as in the microscopic theory of fluid phase behavior is entropy. The established molecular interpretation of entropy is that of a measure of disorder, randomness, and chaos in the molecular state of a system. It is defined by the Boltzmann relation

$$S = k \ln W. \qquad (1.7)$$

The measure of randomness, W (German: *Wahrscheinlichkeit*), is defined as the number of different molecular states of a system that are consistent with its macroscopic state as defined by its temperature, its volume, and its composition. Here, molecular states refer to position coordinates as well as to momenta. Thus a billiard ball system with a broad distribution of its billiard balls over the available position and momentum values is more random than one in which all billiard balls are confined to a narrow subspace with a uniform momentum and will correspondingly have a higher entropy. To calculate a numerical value for the entropy of a billiard ball gas it is necessary to derive the number of different molecular states of the billiard balls for a fixed thermodynamic state. The number of position states will clearly be proportional to the volume of the system, i.e., the third power of a position coordinate. Analogously, the number of momentum states will be proportional to the third power of momentum and thus to $(mkT)^{3/2}$. Also, the entropy as an extensive quantity will be proportional to the number of billiard balls. So we have for the entropy of a billiard ball gas

$$S = k \ln W \sim Nk(3/2 \ln T + 3/2 \ln m + \ln V). \qquad (1.8)$$

Again, this result is fully consistent with that for a noble gas under conditions of room temperature and pressure.

Studying the properties of a billiard ball system by simple kinetic and statistical considerations thus provides closed formulae for its thermodynamic properties. It turns out that the billiard ball system is a useful molecular model for the properties of the noble gases under conditions of room temperature and pressure. Clearly, it is generally far away from the true molecular behavior of a fluid. So, when the temperature is raised to very high values, say 10,000 K, even a simple monatomic gas at normal pressure will deviate from the billiard ball gas model by ionization. Also, at higher pressures, entirely different fluid phase behavior will be observed, associated with the interactions, repulsive and attractive, between the molecules. Further deviations from the simple billiard ball model are due to the fact that molecules are not monatomic, but consist of various atoms. To describe more general aspects of fluid phase behavior, we thus must turn to more realistic models for the molecules and, in particular, the interactions between them. We then find that the simple kinetic arguments used above to provide the macroscopic properties become insufficient. We will have to turn to more involved statistical methods. Also, the results will no longer be presentable in simple closed formulae, but rather, complicated expressions will appear. Still, there is no unique molecular model of fluid phase behavior, but rather an enormous plenitude, reflecting the many different fields of application and the complexity considered to be adequate. As we shall see, even models that are highly simplified in comparison with the true molecular situation can be very valuable tools for fluid behavior prediction if properly applied.

1.4 Summary

The properties of fluids in thermodynamic equilibrium control energy and material conversion processes in manifold technical equipment. Fluid systems display a large variety of phenomena that are encountered in technical processes, such as heat effects, vapor–liquid equilibria, solution of gases in liquid phases, liquid–liquid equilibria, and high-temperature dissociation effects. Quantitative information on the phase behavior of fluid systems can be obtained from the general equations of classical thermodynamics as soon as adequate numerical models for the systems are available. Such models can basically be generated from experimental data. However, this route is only sensible for a small number of particularly important fluids.

More generally, the route to such models has to be based on knowledge about the properties of atoms and molecules; i.e., it originates in the microscopic world underlying fluid systems. Molecules have complicated geometrical structures resulting from the bonds between the atoms. The energy of a molecular system can be formulated as a sum of two contributions, a kinetic and a potential term. The kinetic term is related to the motion of the single molecules, whereas the potential term results from interactions of the molecules with each other. Depending on the complexity and largeness of the molecules, the separability of the kinetic and potential energy can be quite different. The molecular motions related to internal flexibility, such as vibration and internal rotation, are purely kinetic for small molecules. For large molecules, in contrast, this is not so, and all types of motion have a potential as well as a kinetic contribution. Because the true molecular situation in a fluid is far too complicated to be treated mathematically, it becomes necessary to design molecular models. A first and highly simplified molecular model of a fluid is provided by the billiard ball gas. Applying simple kinetic and statistical considerations to it provides molecular interpretations and closed formulae for its thermodynamic functions. These results are fully adequate for a noble gas under conditions of room temperature and pressure as a particularly simple application. Much more complex approaches are required to set up molecular models for more general fluid systems. No generally applicable model has yet emerged. Rather, we are faced with a plenitude of molecular models for the many different types of application and the complexity considered adequate.

1.5 References to Chapter 1

[1] K. Lucas, V. Buß, U. Delfs, and M. Speis. *Int. J. Thermophys.*, 14:291, 1993.
[2] N. Peters, G. Paczo, R. Seiser, and K. Seshadre. *Combust. and Flame*, 128:38, 2002.
[3] W. E. Moeckel and K. C. Weston. *NACA*, TN 4265:466, 1958.
[4] J. Gmehling and U. Onken. *DECHEMA Chemistry Data Series*. DECHEMA, Frankfurt, 1977. Vol. 1. Parts 1, 2a and 2b.
[5] M. J. Paz-Andrade. *International Data Series A*. Thermodynamic Research Center, A & M University Texas, 1973.
[6] M. Diaz-Penar and C. Menduina. *J. Chem. Thermodyn.*, 6:1097, 1974.
[7] V. N. Nguyen and F. Kohler. *Fluid Phase Equilibria*, 50:267, 1989.
[8] A. H. Harvey. *AIChE J.*, 42:1491, 1996.
[9] J. Krissmann, M. A. Siddiqi, and K. Lucas. *Fluid Phase Equilibria*, 141:221, 1997.
[10] A. Schäfer. BASF AG. Personal communication.
[11] P. W. Atkins. *Molecules*. Scientific American Library, New York, 1987.
[12] P. W. Atkins. *The Second Law*. Scientific American Library, New York, 1984.

2 Foundations

The book deals with the prediction of the macroscopic behavior of fluids from the properties of their molecular constituents. The basis of such prediction is the availability of molecular models. Designing molecular models for fluids has an interdisciplinary background of foundations. Their thorough understanding is the basis for developing new models and appreciating the promise as well as the limitations of those that are established.

Molecular models are formulated in terms of the energy of a system of three-dimensional flexible bodies, i.e., the molecules. This molecular energy depends on the geometrical structures of the molecules and the force field they are moving in. The relation between the geometrical structure of a body and its energy is defined in mechanical terms. The force field results from the electrical properties of the molecules. On this level the models are thus based on classical mechanics and electrostatics. Classical theory, although most powerful even on the molecular scale, is, however, incomplete in the sense that it does not provide information on the geometry of the molecules or on their charge distributions as the origin of the electrical force field. This gap is closed by quantum mechanics, which also gives important corrections to the classical results in order to make them ultimately applicable to molecules. The link between the molecular energy of a system and its macroscopic thermodynamic functions is provided by statistical mechanics and by computer simulation. By using this link the molecular model leads to numerical data for the thermodynamic functions, from which the macroscopic behavior of a fluid can be calculated by the laws of classical thermodynamics.

In this chapter we shall recapitulate and summarize the most important results of these various disciplines, to the extent that they are used in the later applications.

2.1 The Macroscopic Framework: Classical Thermodynamics

In classical thermodynamics the equilibrium behavior of fluids is described in terms of state quantities. These state quantities are not independent of each other. Rather, they are interrelated and, as a consequence, the various aspects of thermodynamic behavior are also interrelated. As an example, the temperature dependence of phase equilibrium can be shown to be related to the effect of heat upon the mixing of the pure components. Classical thermodynamics provides the general network of equations for these interrelations and for macroscopic fluid phase behavior in terms of them. It is generally valid, i.e., not restricted to particular substances, and it is the framework in which molecular models for fluids are applied [1,2].

2.1.1 General Relations

According to the first law, the internal energy of a system changes to the extent that energy in its various forms is transferred to or from the system across its boundaries. On the basis of this law and limiting consideration to fluids without electromagnetic fields, appreciable gravity and surface effects, the differential of the internal energy $U = U(S, V, \{N_j\})$ can be represented in a basic and general way by the relation

$$dU = TdS - pdV + \sum \mu_j dN_j. \tag{2.1}$$

This is the fundamental relation for simple systems. Most systems in applications of energy and material conversion belong to this class. The basic variables of the energy function U are the entropy S, the volume V, and all the numbers of moles or molecules of the various components $\{N_j\}$. We here use the notation $\{N_j\}$ interchangeably for all the numbers of moles as well as of molecules. The number of molecules is derived from the number of moles by multiplication by the Avogadro number; cf. App. 1. These basic variables are the coordinates by which a simple system responds in a fundamental way to energy transfer across the system boundaries. The related basic intensity variables are the temperature $T = (\partial U/\partial S)_{V,\{N_j\}}$, the pressure $p = -(\partial U/\partial V)_{S,\{N_j\}}$, and the chemical potential $\mu_j = (\partial U/\partial N_j)_{S,V,N_j^*}$, where N_j^* means constant numbers of moles or molecules except N_j. These quantities measure the intensity of the energy change due to a change of entropy, volume, and the amount of substance of one component, respectively. The fundamental relation is a basic and general relationship in the sense that no substance-specific limitations have gone into its derivation. It therefore produces general relationships for the equilibrium behavior of fluids without reference or limitation to particular substances.

Unfortunately, the independent variables of this energy function, i.e., the state quantities S, V, and $\{N_j\}$, are quite inconvenient for practical calculations. It is possible, however, to translate the energy function $U(S, V, \{N_j\})$ into equivalent functions with more convenient associated variables by Legendre transformations. The most important thermodynamic potentials emerging from this procedure are the Helmholtz free energy or Helmholtz potential,

$$A(T, V, \{N_j\}) = U - TS, \tag{2.2}$$

with the differential

$$dA = -SdT - pdV + \sum \mu_j dN_j \tag{2.3}$$

and the Gibbs free energy or Gibbs potential,

$$G(T, p, \{N_j\}) = U + pV - TS = H - TS, \tag{2.4}$$

with the differential

$$dG = -SdT + Vdp + \sum \mu_j dN_j. \tag{2.5}$$

The differential of the Helmholtz free energy in terms of T, V, and $\{N_j\}$ and the differential of the Gibbs free energy in terms of T, p, and $\{N_j\}$ are fundamental relations of the same basic significance as the fundamental relation (2.1). Thus, all thermodynamic state quantities may be calculated from them in terms of the associated variables by working out the appropriate derivatives. The enthalpy $H = U + pV$, for example, as a function of T, p, and $\{N_j\}$, can be calculated from

$$H(T, p, \{N_j\}) = G - T\left(\frac{\partial G}{\partial T}\right)_{p,\{N_j\}} = \left(\frac{\partial(G/T)}{\partial(1/T)}\right)_{p,\{N_j\}}, \tag{2.6}$$

which is the Gibbs–Helmholtz equation. For the pressure in terms of T, V, $\{N_j\}$, i.e., the equation of state, (2.3) immediately gives

$$p(T, V, \{N_j\}) = -\left(\frac{\partial A}{\partial V}\right)_{T,\{N_j\}}. \tag{2.7}$$

Finally, the chemical potential, the most important quantity determining the equilibrium behavior of systems, can be computed from

$$\mu_j(T, V, \{N_j\}) = \left(\frac{\partial A}{\partial N_j}\right)_{T,V,N_j^*} \tag{2.8}$$

as a function of T, V, and $\{N_j\}$, or from

$$\mu_j(T, p, \{x_j\}) = \left(\frac{\partial G}{\partial N_j}\right)_{T,p,N_j^*} = h_j(T, p, \{x_j\}) - Ts_j(T, p, \{x_j\}) \tag{2.9}$$

as a function of T, p, and $\{x_j\}$, with $\{x_j\}$ as the total set of mole fractions of the components. Because differences of the Helmholtz free energy and the Gibbs free energy are known to represent the minimum work required for an isothermal change of state, we note from (2.8) and (2.9) that the chemical potential μ_i can be interpreted as the reversible work required to introduce one molecule of component i at constant temperature and volume or pressure, respectively. Eq. (2.9) provides an example of a partial molar quantity, here the partial molar Gibbs free energy, which is identical with the chemical potential in terms of T, p, and $\{x_j\}$. From (2.4) it is seen that the chemical potential can be obtained from the partial molar enthalpy and the partial molar entropy, as explicitly shown in (2.9). Generally, a partial molar quantity z_j is defined as the partial derivative of the associated extensive variable Z with respect to the mole number N_j of a component j, where T, p, and all other mole numbers are held constant. It is an intensive property, i.e., it does not depend on the total amount of substance, and it provides a measure of how much of the extensive property is to be ascribed to a particular component in a mixture, i.e., $Z = \sum N_j z_j$.

According to the second law the equilibrium state in an isolated system is defined by the maximum of entropy. This condition can be translated into that of the minimum of Helmholtz free energy for given values of T, V, and $\{N_j\}$ or the minimum of Gibbs free energy for given values of T, p, and $\{x_j\}$. In heterogeneous systems with K components, the following conditions for equilibrium between different phases (α), (β), $\cdots$ result,

$$T^{(a)} = T^{(\beta)} = \cdots \tag{2.10}$$

$$p^{(a)} = p^{(\beta)} = \cdots \tag{2.11}$$

$$\mu_i^{(a)} = \mu_i^{(\beta)} = \cdots \quad i = 1, 2, \ldots, K, \tag{2.12}$$

where the condition of equal chemical potentials controls the distribution of the components among the phases. In homogeneous systems with r chemical reactions the isothermal–isobaric equilibrium composition can be obtained from

$$0 = \sum_i v_{i,j} \mu_i \quad j = 1, 2, \ldots, r, \tag{2.13}$$

where j denotes the particular reaction considered and $v_{i,j}$ is the stoichiometric coefficient of component i in reaction j. These equations are to be solved under the condition of conservation of elements. In the general case, phase and reaction equilibria arise simultaneously and many reactions may have to be considered. Such complex equilibria, too, are defined by the set of equations (2.10) to (2.13), and the distribution of the various components over the various phases can be calculated from them. A sometimes more convenient alternative starting point for such calculations is the minimum condition for the Gibbs free energy, i.e.,

$$\sum_p \sum_i N_i^{(p)} \mu_i^{(p)} = \text{Minimum}, \tag{2.14}$$

where p denotes the phases over which the components i may be distributed.

Summarizing, we note that a relation of the Helmholtz free energy as a function of temperature, volume, and the mole numbers or, alternatively, the Gibbs free energy as a function of temperature, pressure, and the mole numbers completely defines the thermodynamic behavior of a system. Combining this information with the laws of classical thermodynamics allows all process calculations to be executed on the thermodynamic level. Unfortunately, however, classical thermodynamics does not provide numerical values for the thermodynamic potentials A and G as a function of the practical variables such as temperature, pressure or volume, and composition. Therefore, within the framework of classical thermodynamics, such values must be obtained from experiments on a particular system. The thermodynamic potentials $A(T, V, \{N_j\})$ and $G(T, p, \{N_j\})$ cannot be directly measured for a given fluid. However, they can be calculated from other functions that can be measured. Typical properties accessible by experiment are heat effects such as heat capacities, heats of mixing, and heats of reaction. Also measurable are relationships between temperature, pressure, and volume as well as data on the phase and reaction equilibrium.

2.1.2 Heat Capacity

The temperature dependence of thermodynamic functions can be determined in terms of heat capacities, defined by

$$C_V = \left(\frac{\partial U}{\partial T} \right)_{V, \{N_j\}} \tag{2.15}$$

and

$$C_p = \left(\frac{\partial H}{\partial T} \right)_{p, \{N_j\}}. \tag{2.16}$$

These heat capacities are state quantities; i.e., they depend on temperature, on volume resp. pressure, and on the composition of the system. Of particular significance is the heat capacity in the ideal gas state, which depends on temperature and, trivially, on composition. Heat capacities can be measured by caloric experiments. Such measurements, when done with the accuracy required for practical process calculations, are rather demanding. Although many data have been measured for simple pure fluids in standard regimes of temperature and pressure, this is not so for complex fluids, for other regions, and, in particular, for mixtures.

2.1.3 Equation of State

A particularly elegant and convenient formulation of the thermodynamic behavior of a system can be obtained in terms of its equation of state. Calculation of thermodynamic functions from an equation of state is most effectively done by introducing residual state quantities according to

$$Z^{\text{res}}(T, V, \{N_j\}) = Z(T, V, \{N_j\}) - Z^{\text{ig}}(T, V, \{N_j\}) \tag{2.17}$$

or

$$Z^{\text{res}}(T, p, \{N_j\}) = Z(T, p, \{N_j\}) - Z^{\text{ig}}(T, p, \{N_j\}). \qquad (2.18)$$

Here Z^{ig} denotes the state quantity Z at $T, V, \{N_j\}$ or $T, p, \{N_j\}$ evaluated using the equation of state for the ideal gas; i.e.,

$$(pV)^{\text{ig}} = NRT \qquad (2.19)$$

with N as the number of moles, or

$$(pV)^{\text{ig}} = NkT \qquad (2.20)$$

with N as the number of molecules. Here R is the universal gas constant and k the Boltzmann constant, cf. App. 1, which are related by

$$R = kN_A, \qquad (2.21)$$

with N_A as the Avogadro number. Thus, one finds for the residual Helmholtz free energy with (2.3)

$$A^{\text{res}}(T, V, \{N_j\}) = -\int_{\infty}^{V} \left[p(T, V, \{N_j\}) - \frac{NRT}{V} \right] dV \qquad (2.22)$$

and for the residual Gibbs free energy with (2.5)

$$G^{\text{res}}(T, p, \{N_j\}) = \int_{0}^{p} \left[V(T, p, \{N_j\}) - \frac{NRT}{p} \right] dp. \qquad (2.23)$$

The volume dependence of the Helmholtz free energy at constant temperature and constant composition is thus given by

$$A(T, V, \{N_j\}) - A(T, V^0, \{N_j\})$$
$$= A^{\text{res}}(T, V, \{N_j\}) - A^{\text{res}}(T, V^0, \{N_j\}) - NRT \ln \frac{V}{V^0}, \qquad (2.24)$$

whereas the pressure dependence of the Gibbs free energy at constant temperature and constant composition can be expressed as

$$G(T, p, \{N_j\}) - G(T, p^0, \{N_j\})$$
$$= G^{\text{res}}(T, p, \{N_j\}) - G^{\text{res}}(T, p^0, \{N_j\}) + NRT \ln \frac{p}{p^0}. \qquad (2.25)$$

The pressure and volume dependence of the thermodynamic functions can be generally formulated in terms of A^{res} or G^{res} and is thus determined by the equation of state. The difference of two state quantities does not depend on the mathematical route taken from one state point to the other. Therefore it is possible and frequently convenient to consider the temperature variation exclusively in the ideal gas state. When this is done, the dependence of the thermodynamic functions on temperature and density or pressure can be calculated from heat capacities in the ideal gas state and the equation of state. Thus, as an example of the resulting equations, one gets for the entropy difference between the states $(T, V, \{N_j\})$ and $(T^0, V^0, \{N_j\})$

$$S(T, V, \{N_j\}) - S(T^0, V^0, \{N_j\})$$

$$= \int_{T^0}^{T} \frac{C_V^{\text{ig}}}{T} dT + S^{\text{res}}(T, V, \{N_j\}) - S^{\text{res}}(T^0, V^0, \{N_j\}) + NR \ln \frac{V}{V^0}, \quad (2.26)$$

where $S^{\text{res}} = -(\partial A^{\text{res}}/\partial T)_{V,N_j}$ and A^{res} is calculated from the equation of state via (2.22). Finally, the residual chemical potential of a component j in a mixture is related to the equation of state by

$$\mu_j^{\text{res}}(T, n, \{x_i\}) = a^{\text{res}}(T, n, \{x_i\}) + \frac{p^{\text{res}}(T, n, \{x_i\})}{n}$$

$$+ \left(\frac{\partial a^{\text{res}}(T, n, \{x_i\})}{\partial x_j} \right)_{T,n,x_j^*}$$

$$- \sum_{i=1}^{k} x_i \left(\frac{\partial a^{\text{res}}(T, n, \{x_i\})}{\partial x_j} \right)_{T,n,x_j^*}. \quad (2.27)$$

Here $\{x_i\}$ denotes all mole fractions of the components and $n = (N/V)$ is the molar density. The derivative of the molar residual free energy a^{res} with respect to the mole fraction of one component in (2.27) is taken formally under the condition of all other mole fractions being held constant. Its connection with the equation of state is given again by (2.22).

It is thus possible to formulate the equilibrium behavior of a fluid system in terms of its equation of state and the ideal gas heat capacity. Phase equilibria are determined exclusively by the equation of state. When chemical reactions are considered we further need data for the enthalpy of formation and for the absolute entropy of a component in the pure ideal gas state, which can be assumed to be available from literature tables. The formulation of the thermodynamic behavior of a system in terms of an equation of state is particularly convenient in applications where large differences in density, i.e., from gas to liquid, have to be considered. Typical areas of application are high-pressure equilibria in the air products or, more generally, the gas industries. Also, liquid–liquid equilibria in mixtures containing large and small molecules are preferably described in terms of an equation of state. Setting up an equation of state requires accurate p, v, T data, which are usually augmented by other types of data, and is a very demanding task, even for pure fluids. For mixtures the effort rises exponentially with the number of components.

2.1.4 Fugacity, Activity, Excess Functions

Systems which are exclusively considered in their liquid state should not necessarily be treated by an equation of state, since this requires data over the whole region of states, i.e., from low density to liquid density. In such cases other experimental information is more practical. Thus, for the chemical potential of a component i in a liquid mixture, we generally write

$$\mu_i(T, n, \{x_j\}) = \mu_i^{\mathrm{r}}(T, p) + RT \ln \frac{f_i(T, p, \{x_j\})}{f_i^{\mathrm{r}}(T, p)}$$
$$= \mu_i^{\mathrm{r}}(T, p) + RT \ln a_i^{\mathrm{r}}. \tag{2.28}$$

Here $\mu_i^{\mathrm{r}}(T, p)$ is the chemical potential of component i in a suitable reference state at a fixed composition. The f-functions are referred to as fugacities, where f_i^{r} is the fugacity associated with the reference state and $f_i(T, p, \{x_j\})$ the fugacity of component i in the state considered. The ratio of the fugacities is referred to as activity a_i^{r}. The activity is practically replaced by the activity coefficient $\gamma_i^{\mathrm{r}} = a_i^{\mathrm{r}}/x_i$. Fugacities and activities can be calculated from the equation of state. In liquid systems they are normally obtained from data of the phase equilibrium.

Different reference states have been defined for different applications. For liquid mixtures in which all components at the temperature and the pressure of the mixture (or a slightly elevated pressure) exist as pure liquid substances, this pure liquid state is a suitable reference state for all components and we have

$$\mu_i(T, p, \{x_j\}) = \mu_{0i}(T, p) + RT \ln(\gamma_i^0 x_i). \tag{2.29}$$

Here $\mu_{0i}(T, p)$ is the chemical potential of pure liquid component i at temperature and pressure of the system, which can be obtained from the enthalpy of formation and the absolute entropy in the standard state of pure liquid by (2.9). Such pure component data are usually available. Further, γ_i^0 is the associated activity coefficient. The particular case of an ideal solution (is) is given for $\gamma_i^0 = 1$; i.e.,

$$\mu_i^{\mathrm{is}}(T, p, x_i) = \mu_{0i}(T, p) + RT \ln x_i. \tag{2.30}$$

From the general thermodynamic relations we then find, e.g.,

$$h^{\mathrm{is}}(T, p, \{x_j\}) = \sum x_i h_{0i}(T, p) \tag{2.31}$$

and

$$s^{\mathrm{is}}(T, p, \{x_j\}) = \sum x_i h_{0i}(T, p) - R \sum x_i \ln x_i. \tag{2.32}$$

Because real liquid mixtures do not form ideal solutions, the ideal solution laws are corrected by so-called excess functions, defined for an arbitrary molar state quantity z at constant pressure by

$$z^{\mathrm{E}}(T, p, \{x_j\}) = z(T, p, \{x_j\}) - z^{\mathrm{is}}(T, p, \{x_j\}). \tag{2.33}$$

Again, from the general thermodynamic relations, we find

$$\frac{\partial(g^{\mathrm{E}}/T)}{\partial(1/T)} = h^{\mathrm{E}} \tag{2.34}$$

and

$$\left(\frac{\partial(Ng^{\mathrm{E}})}{\partial N_i}\right)_{T,p,n_i^*} = RT\ln\gamma_i^0.$$ (2.35)

The equilibrium condition for vapor–liquid equilibrium can then be most conveniently written in terms of fugacity and activity coefficients as

$$x_i''p = x_i'\gamma_i^{0i}\left(T, p, \{x_i\}\right)p_{s0i}F_i$$ (2.36)

with

$$F_i = \frac{\phi_{0i}''(T, p_{s0i})}{\phi_i''(T, p, \{x_i\})}\exp\int_{p_{s0i}}^{p}\frac{v_{0i}}{RT}\,\mathrm{d}p.$$ (2.37)

Here x_i'' and x_i' denote the mole fractions of component i in the vapor and in the liquid phase, respectively, ϕ_{0i}'' the fugacity coefficient of pure vapor i, ϕ_i'' the fugacity coefficient of component i in the vapor mixture, v_{0i} the molar volume of pure liquid component i, and p_{s0i} the vapor pressure of pure component i. The fugacity coefficients are just the fugacities divided by their ideal gas values. At low pressures we usually can set $F_i = 1$. Analogous equations can be formulated for the liquid–liquid equilibrium. Data for the activity coefficient $\gamma_i^0(T, p, \{x_i\})$ can be calculated from phase equilibrium measurements, in particular vapor–liquid and liquid–liquid equilibria, and used to provide functional relationships for the excess Gibbs free energy g^{E}. Independent information on the temperature dependence is available from experimental data of the excess enthalpy h^{E}. We generally assume that experimental data for the pure fluid vapor pressure are available.

Other types of mixtures, such as solutions of gases and solids in liquids or electrolyte solutions, are described analogously but with other reference states. Frequently, e.g., in gas solubility problems, the pure state of a dissolved component i at the temperature and the pressure of the mixture is not liquid, but rather gaseous. Then the pure real state is not a suitable reference state for this component. Instead, a hypothetical pure liquid state with reference to an ideal dilute solution (ids) is frequently used, and the fugacity of component i in this state, denoted by an asterisk $*$, is given by

$$f_i^*(T, p) = \lim_{x_i\to 1} f_i^{\mathrm{ids}} = \lim_{x_i\to 1}[x_i H_i^*(T, p)] = H_i^*(T, p).$$ (2.38)

Here $H_i^*(T, p)$ is the Henry coefficient of solute i in the solution. It is a hypothetical pure fluid property in the sense of the limit $x_i \to 1$. Its relation to the ideal dilute solution fugacity is $f_i^{\mathrm{ids}} = x_i H_i^*$, and its definition in terms of the fugacity f_i is $H_i^* = \lim_{x_i\to 0}(f_i/x_i)$, since $f_i(x_i = 0) = 0$. Clearly its numerical value depends on the nature and on the composition of the solute-free solvent mixture, and this is to be understood without explicitly introducing the composition variables in addition to temperature and pressure in $H_i^*(T, p)$. For the chemical potential of component i in this hypothetical pure fluid reference state, one can thus write, with $f_{0i}^{ig} = p$, from (2.28),

$$\mu_i^*(T, p) = \mu_{0i}^{ig}(T, p) + RT \ln \frac{H_i^*(T, p)}{p}, \qquad (2.39)$$

where the same remarks as above apply to the dependence of $\mu_i^*(T, p)$ and $H_i^*(T, p)$ on the composition of the solvent. The chemical potential of component i in a liquid mixture then reads

$$\mu_i(T, p, \{x_i\}) = \mu_i^*(T, p) + RT \ln \frac{f_i(T, p, \{x_i\})}{H_i^*(T, p)}. \qquad (2.40)$$

The ratio of the fugacity of solute i in the mixture to its Henry coefficient is again denoted as activity and can again be expressed in terms of an activity coefficient, although this activity coefficient is obviously different from that defined in (2.28). We denote it by an asterisk $*$ and thus have

$$a_i^*(T, p, \{x_i\}) = \frac{f_i(T, p, \{x_i\})}{H_i^*(T, p)} \qquad (2.41)$$

and

$$\gamma_i^*(T, p, \{x_i\}) = a_i^*(T, p, \{x_j\})/x_i, \qquad (2.42)$$

where in both definitions the superscript $*$ denotes the reference state related to the properties of an ideal dilute solution. For the chemical potential one then has from (2.40)

$$\mu_i(T, p, \{x_i\}) = \mu_i^*(T, p) + RT \ln x_i + RT \ln \gamma_i^*(T, p, \{x_j\}), \qquad (2.43)$$

and the condition for phase equilibrium for a solute i in the gas phase (G) and the liquid phase (L) at normal pressure is given by

$$x_i^G p = x_i^L H_i^*(T, p) \gamma_i^*(T, p, \{x_i\}). \qquad (2.44)$$

In aqueous solutions we usually describe the composition of the solute i in terms of molality m_i, i.e., the number of moles of component i per kg of water. The reference state is then defined as a (hypothetical) ideal dilute solution with $m_i = 1$ and denoted by $\square$. We then have

$$\mu(T, p, \{m_i\}) = \mu_i^\square(T, p) + RT \ln \frac{f_i(T, p, \{m_i\})}{H_i^\square(T, p)}$$
$$= \mu_i^\square(T, p) + RT \ln m_i + RT \ln \gamma_i^\square(T, p, \{m_i\}), \qquad (2.45)$$

where m_i is made dimensionless by dividing the actual molality by the unit molality $m_i^\square = 1 \text{mol/kg}$.

According to (2.39), the Henry coefficient describes the change of chemical potential of the solute i during the isothermal–isobaric transformation from the pure ideal gas state to the reference state of an ideal dilute solution. The pure ideal gas data are assumed to be known. Data for the Gibbs free energy of formation and the absolute entropy in the associated standard state of the ideal dilute solution, i.e., at $t° = 25°C$ and $p° = 1$ bar, are available in tables for many components in water and some other solvents. Generally, they have to be considered as unknown, as all mixture properties. They are experimentally accessible by solution data at low concentrations of the solute, which can be used to determine the Henry coefficient from (2.44) and then give the standard state properties. The activity coefficient corrects for real solution behavior. It must be obtained from data of the solution under consideration. Chemical reaction equilibria can be calculated from the above equations for the chemical potentials, when the standard state values as well as the activity coefficients are available.

In summary, classical thermodynamics tells us that there is one particular function from which all information about fluid phase behavior can be obtained by established computational procedures. This particular function is the Helmholtz free energy $A(T, V, \{N_j\})$ or an equivalent fundamental function. Practically, one uses either of two well-established approaches, the equation of state approach or the excess function/activity coefficient approach. The equation of state approach is more general, covering the whole range of density, but use of an equation of state can be computationally demanding. On the contrary, the excess function/activity coefficient approach is much simpler to use, but is restricted to applications in which the fluid phase behavior is controlled by mixing effects at constant density. In this book, we shall develop molecular models for fluids, which are either presented in the equation of state form or in the excess function form. This will be augmented by models for the ideal gas heat capacities and for standard state properties in the various standard states and thus, in combination with the equations of classical thermodynamics, will provide a complete set of information for the prediction of the phase behavior of fluids.

2.2 From the Microscopic to the Macroscopic World: Statistical Mechanics

The basis of any molecular model for fluids is a link between the molecular properties and the fundamental thermodynamic function of a system. This is provided by statistical mechanics in form of relationships between the energy of a molecular system and its macroscopic thermodynamic functions. Statistical mechanics is founded on some postulates in combination with statistical and combinatorial results for systems consisting of a large number of elements [3].

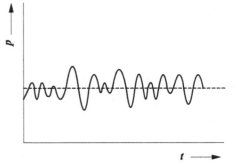

Fig. 2.1. Microscopic pressure fluctuations for a fixed macrostate in the classical view.

2.2.1 Macrostate and Microstate

The macroscopic thermodynamic state of a system, which we refer to as its macrostate, is defined in terms of a few macroscopic quantities. In simple systems, $k + 2$ state quantities determine each extensive state quantity entirely, where k is the number of components; cf. Section 2.1. Thus one can prescribe U, V, and all mole numbers $\{N_j\}$ or, alternatively, T, V, and $\{N_j\}$ in order to define the macrostate of the system.

For each macrostate very many different microstates are possible. Here a microstate is a state of the system that can be identified by observations on the atomic scale. According to the classical point of view, a microstate is defined in terms of fixed values of the coordinates and momenta of all atoms in the system; cf. Section 2.3. While the system is kept in a fixed macrostate, the molecular motions will lead to very many different microstates during the time of observation. Some of the macroscopic state quantities of the system can also be defined in each individual microstate, e.g., the dynamical variables pressure and energy and also the density. The pressure results from the forces that the molecules transfer to the wall on collisions with it, whereas the energy is the sum of the kinetic and potential energies of the molecules. Both quantities will in general fluctuate while the system, being in a fixed macrostate, runs through its very many different microstates. According to the classical view, all dynamical variables fluctuate continuously with time t. Thus, if we had a measuring device of atomic resolving capability, we would observe pressure fluctuations about a mean value as shown in Fig. 2.1. However, because the measuring device actually has macroscopic dimensions with an associated inertia, it provides an average pressure value given by

$$p = \lim_{\tau \to \infty} \frac{1}{\tau} \int p(t)\mathrm{d}t, \tag{2.46}$$

where τ is the time of measurement.

We know that the properties of atomic systems must basically be described in terms of quantum mechanics. Quantum mechanics tells us that the energy of a system cannot assume arbitrary but only discrete values. The energy is

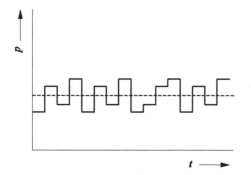

Fig. 2.2. Microscopic pressure fluctuations for a fixed macrostate in the quantum mechanical view.

quantized; cf. Section 2.5. The same is true for the positions and the momenta of the molecules. The continuously varying microstates in classical mechanics have their counterparts in the quantum states of quantum mechanics, in which the dynamical variables in the different quantum states may only assume discrete values in subsequent time steps Δt_i. Thus, in the quantum mechanical view, we find the microscopic interpretation of a macroscopic pressure measurement,

$$p = \lim_{\tau \to \infty} \frac{1}{\tau} \sum p_i \Delta t_i, \qquad (2.47)$$

where the pressure fluctuations may look as shown in Fig. 2.2.

Thus, macroscopic thermodynamic properties can statistically be interpreted as time averages over the corresponding values in the various microstates. We are now looking for a practical procedure to evaluate these time averages. This requires turning from time averages to ensemble averages.

2.2.2 Ensemble Averages

An ensemble is a collection of very many systems in the same macrostate. If the macrostate is defined in terms of U, V, and $\{N_j\}$ (isolated system), each system of the ensemble has these values U, V, $\{N_j\}$ and the ensemble is referred to as a microcanonical ensemble. If, alternatively, the macrostate is defined in terms of T, V, $\{N_j\}$ (closed system in contact with a heat bath), then each system of the ensemble has these values of T, V, $\{N_j\}$ and the ensemble is referred to as a canonical ensemble. If, finally, the system is defined in terms of T, V, $\{\mu_j\}$ (open system in contact with a heat bath and material reservoir), again all systems of the ensemble have these values of T, V, $\{\mu_j\}$ and the ensemble is referred to as a grand canonical ensemble. Other types of ensembles can be defined, but we do not need them here.

Although the systems of an ensemble are entirely identical macroscopically, this is not the case microscopically. Because the number of systems may be thought to be arbitrarily large, all the different microstates compatible with the given macrostate of the system will be represented in the ensemble at each instant of time. Statistical mechanics makes use of this fact and is founded on two postulates about the properties of ensembles.

First postulate of statistical mechanics:
The time average of a dynamical quantity in a macroscopic system equals its ensemble average.

Given a dynamical quantity x, we thus postulate that

$$x = \sum_i P_i x_i. \qquad (2.48)$$

Here x_i is the value of x in quantum state i and P_i the relative probability of this state. This relative probability equals the number of systems in quantum state i divided by the total number of systems. Thus (2.48) is the usual averaging recipe, just like the one used, e.g., to compute the average age of a population. The first postulate is immediately plausible, as demonstrated by the following example. We consider a box containing three equal-sized balls, one being white and two being red. When one ball is taken out of the box blindly and put back and this experiment is repeated a total of 999 times, the ratio of white balls to red balls taken will be 1:2; i.e., a white ball will have been taken 333 times, a red ball 666 times. The result of this experiment, which is repeated many times, is a time average and corresponds to the pressure measurement with a macroscopic measuring instrument. The same average would clearly emerge by ensemble averaging, i.e., if 999 hands took one ball from each of 999 boxes of the ensemble, each containing one white ball and two red balls. Thus, we find that the time average equals the ensemble average.

Second postulate of statistical mechanics:
In a microcanonical ensemble all possible states are equally probable.

Thus, for two microscopic states i and j in a system with fixed values of U, V, and $\{N_j\}$, i.e., with $E_i = E_j$, we find for the associated probabilities

$$P_i = P_j. \qquad (2.49)$$

Although this postulate is restricted to the microcanonical ensemble, we shall see below that it also determines the probability of a microstate in the canonical as well as in the grand canonical ensemble.

The two postulates of statistical mechanics cannot be proven rigorously. While the first postulate is immediately plausible the second may look somewhat obscure at first sight. It is sufficient to say here that these postulates may be considered as basic axioms, like the laws of classical thermodynamics, which have been abstracted from a comparison of the conclusions drawn from them with observation.

2.2.3 Relative Probability of a Microstate

To calculate any dynamical property x from (2.48) we need the relative probabilities of the various microstates, i.e., $\{P_i\}$. From the second postulate we know that these probabilities will depend on the energy E_i. For a system at constant values of N, V, T, i.e., a system of a canonical ensemble, this energy will fluctuate over the many possible values $\{E_i\}$ of the system. It can be derived by statistical arguments, cf. App. 3, that the probability of a microstate i in such a system is given by

$$P_i = \frac{e^{-E_i/kT}}{Q}. \tag{2.50}$$

The term $e^{-E_i/kT}$ is referred to as the Boltzmann factor, and

$$Q = \sum_i e^{-E_i/kT} \tag{2.51}$$

is the so-called canonical partition function. In a system with constant values of T, V, μ not only the energy, but also the number of molecules will fluctuate. Thus, the energy of a microstate will carry two indices i and j, E_{ij}, referring to the energy state j associated with a system of i molecules. We thus expect a statistical formulation of the ensemble averages containing a sum not only over energy states but also over the number of molecules. Thus, the sum in (2.51) will be replaced by a double sum over i and j. In analogy to the procedure of the canonical ensemble we find

$$P_{ij} = \frac{e^{-E_{ij}/kT} e^{i\mu/kT}}{\Xi} \tag{2.52}$$

with the so-called grand canonical partition function

$$\begin{aligned}
\Xi &= \sum_i \sum_j e^{-E_{ij}/kT} e^{i\mu/kT} \\
&= \sum_N e^{N\mu/kT} \left(\sum_j e^{-E_{Nj}/kT} \right) \\
&= \sum_N Q_N e^{N\mu/kT}.
\end{aligned} \tag{2.53}$$

The grand canonical partition function can thus be expressed in terms of the canonical partition function and consists of an expansion in successive terms with 1 molecule, 2 molecules, 3 molecules, etc., up to N molecules. For a mixture with components α, β, ... the grand canonical partition function reads

$$\begin{aligned}
\Xi &= \sum_{N_\alpha} \sum_{N_\beta} \cdots \sum_j e^{N_\alpha \mu_\alpha/kT} e^{N_\beta \mu_\beta/kT} \cdots e^{-E_{N_\alpha N_\beta \cdots j}/kT} \\
&= \sum_{N_\alpha} \sum_{N_\beta} e^{N_\alpha \mu_\alpha/kT} e^{N_\beta \mu_\beta/kT} \cdots Q_{N_\alpha, N_\beta \cdots}
\end{aligned} \tag{2.54}$$

with $E_{N_\alpha N_\beta \dots j}$ being the molecular energy of a system with N_α molecules of component α, N_β molecules of component β, etc., in the quantum state j associated with these numbers of molecules and $Q_{N_\alpha, N_\beta} \dots$ the canonical partition function of such a system.

2.2.4 Thermodynamic Functions

We first derive relationships between the thermodynamic functions and the canonical partition function. According to the first postulate of statistical mechanics, we can equate the macroscopic dynamical quantities of a system to the corresponding ensemble averages. This can be directly evaluated for the internal energy, with the result

$$U = \Sigma P_i E_i = \frac{1}{Q} \sum E_i \, e^{-E_i/kT}$$
$$= kT^2 \left(\frac{\partial \ln Q}{\partial T} \right)_{V, \{N_\alpha\}}. \tag{2.55}$$

The internal energy is thus related to the temperature derivative of the canonical partition function. The entropy is statistically interpreted through a comparison with the fundamental relation (2.1). The total differential of internal energy for a system with constant volume and mole numbers reads

$$dU_{V,N} = \sum_i E_i \, dP_i. \tag{2.56}$$

We here anticipate the quantum mechanical fact, to be derived in Section 2.5, that the $\{E_i\}$ are determined by V, N, so that dE_i at $V, N =$ const. is zero. Further, we ignore the discrete nature of the P_i, because they are so closely spaced that standard differential calculus is adequate. With

$$P_i = \frac{1}{Q} \, e^{-E_i/kT}$$

we find

$$E_i = -kT(\ln P_i + \ln Q). \tag{2.57}$$

Introducing this into (2.56), we get

$$dU_{V,N} = -kT \sum_i (\ln P_i + \ln Q) dP_i$$
$$= -kT d \left(\sum_i P_i \ln P_i \right),$$

where use was made of $\sum dP_i = 0$. We compare this statistical formulation of dU with the fundamental relation; cf. (2.1),

$$dU_{V,N} = T dS.$$

For the product of the thermodynamic temperature and the differential of entropy we thus find

$$T\mathrm{d}S = -kT\mathrm{d}\left(\sum_i P_i \ln P_i\right),$$

which gives, by integration,

$$S = -k\sum_i P_i \ln P_i + C.$$

From the third law we know that the entropy becomes zero at the absolute zero of temperature for a perfect crystal; i.e.,

$$\lim_{T\to 0} S = 0 \quad \text{perfect crystal.}$$

A perfect crystal is a system that has a well-defined microstate at $T = 0$, i.e., at its state of lowest energy. This means that

$$\lim_{T\to 0} P_1 = 1, \quad \text{for } P_2 = P_3 = \cdots = 0.$$

Agreement of the statistical equation for entropy with the third law thus requires that

$$C = 0,$$

leading to

$$S = -k\sum_i P_i \ln P_i. \tag{2.58}$$

Because of

$$\ln P_i = -E_i/kT - \ln Q$$

we have

$$\sum_i P_i \ln P_i = -\ln Q - \sum_i P_i \frac{1}{kT} E_i$$

and thus

$$S = k\ln Q + \frac{U}{T} = k\ln Q + kT\left(\frac{\partial \ln Q}{\partial T}\right)_{V,\{N_\alpha\}}. \tag{2.59}$$

This gives the entropy in terms of the canonical partition function. With U and S we can express the Helmholtz free energy A in terms of the canonical partition function, cf. (2.2),

$$A = U - TS = U - kT\ln Q - U = -kT\ln Q. \tag{2.60}$$

From the general thermodynamic relations we then find for the pressure

$$p = -\left(\frac{\partial A}{\partial V}\right)_{T,\{N_\alpha\}} = kT\left(\frac{\partial \ln Q}{\partial V}\right)_{T,\{N_\alpha\}} \tag{2.61}$$

and for the chemical potential of component α

$$\mu_\alpha = \left(\frac{\partial A}{\partial N_\alpha}\right)_{T,V,N_\alpha^*} = -kT \left(\frac{\partial \ln Q}{\partial N_\alpha}\right)_{T,V,N_\alpha^*}. \qquad (2.62)$$

These statistical mechanical expressions for the thermodynamic functions in terms of the canonical partition function have ultimately been derived by an appeal to classical thermodynamics and thus are consistent with that theory. They reach, however, far beyond the limit of classical thermodynamics in the sense that they provide connections to the molecular properties of a system, namely its spectrum of possible energy values. They are the fundamental molecular equations for the thermodynamic functions in terms of the independent variables T, V, $\{N_\alpha\}$. Particular systems differ in their particular expressions for the energy values E_i.

In the grand canonical ensemble the internal energy and the number of molecules fluctuate. They are thus calculated from

$$U = \sum_i \sum_j P_{ij} E_{ij} = \frac{\sum_i \sum_j E_{ij} \, e^{-E_{ij}/kT} \, e^{i\mu/kT}}{\varXi} \qquad (2.63)$$

and

$$N = \sum_i \sum_j P_{ij} \, i = \frac{\sum_i \sum_j i e^{-E_{ij}/kT} \, e^{i\mu/kT}}{\varXi}. \qquad (2.64)$$

It can immediately be seen that the statistical analogs for the number of molecules and the internal energy in the grand canonical ensemble are

$$N = kT \left(\frac{\partial \ln \varXi}{\partial \mu}\right)_{T,V} \qquad (2.65)$$

and

$$U = kT^2 \left(\frac{\partial \ln \varXi}{\partial T}\right)_{V,\mu} + \mu N. \qquad (2.66)$$

The total differential of the internal energy in a system of the grand canonical ensemble at constant V and N is in statistical and thermodynamic terms

$$dU_{V,N} = \sum_i \sum_j E_{ij} dP_{ij} = TdS,$$

because then the energy values E_{ij} and also the molecule numbers i are constant. With

$$\frac{E_{ij}}{kT} = -\ln P_{ij} - \ln \varXi + \frac{\mu \cdot i}{kT}$$

from (2.52) and hence

$$
\begin{aligned}
dU &= \sum_i \sum_j E_{ij} dP_{ij} = -kT \sum_i \sum_j \ln P_{ij} dP_{ij} \\
&\quad - kT \sum_i \sum_j \ln \varXi \, dP_{ij} + \mu \sum_i \sum_j i \, dP_{ij} \\
&= -kT d\left[-\sum_i \sum_j P_{ij} \ln P_{ij} \right] = T dS,
\end{aligned}
$$

where $\sum_i \sum_j dP_{ij} = 0$ and $dN = 0$ have been used, we arrive at

$$
S = -k \sum_i \sum_j P_{ij} \ln P_{ij}.
$$

Here the third law has been taken into account. Replacing $\ln P_{ij}$ by introducing the expression (2.52), we get

$$
-k \sum_i \sum_j P_{ij} \ln P_{ij} = \frac{1}{T} \sum_i \sum_j (E_{ij} - \mu \cdot i) P_{ij} + k \ln \varXi,
$$

leading to the statistical analog for the entropy in the grand canonical ensemble, as

$$
S = \frac{1}{T}(U - N\mu) + k \ln \varXi = kT \left(\frac{\partial \ln \varXi}{\partial T} \right)_{V,\mu} + k \ln \varXi. \tag{2.67}
$$

Introducing here the definition of the Gibbs potential,

$$
G = N\mu = U + pV - TS, \tag{2.68}
$$

gives

$$
pV = kT \ln \varXi \tag{2.69}
$$

as the statistical analog for the pressure, i.e., the equation of state, in the grand canonical ensemble. The product pV is referred to as the grand canonical potential, another thermodynamic potential derivable from the fundamental relation by Legendre transformation to the variables T, V, μ.

EXERCISE 2.1

Use (2.67) for the entropy analog in the grand canonical ensemble to derive Boltzmann's relationship between entropy and thermodynamic probability, $S = k \ln W$, for an isolated system.

Solution

The entropy in the grand canonical ensemble is given by

$$
S = \frac{1}{T}(U - N\mu) + k \ln \varXi.
$$

In an isolated system the number of molecules as well as the energy will have a fixed value. The grand canonical partition function then reads

$$\varXi = \varOmega(N, V, E)e^{-E/kT}e^{N\mu/kT},$$

where $\varOmega(N, V, E)$ is the number of microstates associated with the energy value $E = U$ and the number of molecules N. Introducing this into the above equation for the entropy gives

$$S = k \ln \varOmega(N, V, U).$$

This is Boltzmann's relationship between entropy and probability. It is the fundamental equation of the microcanonical ensemble, whose systems are characterized by constant values of U, V, and N and thus have $S(U, V, N)$ as the associated thermodynamic potential. Also, since a large number of microstates implies a great deal of indeterminacy in the atomic level, this equation also serves as the basis for qualitative statements concerning entropy and disorder, entropy and randomness, entropy and loss of information, etc. The way this equation is usually written is $S = k \ln W$, where W refers to the German word for probability (*Wahrscheinlichkeit*); cf. (1.7). We note in particular that this atomic interpretation of entropy is restricted to an isolated system.

2.2.5 The Semiclassical Approximation

The canonical partition function

$$Q = \sum_i e^{-E_i/kT}$$

plays a prominent role in the theory of molecular models. All thermodynamic properties of a system can be derived from it. It was obtained on the basis of the anticipated quantum mechanical fact that the energy of a system can only assume discrete values. The sum extends over all distinguishable states that are accessible to the system. To evaluate the partition function, and hence the thermodynamic functions, we need the energy values in all accessible states and a recipe to evaluate the summation. Information about the energy states is not available in the general case, and also, carrying out the summation is practically difficult in general. However, in many important applications, approximations can be made that lead to a form of the canonical partition function that can practically be worked out.

One particularly important step in this direction is the semiclassical approximation. In this approximation we return to the classical concept of an atomic microstate as being determined by the continuously variable position and momentum coordinates describing the dynamical state of the molecular system. As we can imagine, this approximation is not adequate generally for all modes of molecular energy. But frequently the quantization of energy is so small, i.e., the energy values are so closely spaced, that the summation can be replaced by an integral, referred to as the phase integral, where the position and momentum coordinates define the phase space. In many applications, advantage can then be taken of established closed solutions for the integrals instead of performing a tedious summation. In the phase integral the quantum mechanical energy values are replaced by their classical counterpart, i.e., the so-called Hamilton function [4], which is the total energy of the molecular system. In contrast to E_i, which requires solution of the equations of quantum mechanics, the Hamilton function is based on classical mechanics and thus is easily derived for specific molecular models, as we will see in Section 2.3. In a system consisting of monatomic molecules the position coordinates and momenta in the Hamilton function refer to the atomic nuclei, because the motion of the electrons relative to the atomic nucleus can be considered separately due to the Born–Oppenheimer approximation; cf. Sections 2.3 and 2.5. In polyatomic molecules they refer to the individual nuclei in a molecule because their relative arrangement is not rigid due to vibrations and internal rotations. In general, the classical microstate of a molecular system thus is determined by the position coordinates and momenta of all its nuclei. We denote the total number of these position coordinates and momenta for N molecules by Γ^N. The Hamilton function $H(\Gamma^N)$, i.e., the total classical energy of a system in a microstate defined by Γ^N, is a continuous function of all the position coordinates and momenta of the molecular system under consideration. The summation over all quantum states is accordingly replaced by an integration over Γ^N. The canonical partition function is thus written classically as

$$Q = C \int_{\Gamma^N} e^{-H(\Gamma^N)/kT} d\Gamma^N,$$

where the integral is the phase integral and

$$C = \frac{\sum_i e^{-E_i/kT}}{\int_{\Gamma^N} e^{-H(\Gamma^N)/kT} d\Gamma^N}.$$

Even at sufficiently high temperature, where quantization effects tend to become negligible for all types of energy modes, the constant differs from unity because of two principal defects of the phase integral relative to the quantum mechanical partition function, which are independent of the energy quantization.

One of these principal defects is related to the indistinguishability of identical molecules. In the quantum mechanical partition function the summation goes over all distinguishable states accessible to the system. Thus the indistinguishability is taken into account. In the phase integral, however, identical molecules are treated as if they were distinguishable. Consider a system of monatomic molecules. When we exchange the center-of-mass coordinates of two identical molecules we find two identical contributions to the integral that only arise once in the quantum mechanical partition function. With N molecules each configuration is thus counted $N!$ times instead of only once in the partition function and we have to divide by $N!$ to correct for this. In the case of polyatomic molecules this argument has to be extended to indistinguishability with respect to further coordinates. The associated quantum mechanical correction factors will be introduced in Chapter 3, when the semiclassical approximation to the partition function for single-molecule motions will be discussed.

The other principal defect of the phase integral is related to the Heisenberg uncertainty principle. We shall see in Section 2.5 that due to this quantum mechanical principle there is a minimum combined uncertainty in the values for a position coordinate and its conjugate momentum, say x and p_x, of $(\Delta x)(\Delta p_x) \sim h$, where h is Planck's constant. From a classical point of view the coordinates and momenta are continuous variables and any combination of them can be specified without limitation of precision. In quantum mechanics, on the other hand, the space of position coordinates and momenta is subdivided into cells with a volume on the order of h. The classical differential $dpdr$ for a single molecule thus cannot be considered as an infinitesimal small volume element in phase space but rather corresponds to a volume of h^f. Here, f is the number of pairs of position coordinates and momenta per molecule, to which we refer as degrees of freedom; cf. Section 2.3. We have $f = 3$ per atom in the molecule, corresponding to the three directions in space. So, for a diatomic molecule, we would have $f = 6$. For N molecules we thus have a cell volume of h^{Nf}. The phase integral must be divided by this volume in the position and momentum space to make it consistent with the quantum mechanical partition function in the classical limit. This also eliminates the dimensional factor that is introduced into the phase integral by $d\Gamma^N$.

Clearly, these arguments are far from being rigorous and may only serve to make the semiclassical approximation to the canonical partition function plausible. However, in the particular case of an ideal gas, we will investigate its validity by comparing with exact evaluations of the quantum mechanical partition function in Chapter 3. It will be confirmed in such cases, where the energy values are so closely spaced that the summation can be replaced by an integration. We thus arrive at the following expression for C,

$$C = \frac{1}{N!h^{Nf}}, \tag{2.70}$$

and the semiclassical formulation of the canonical partition function reads

$$Q = \frac{1}{N!h^{Nf}} \int_{\Gamma^N} e^{-H(\Gamma^N)/kT} d\Gamma^N. \tag{2.71}$$

Here we reiterate that further correction factors due to indistinguishability of molecular orientations, in addition to $N!$, may have to be introduced, as shown in Chapter 3. The term "semi" in "semiclassical" refers to the two quantum mechanical corrections that have been applied to the phase integral in order to make it consistent with the quantum mechanical partition function. When the system is a mixture of various components α, the semiclassical formulation of the canonical partition function is

$$Q = \frac{1}{\prod_\alpha (N_\alpha! \, h^{N_\alpha f_\alpha})} \int\limits_{\Gamma^N} e^{-H(\Gamma^N)/kT} \mathrm{d}\Gamma^N, \tag{2.72}$$

because now the Hamilton function is no longer invariant to permutations of all molecules but only to permutations of molecules of a particular component.

The Hamilton function of a classical molecular system is identical with its total energy as manifested by the motions and positions of its molecules. It consists of contributions from the translational motions of the centers of mass, from external rotations of the molecules around axes through the center of mass, and from internal molecular motions such as internal rotations and vibrations. According to classical mechanics, cf. Section 2.3, all these contributions can formally be split into a kinetic term and a potential term and we get for the Hamilton function

$$H = H^{\mathrm{kin}} + H^{\mathrm{pot}}. \tag{2.73}$$

The kinetic part of the Hamilton function collects all contributions to the total energy in the molecular system that do not depend on molecular environment. Thus, it includes the kinetic parts of translation, of external and internal rotations, and of vibrations. These are the contributions of the single-molecule energy modes to be considered in Section 2.3. Generally, we therefore have $H^{\mathrm{kin}} = H^{\mathrm{kin}}(\Gamma^{N,\mathrm{kin}})$, where $\Gamma^{N,\mathrm{kin}}$ is shorthand for all coordinates referring to the motions of the single molecules. The potential part summarizes those contributions that depend on the molecular environment. Generally, all energy modes of the molecules are influenced by their neighbors, including the internal motions such as vibration and internal rotation. So they all will have a potential part in addition to their kinetic part. Consequently, the potential part of the Hamiltonian, referred to as the intermolecular potential energy U, will in general depend on all coordinates describing the influence of molecular environment on the total energy of the molecular system. If we use $\Gamma^{N,\mathrm{pot}}$ as a shorthand for these coordinates we have $U = U(\Gamma^{N,\mathrm{pot}})$. Then, in the spirit of (2.73), we can factor the canonical partition function in the semiclassical approximation as

$$Q = \frac{1}{\prod_\alpha (N_\alpha! h^{N_\alpha f_\alpha})} \int\limits_{\Gamma^{N,\mathrm{kin}}} e^{-H^{\mathrm{kin}}(\Gamma^{N,\mathrm{kin}})/kT} \mathrm{d}\Gamma^{N,\mathrm{kin}}$$

$$\cdot \int\limits_{\Gamma^{N,\mathrm{pot}}} e^{-H^{\mathrm{pot}}(\Gamma^{N,\mathrm{pot}})/kT} \mathrm{d}\Gamma^{N,\mathrm{pot}} = \frac{1}{\prod_\alpha (N_\alpha! h^{N_\alpha f_\alpha})} Q^{\mathrm{kin}} Q^{\mathrm{C}}, \tag{2.74}$$

where

$$Q^{\mathrm{C}} = \int\limits_{\Gamma^{N,\mathrm{pot}}} e^{-U(\Gamma^{N,\mathrm{pot}})/kT} \mathrm{d}\Gamma^{N,\mathrm{pot}} \tag{2.75}$$

is referred to as the configurational partition function or the configuration integral.

EXERCISE 2.2

Evaluate the canonical partition function due to translational motion with zero potential energy in the semiclassical approximation.

Solution

The kinetic contribution of translational motion to the Hamilton function of a N molecule system is just the sum of all translational kinetic energies of the N molecules in the three directions of space; i.e.,

$$H_{\text{tr}}^{\text{kin}} = \sum_{i}^{3N} \frac{p_i^2}{2m_i},$$

where p_i are the momenta of the centers of mass. Since no other contribution to the total Hamilton function depends on the momenta, they can be integrated over separately to give, cf. App. 6 for the integral,

$$Q_{\text{tr}} = \frac{1}{N! h^{N f_{\text{tr}}}} \int_{-\infty}^{+\infty} e^{-\sum_i^{3N}(p_i^2/2m_i)/kT} \, \mathrm{d}p^N \mathrm{d}r^N$$

$$= \frac{1}{N! h^{3N}} \left[\int_{-\infty}^{+\infty} e^{-\frac{p^2}{2mkT}} \, \mathrm{d}p \right]^{3N} V^N$$

$$= \frac{1}{N!} \left(\frac{2\pi mkT}{h^2} \right)^{3N/2} V^N = \frac{1}{N!} \Lambda^{-3N} V^N,$$

where

$$\Lambda = \sqrt{\frac{h^2}{2\pi mkT}}$$

is the so-called thermal de Broglie wavelength. Because $U = 0$ the configurational integral is simply V^N.

2.3 Kinetic Energy of a Molecular System: Classical Mechanics

We learn from statistical mechanics that it is the energy of a molecular system that determines the partition function and thus the macroscopic behavior of a system. This energy consists of a kinetic and a potential contribution. The kinetic energy results from the motions of the molecules and can be formulated independent of the configuration of the molecular environment. The potential energy, in contrast, is associated with the electrical force field, in which the molecules move. This force field is generated by the other molecules in the system and so the potential energy depends on their position coordinates, i.e., on the configuration of the molecular environment. In this section, we derive explicit formulae for the kinetic energy associated with the motion of the molecules.

In referring to molecules as geometrical objects moving about in an intermolecular force field, an obvious approach is to describe their kinetic energy in terms of classical mechanics. We know, however, that classical mechanics is in conflict with experimental evidence on the molecular level. It has been established since the beginning of the last century that a different kind of mechanics, quantum mechanics, is required to describe the energy of objects as small as molecules. Still, it turns out that classical mechanics is not in general contradiction with quantum mechanics, but rather the two theories converge, under suitable circumstances, to the same results. Many aspects of molecular behavior can in fact be formulated in terms of the familiar concepts of classical mechanics and corrected in a simple way to reproduce the experimental facts. Because classical mechanics is conceptually and computationally much simpler than quantum mechanics, it makes sense to base the molecular models for fluids as far as possible on classical mechanical concepts [3,4]. This was anticipated in formulating the canonical partition function in terms of the classical mechanical energy of a molecular system in the semiclassical approximation.

2.3.1 Basic Equations of Classical Mechanics

The basic law of classical mechanics is Newton's second law. This law expresses the relation between the acceleration of a mass point of mass m, e.g., a molecular particle, say in the x direction, and the force $F(x)$ it experiences on the way,

$$m\frac{d^2x}{dt^2} = F(x) \tag{2.76}$$

or

$$\frac{dp}{dt} = F(x), \tag{2.77}$$

where $p = m\,dx/dt$ is the momentum of the particle in the x direction and x is a function of time t. Analogous equations are valid for the y and z directions. In our applications the force $F(x)$ can always be expressed in terms of the negative derivative of the potential energy $U(x)$ with respect to the position coordinate x according to $F(x) = -dU/dx$, where $U(x)$ results from the electrical interactions of the molecular charges. We refer to such systems as conservative systems, for reasons that will become immediately clear. For, after multiplying both sides of (2.77) by $p/m = dx/dt$ and integrating over time, we find that

$$E_1^{\text{kin}} + U_1 = E_2^{\text{kin}} + U_2 = E^{\text{kin}} + U = \text{const.} \tag{2.78}$$

This is the familiar energy conservation equation of classical mechanics, with

$$E^{\text{kin}} = \frac{p^2}{2m} \tag{2.79}$$

as the kinetic energy of the particle in terms of its momentum p. The total energy, i.e., the sum of the kinetic and the potential energy of the particle, is constant, i.e., independent of time t, and can be prescribed as a boundary condition. We then find that (2.78) is a differential equation for x as a function of time. Solving it gives the position and momentum of the particle as functions of time. The same information can be obtained by solving (2.76) and (2.77) from knowledge of the force acting on the particle at any location. A statement of both $x(t)$ and $p(t)$ is called the trajectory of the particle. So, given the location and the momentum of the particle at some moment $t = 0$, the whole future behavior of the particle can be predicted.

The above formulation of the classical mechanical equations is adequate for a single particle moving along a trajectory described in a cartesian system of coordinates in a conservative force field. In the context of molecular models a somewhat different formulation is more suitable. We frequently find that the particles of a molecular model cannot move freely, but are subject to certain restrictions. An example is a nucleus bonded to a second nucleus in a rotating diatomic molecule. The energy of such a system is not easily described in terms of single-particle Newtonian equations in cartesian coordinates. Instead, it is more convenient to formulate the classical mechanical equations in terms of so-called degrees of freedom, characterizing the molecule as a whole, to which a set of independently variable generalized position coordinates $\{q_k\}$ is associated. The notion of degrees of freedom and their relation to the motion of molecules will be discussed below. Here it is sufficient to anticipate that generalized position coordinates $\{q_k\}$, with $k = 1, 2, \ldots, f$, where f is the total number of degrees of freedom, can be defined, which adequately describe the configuration space of a molecule. The associated velocity $\dot{q}_i$ to a coordinate q_i is, as usual, the time derivative $\dot{q}_i = \mathrm{d}q_i/\mathrm{d}t$, and $\{\dot{q}_k\}$ are referred to as generalized velocities. We now introduce the Lagrange function $L(\{q_k\}, \{\dot{q}_k\}, t)$,

$$L(\{q_k\}, \{\dot{q}_k\}) = E^{\mathrm{kin}}(\{\dot{q}_k\}) - E^{\mathrm{pot}}(\{q_k\}), \qquad (2.80)$$

where $E^{\mathrm{kin}}(\{\dot{q}_k\})$ is the kinetic energy of the molecular system in terms of the generalized velocities and $E^{\mathrm{pot}}(\{q_k\})$ its potential energy in terms of the generalized position coordinates. Evaluating this for a single particle moving in a cartesian coordinate system shows that the Newtonian equation of motion takes the form

$$\frac{\partial L}{\partial x} - \frac{\mathrm{d}}{\mathrm{d}t}\left(\frac{\partial L}{\partial \dot{x}}\right) = 0$$

and we have $p_x = \partial L/\partial \dot{x}$ and $F_x = \partial L/\partial x$, as immediately follows from the above definitions. It can be shown that the above equation of motion in the Langrangian form is independent of the coordinate system. We therefore define generalized momenta generally in terms of the generalized velocities by

$$p_i = \frac{\partial L}{\partial \dot{q}_i} = \frac{\partial E^{\text{kin}}}{\partial \dot{q}_i} \tag{2.81}$$

and generalized forces by

$$F_i = \frac{\partial L}{\partial q_i} = -\frac{\partial E^{\text{pot}}}{\partial \dot{q}_i}. \tag{2.82}$$

The equation of motion for a degree of freedom i in the Lagrangian form is then

$$\frac{\partial L}{\partial q_i} - \frac{\mathrm{d}}{\mathrm{d}t}\left(\frac{\partial L}{\partial \dot{q}_i}\right) = 0. \tag{2.83}$$

Finally, we wish to formulate the total energy, i.e., the sum of the kinetic and the potential energy, in terms of generalized coordinates, since this is the quantity needed in the partition function; cf. Section 2.2. This then leads to the Hamilton function

$$H(\{q_k\}, \{p_k\}) = \sum_i \dot{q}_i p_i - L(\{q_k\}, \{\dot{q}_k\}), \tag{2.84}$$

which has been introduced before in an ad hoc manner. It is here seen to result from the Lagrange function by a Legendre transformation replacing the generalized velocities $\{\dot{q}_k\}$ by generalized momenta $\{p_k = \partial L/\partial \dot{q}_k\}$ and thus has the same information content as L. From (2.80) one immediately sees, as required, that the Hamilton function is the sum of the kinetic and the potential energy; i.e.,

$$H(\{q_k\}, \{p_k\}) = E^{\text{kin}}(\{p_k\}) + E^{\text{pot}}(\{q_k\}) = H^{\text{kin}}(\{p_k\}) + U(\{q_k\}). \tag{2.85}$$

Thus, the Newtonian equations of motion in the Hamiltonian form for degree of freedom i are

$$\frac{\partial H}{\partial p_i} = \dot{q}_i \tag{2.86}$$

and

$$\frac{\partial H}{\partial q_i} = -\dot{p}_i. \tag{2.87}$$

The Hamilton function expresses the total classical energy of a degree of freedom in terms of generalized coordinates and momenta. It thus has a well-defined significance and will be shown to be worked out relatively easily for a molecular system. Because it determines the semiclassical formulation of the canonical partition function, extensive use will be made of the Hamilton function in later applications.

2.3.2 Molecular Degrees of Freedom

The energy displayed by a single molecule can be described classically in terms of the dynamics of n mass points, where n is the number of atoms in the molecule. Generally, the dynamics of a body consisting of n mass points is determined by $3n$ pairs of position and momentum coordinates. Thus a molecule containing n atoms has $3n$ momentum coordinates to describe its kinetic energy. In this model of a molecule, the electrons do not appear, because it is assumed that all the mass of a molecule is concentrated in the atomic nuclei and thus the electrons do not contribute to the molecular dynamics. Also, because the electrons have negligible mass compared to the nuclei, their motions are on a different time scale. We can, therefore, consider the motion of the electrons to occur at fixed positions of the nuclei. This approximation yields an intramolecular potential energy that is defined by the positions of the nuclei and that controls the internal motions of the molecule. This separation of electronic and nuclear motions is referred to as the Born–Oppenheimer approximation. It will be discussed more formally in Section 2.5 and is essentially exact. As discussed above, the $3n$ momentum coordinates of the n nuclei are subject to constraints represented by the intramolecular bonds. Therefore, it is not convenient to describe them individually, but rather to attribute them to an equal number of modes of motion, the so-called degrees of freedom; cf. Fig. 2.3.

A monatomic molecule has three modes of motion without constraints, i.e., translation in the three directions of space. Its Hamilton function is formulated in terms of the associated cartesian momenta and coordinates.

The dynamic energy state of a diatomic molecule with $3n = 6$ momenta is characterized by three modes of translational motion of the center of mass CM in three directions, two rotational motions defined by the orientational angles θ, ϕ around two axes through the center of mass (the rotation around the axis connecting the two atoms does not have any energy because the associated moment of inertia is zero), and one vibrational motion of the atomic nuclei in the direction of the main axis r that leaves the center of mass in its position. Although the translational and rotational motion is entirely free for single molecules, the vibrational motion will be influenced by the charge distribution within the molecules because the nuclei are moved against each other in the intramolecular force field produced by the electrons. The vibrational energy of a single molecule will thus contain an intramolecular potential contribution, not to be confused with the potential energy due to the molecular environment. In the interpretation of the molecular motions of a diatomic molecule in terms of degrees of freedom the six momenta p_{1x}, p_{1y}, p_{1z}, p_{2x}, p_{2y}, p_{2z} of the two atoms are thus replaced by p_{xCM}, p_{yCM}, p_{zCM}, p_θ, p_ϕ, p_r, i.e., the three momentum coordinates of the center of mass translation, the two rotational momenta, and one momentum of vibration; cf. Fig. 2.3a. We thus have three translational degrees of freedom, two rotational degrees of freedom, and one degree of

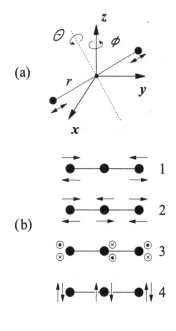

(a)

(b)

(c)

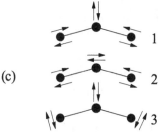

Fig. 2.3. Modes of motion of simple molecules. **(a)** diatomic molecule; **(b)** linear triatomic molecule: 1 stretching mode 1, 2 stretching mode 2, 3 bending mode 1, 4 bending mode 2; **(c)** nonlinear triatomic molecule: 1 stretching mode 1, 2 stretching mode 2, 3 bending mode.

freedom associated with vibration. The Hamilton function will be formulated in terms of these momenta and the associated coordinates.

The visualization of the dynamical energy state of triatomic molecules, which has $3n = 9$ momentum coordinates, is considerably more complicated. Linear triatomic molecules have three translational degrees of freedom of the center of mass, two rotational degrees of freedom around axes through the center of mass, and a further $3 \cdot 3 - 5 = 4$ degrees of freedom for internal molecular motions, i.e., those motions that are associated with internal deformations of the molecular geometry. These internal modes do not change the position of the center of mass. For a linear triatomic molecule the only internal modes are the four vibrational movements shown in Fig. 2.3b, i.e., two stretching modes and two bending modes. A bending mode can occur in any plane in space, which amounts to considering it in two mutually vertical planes. Nonlinear triatomic molecules have three translational degrees of freedom of the center of mass, three rotational degrees of freedom around axes through the center of mass, and a further $3 \cdot 3 - 6 = 3$ degrees of freedom of internal modes with a fixed center of mass, i.e., the three vibrations shown in Fig. 2.3c. Here the bending vibration is counted only once because its various planes are taken into account by the molecular rotation around the third axis not present in linear molecules. Again, translation and rotational modes are entirely free while the vibrational modes are controlled by the intramolecular charge distribution. The Hamilton function will be formulated in terms of the momenta and position coordinates associated with the various degrees of freedom.

Molecules with four and more atoms have three center-of-mass translational degrees of freedom, three rotational degrees of freedom around axes through the center of mass, and a further $3n - 6$ degrees of freedom for internal modes. A large number of internal modes is difficult to visualize because besides the vibrations a further energy mode referred to as internal rotation may occur. This is often associated with rotations of molecular groups around a single carbon-carbon bond. Like vibration, it will generally be influenced by the intramolecular charge distribution. All these internal motions have in common that the center of mass of the molecule remains fixed. Again, the Hamilton function will be formulated in terms of proper momenta and position coordinates.

We summarize that the kinetic energy of a molecular system arises from contributions of translation, of rotation, of vibration, and of internal rotation of the single molecules. In general, we have three degrees of freedom for translation, three degrees of freedom for rotation, and $3n - 6$ degrees of freedom associated with internal motions of the molecules. Since the energy is the crucial property of any molecular model it is essential to formulate the various contributions properly. This will be done below by simplified mechanical models of molecular motion. These models express the molecular energy in terms of mechanical properties, such as the mass and the moments of inertia, and appropriate momenta and position coordinates.

2.3.3 A Model for Vibration: The Harmonic Oscillator

The linear harmonic oscillator serves to model the vibrational motion of a diatomic molecule. In this model of vibration, a linear restoring force is assumed to act between the nuclei, according to

$$F = -\kappa(r - r_e) = -\kappa x \qquad (2.88)$$

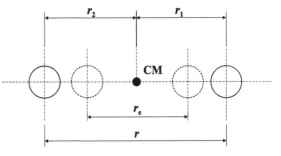

Fig. 2.4. Vibration of a diatomic molecule. CM = center of mass.

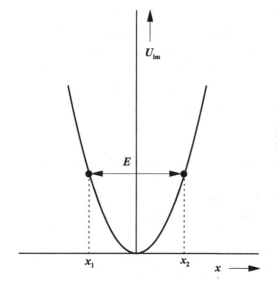

Fig. 2.5. Intramolecular potential energy of the harmonic oscillator.

with κ as the force constant, r_e as the distance between the two atoms at equilibrium, and $x = r - r_e$. Fig. 2.4 shows the configuration of the nuclei and the notation. The associated intramolecular potential energy is given by

$$U_{\mathrm{im}} = \int \kappa(r - r_e)\mathrm{d}r = \frac{1}{2}\kappa x^2. \tag{2.89}$$

This intramolecular potential energy of a harmonic oscillator is shown in Fig. 2.5. In a real molecule it will be given by the force field due to the electrostatic interactions of the electrons, cf. Section 2.5, and will not be a symmetric parabola. Still, as will be shown later, the harmonic oscillator is a good model for many applications. In Fig. 2.5 the intramolecular potential energy is seen to limit the motion in the sense that vibration with a total energy of E can only extend up to the location x_1, x_2 in Fig. 2.5, because there $E = U_{\mathrm{im}}$, which defines the turning point. The kinetic energy of the oscillator is given by

$$E^{\mathrm{kin}} = \frac{p_1^2}{2m_1} + \frac{p_2^2}{2m_2} = \frac{1}{2}m_1\dot{r}_1^2 + \frac{1}{2}m_2\dot{r}_2^2. \tag{2.90}$$

Transformation to the vibrational degree of freedom involves introducing an appropriate generalized coordinate q, which obviously is $r = r_1 + r_2$. Using the condition $m_1 r_1 = m_2 r_2$ for the definition of the center of mass we find that

$$r_1 = \frac{m_2}{m_1 + m_2}r$$

and

$$r_2 = \frac{m_1}{m_1 + m_2}r,$$

leading to

$$E^{\mathrm{kin}} = \frac{1}{2}m_r\dot{r}^2, \tag{2.91}$$

where m_r is the reduced mass, given by

$$m_r = \frac{m_1 m_2}{m_1 + m_2}. \tag{2.92}$$

The associated generalized momentum is

$$p_r = \frac{\partial E^{kin}}{\partial \dot{r}} = m_r \dot{r},$$

which gives the total Hamilton function as

$$H_v = \frac{1}{2m_r} p_r^2 + \frac{1}{2}\kappa(r - r_e)^2 = H(p_r, r). \tag{2.93}$$

The force constant κ can be related to the frequency of vibration. Starting from the equation of motion for atom number 1,

$$-\kappa(r - r_e) = m_1 \frac{d^2 r_1}{dt^2}, \tag{2.94}$$

and transforming r_1 to the generalized coordinate r gives

$$m_r \frac{d^2 r}{dt^2} + \kappa(r - r_e) = m_r \frac{d^2 x}{dt^2} + \kappa x = 0.$$

This is the differential equation for a harmonic oscillator written in terms of a mass point of mass m_r undergoing a harmonic vibration along the coordinate x. Solution gives x as a periodic function of time with the frequency

$$v_0 = \frac{1}{2\pi} \sqrt{\frac{\kappa}{m_r}},$$

which shows that the force constant κ is related to the frequency by

$$\kappa = 4\pi^2 v_0^2 m_r. \tag{2.95}$$

For polyatomic molecules it can be shown that the total intramolecular potential energy of the harmonic oscillator can be formulated as a sum of quadratic terms such as

$$U_{im} = U_1 + U_2 + \cdots = \sum_{l=1}^{\substack{3n-5 \\ 3n-6}} U_l \tag{2.96}$$

with

$$U_l = \frac{1}{2}\kappa_l(q_l - q_{l,e})^2, \tag{2.97}$$

where q_l are so-called normal coordinates. In normal coordinates each vibrational degree of freedom thus has an individual potential energy. The number of independent vibrations is $(3n - 6)$ for nonlinear and $(3n - 5)$ for linear molecules, with n as the number of atoms. They are called normal modes. With κ_l we denote the force constant of the lth normal mode, which can be expressed in terms of a vibration frequency. All nuclei oscillate with the same frequency and pass through the equilibrium position simultaneously.

2.3.4 A Model for Rotation: The Rigid Rotator

To model the rotational motion of a diatomic molecule we study the linear rigid rotator. Fig. 2.6 shows the configuration of the atoms and the notation. The kinetic energy of the rotator is, in cartesian coordinates,

$$E_{\mathrm{r}}^{\mathrm{kin}} = \frac{p_1^2}{2m_1} + \frac{p_2^2}{2m_2} = \frac{1}{2}m_1 \left(\dot{x}_1^2 + \dot{y}_1^2 + \dot{z}_1^2 \right) + \frac{1}{2}m_2 \left(\dot{x}_2^2 + \dot{y}_2^2 + \dot{z}_2^2 \right).$$

In this formulation the motion of the rotator is not yet constrained to be a rigid rotation with a fixed interatomic distance. To formulate the energy of the rotational degree of freedom with this restriction properly we introduce polar coordinates as adequate generalized coordinates via

$$x = r \sin \theta \cos \phi$$
$$y = r \sin \theta \sin \phi$$
$$z = r \cos \theta$$

and obtain for the components of the velocity, for a constant value of r,

$$\dot{x} = \frac{\mathrm{d}x}{\mathrm{d}t} = r\dot{\theta} \cos \theta \cos \phi - r\dot{\phi} \sin \theta \sin \phi$$
$$\dot{y} = \frac{\mathrm{d}y}{\mathrm{d}t} = r\dot{\theta} \cos \theta \sin \phi + r\dot{\phi} \sin \theta \cos \phi$$
$$\dot{z} = \frac{\mathrm{d}z}{\mathrm{d}t} = -r\dot{\theta} \sin \theta.$$

Here the polar angles θ and ϕ are the generalized coordinates that describe the rotation of the molecule as a whole around two perpendicular axes through the center of mass, as shown in Fig. 2.3a, and their time derivatives $\dot{\theta}$ and $\dot{\phi}$ are the generalized velocities. We thus get for the total kinetic energy of the linear rigid rotator

$$E_{\mathrm{r}}^{\mathrm{kin}} = \frac{1}{2} \left(m_1 r_1^2 + m_2 r_2^2 \right) (\dot{\theta}^2 + \dot{\phi}^2 \sin^2 \theta).$$

Introducing the moment of inertia by

$$I = m_1 r_1^2 + m_2 r_2^2,$$

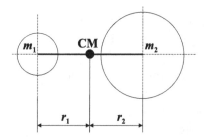

Fig. 2.6. The linear rigid rotator.

the kinetic energy can be written as

$$E_r^{kin} = \frac{I}{2}(\dot{\theta}^2 + \dot{\phi}^2 \sin^2 \theta). \tag{2.98}$$

An inspection of Eq. (2.98) shows that the linear rigid rotator has the same kinetic energy as a mass point of mass I constrained to rotate on a spherical surface at a distance $r = 1$ from the center of mass of the rotator. To find the Hamilton function for the rotational degree of freedom we formulate the kinetic energy in terms of generalized momenta by setting

$$p_\theta = \frac{\partial E^{kin}}{\partial \dot{\theta}} = I\dot{\theta} \tag{2.99}$$

and

$$p_\phi = \frac{\partial E^{kin}}{\partial \dot{\phi}} = I\dot{\phi} \sin^2 \theta, \tag{2.100}$$

leading to

$$E_r^{kin} = \frac{1}{2I} \left(p_\theta^2 + \frac{p_\phi^2}{\sin^2 \theta} \right) = H_r(p_\theta, p_\phi). \tag{2.101}$$

The quantities p_θ and p_ϕ are generalized momenta of dimensions momentum times length, which explains the correct dimensions in this formulation of kinetic energy.

Extension to three dimensions gives the kinetic energy of rotation of a non-linear rigid molecule as [5,6]

$$E_r^{kin} = \frac{1}{2I_A} \left[p_\theta \sin \chi - \frac{\cos \chi}{\sin \theta}(p_\phi - p_\chi \cos \theta) \right]^2$$

$$+ \frac{1}{2I_B} \left[p_\theta \cos \chi + \frac{\sin \chi}{\sin \theta}(p_\phi - p_\chi \cos \theta) \right]^2$$

$$+ \frac{1}{2I_C} p_\chi^2 = H_r(p_\theta, p_\phi, p_\chi). \tag{2.102}$$

Here, I_A, I_B, and I_C are the principal moments of inertia, which can be calculated from the geometry of the molecule by standard methods. The three angles θ, ϕ, and χ are the so-called Euler angles describing the rotational orientation of a rigid body in a space-fixed coordinate system; cf. Fig. 2.7. Here, ϕ is the rotation angle around the space-fixed z-axis (Z), and θ describes the rotation around the new y-axis $(Y'$, not shown in Fig. 2.7) arrived at after the ϕ rotation; cf. also Fig. 2.3a. Finally, χ denotes the rotation around the body-fixed z-axis (z). The p_θ, p_ϕ, and p_χ are the associated generalized momenta. In general, all principal moments of inertia I_A, I_B, and I_C of a molecule are different. We then refer to the molecule as an asymmetrical top (e.g., H_2O). If two principal moments of inertia are equal, the molecule is a symmetrical top (e.g., C_2H_6). If, finally, all principal moments are equal, the molecule is called a spherical top (e.g., CH_4). For a linear molecule with $p_\chi = 0$, $\chi = 0$, and $I_A = I_B = I$ the general expression (2.102) reduces to (2.101).

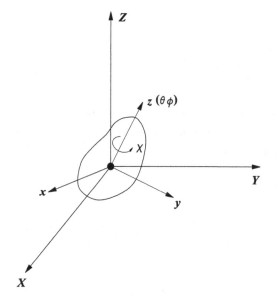

Fig. 2.7. Geometric interpretation of the Euler angles.

2.3.5 A Model for Internal Rotation

In polyatomic nonlinear molecules individual molecular groups may rotate relative to each other. We refer to this degree of freedom as internal rotation. In Fig. 2.8 it is shown for the particularly simple case of ethane that the two CH_3-groups may rotate against each other around the axis of the C-C bond. Clearly, the appropriate generalized coordinate for this degree of freedom is the relative torsional angle ϕ of the two CH_3-groups around the C-C axis. The intramolecular potential energy of this internal rotation depends on the symmetry of the rotating groups. For ethane U_{ir} will be a maximum when the two methyl groups are lined up or "eclipsed" and a minimum when they are "staggered." The minimum can be normalized to zero by including it in the zero of energy. $U_{ir,max}$ is called the potential barrier of internal rotation. The individual C-H bonds are separated by 120°. Thus the minima and maxima of U_{ir} follow each other on rotation by 60°. The potential energy of internal rotation shown in Fig. 2.8 for ethane may be written as

$$U_{ir} = \frac{1}{2} U_{ir,max}(1 - \cos 3\phi). \tag{2.103}$$

The associated kinetic energy of internal rotation can easily be analyzed by classical mechanics. In this simple case of ethane the axis of internal rotation is one of the principal axes of the molecule and the principal moments of inertia thus do not change during a rotation of the two methyl groups around this axis.

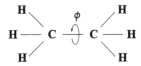

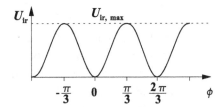

Fig. 2.8. Internal rotation in the ethane molecule.

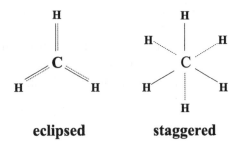

Thus the total kinetic energy of rotation of the molecule around this C-C bond is given as

$$E^{\mathrm{kin}} = \frac{1}{2}I_1(\dot{\alpha}_1)^2 + \frac{1}{2}I_2(\dot{\alpha}_2)^2, \qquad (2.104)$$

where I_1, I_2 are the moments of inertia of both rotational groups with respect to the rotational axis, $\dot{\alpha}_1$, $\dot{\alpha}_2$ are the corresponding angular velocities, and the two rotatory groups are considered to rotate independently. The complete rotational motion of the molecule around this axis splits into a rotation of the rigid rotator around the associated angle θ and an additional internal rotation when the angular velocities of the two rotational groups are different. We thus have

$$\begin{aligned} E^{\mathrm{kin}} &= E_{\mathrm{r}}^{\mathrm{kin}} + E_{\mathrm{ir}}^{\mathrm{kin}} \\ &= \frac{1}{2}(I_1 + I_2)(\dot{\theta})^2 + \frac{1}{2}I_{\mathrm{ir}}(\dot{\phi})^2, \end{aligned}$$

where agreement with (2.104) requires that

$$\begin{aligned} \dot{\theta} &= \frac{I_1\dot{\alpha}_1 + I_2\dot{\alpha}_2}{I_1 + I_2}, \\ \dot{\phi} &= \dot{\alpha}_1 - \dot{\alpha}_2, \end{aligned}$$

and

$$I_{ir} = \frac{I_1 I_2}{I_1 + I_2}. \tag{2.105}$$

We easily realize that for $\dot{\alpha}_1 = \dot{\alpha}_2 = \dot{\alpha}$ and thus $\dot{\theta} = \dot{\alpha}$ and $\dot{\phi} = 0$, we have with

$$E^{kin} = \frac{1}{2}(I_1 + I_2)\dot{\alpha}^2 = \frac{I}{2}\dot{\alpha}^2 = E_r^{kin}$$

the kinetic energy of an external rigid rotation around the axis of internal rotation. This contribution was already considered as the rotation of a rigid rotator; cf. (2.98) with ($\theta = 0$). Correspondingly we find for $I_1\dot{\alpha}_1 = -I_2\dot{\alpha}_2$, i.e., a vanishing total angular momentum ($\dot{\theta} = 0$) and thus a vanishing kinetic energy of external rotation around this axis,

$$E^{kin} = \frac{1}{2}I_{ir}\left(\dot{\phi}\right)^2 = E_{ir}^{kin}$$

as the kinetic energy of the internal rotational degree of freedom. We finally express the kinetic energy of internal rotation in terms of the appropriate generalized momentum coordinate, i.e., by $p_\phi = \partial E_{ir}^{kin}/\partial \dot{\phi} = I_{ir}\dot{\phi}$, and thus get

$$E_{ir}^{kin} = \frac{1}{2}I_{ir}(\dot{\phi})^2 = \frac{1}{2I_{ir}}p_\phi^2. \tag{2.106}$$

The Hamilton function of internal rotation for the ethane molecule is thus given by

$$H_{ir}(p_\phi, \phi) = \frac{1}{2I_{ir}}p_\phi^2 + \frac{1}{2}U_{ir,max}(1 - \cos 3\phi), \tag{2.107}$$

with the generalized coordinate ϕ and the conjugate generalized momentum p_ϕ.

In more general unsymmetrical molecules (2.106) remains valid but a more complicated formulation of the potential energy may be required. Further, in most of the more complicated organic molecules the axis of internal rotation is not a principal axis. The simple formula (2.105) derived above is then a more or less crude approximation and more complicated expressions for I_{ir} to be calculated from the principal moments of inertia and the principal axes have to replace it. If a symmetric rotating group 1 is connected to a rigid unsymmetric molecular group, we find [5]

$$I_{ir} = I_1 \left[1 - I_1 \left(\frac{\cos^2 \alpha}{I_A} + \frac{\cos^2 \beta}{I_B} + \frac{\cos^2 \gamma}{I_C}\right)\right], \tag{2.108}$$

where I_1 is the moment of inertia of the symmetric rotating group with respect to the axis of internal rotation, I_A, I_B, and I_C are the principal moments of inertia of the molecule, and α, β, and γ are the angles between the corresponding principal axes and the axis of internal rotation. The more general expression (2.108) reduces to (2.105) when the internal rotation takes place around a principal axis C because then $\gamma = 0$, α, $\beta = 90°$, and $I_C = I_1 + I_2$. The requirement of a symmetric rotational group ensures the independence of external and internal rotation because then the moments of inertia will be unaffected by the internal rotation. For an unsymmetric rotational group with a small moment of inertia the error will be small. Heavy unsymmetrical rotational groups have to be treated by more complicated methods.

If a molecule possesses several rotating groups $1, 2, \ldots, n$ one can set as an approximation

$$H_{ir} = H_{ir}(1) + H_{ir}(2) + \cdots + H_{ir}(n). \tag{2.109}$$

This approximation becomes exact when the moments of inertia of the rotating groups are sufficiently small so that the internal rotations are independent of each other. This is frequently not accurate enough and we shall address this problem in Section 3.6 on the basis of quantum-chemical calculations.

Whenever a mode of internal rotation is taken into account it replaces a normal mode of vibration. Generally, an internal rotation is expected to occur when a molecular group is connected to some molecular remainder by a single bond.

EXERCISE 2.3

Calculate the principal moments of inertia and the principal axes as well as the moment of inertia adequate for internal rotation for methanol.

Solution

The molecular geometry is shown in Fig. E 2.3.1 as taken from [1]. Note that the projection of the 109° tetrahedral angle between the C-H bonds into the $z - y$ plane gives 120° and that the x-axis is the axis of internal rotation. Only the light hydrogen atoms contribute to the moments of inertia of the rotating CH_3-group. Because the unsymmetric nature of the OH-group is only moderate due to the light hydrogen atom, a first estimate of the moment of inertia for internal rotation may be obtained from (2.105), i.e.,

$$I_{ir} = \frac{I_1 I_2}{I_1 + I_2},$$

with

$$I_1 = 3m_H[1.1\sin(180° - 109°)]^2 = 5.432 \cdot 10^{-40} \, g \; cm^2$$

and

$$I_2 = m_H[0.96\sin(180° - 109°)]^2 = 1.379 \cdot 10^{-40} \, g \; cm^2.$$

We thus obtain

$$I_{ir} = \frac{I_1 I_2}{I_1 + I_2} = 1.100 \cdot 10^{-40} g \; cm^2.$$

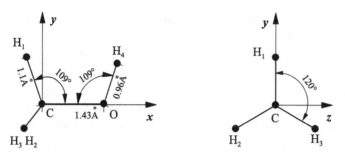

Fig. E 2.3.1. Molecular geometry of the methanol molecule.

More elaborate calculations are required for the principal moments and principal axes of inertia, as needed for the evaluation of (2.108). We start by defining an arbitrary coordinate system and determining the associated coordinates of the atoms. We then determine the coordinates of the center of mass and calculate the coordinates of the atoms in the center of mass coordinate system. With these coordinates we determine the moments of inertia and the products of inertia. Finally we find the principal moments of inertia and the principal axes:

1. Determination of the atom coordinates in an arbitrary coordinate system with the C-atom as the origin, cf. Fig. E 2.3.1:

$$x_C = 0; \quad y_C = 0; \quad z_C = 0$$
$$x_O = 1.43 \text{ Å}; \quad y_O = 0; \quad z_O = 0$$
$$x_{H_1} = -1.1 \cos 71° = -0.358 \text{ Å} = x_{H_2} = x_{H_3}$$
$$x_{H_4} = 1.43 + 0.96 \cos 71° = 1.743 \text{ Å}$$
$$y_{H_1} = 1.1 \sin 71° = 1.04 \text{ Å}$$
$$y_{H_2} = -[\sin(120° - 90°) \cdot 1.1] \cdot \sin 71° = -0.520 \text{ Å}$$
$$y_{H_3} = y_{H_2}$$
$$y_{H_4} = 0.96 \sin 71° = 0.908 \text{ Å}$$
$$z_{H_1} = 0$$
$$z_{H_2} = -1.1 \cos(109° - 90°) \cos(120° - 90°) = -0.9007 \text{ Å}$$
$$z_{H_3} = 0.9007 \text{ Å}$$
$$z_{H_4} = 0.$$

2. Determination of the center of mass and the atom coordinates in the center of mass coordinate system:

In the coordinate system defined in Fig. E 2.3.1 the coordinates of the center of mass are

$$x_{CM} = \frac{\sum m_i x_i}{\sum m_i} = \frac{12.011 \cdot 0 + 15.999 \cdot 1.43 + \{1.008(-0.358)\}3 + 1.008 \cdot 1.743}{12.011 + 15.999 + 4 \cdot 1.008}$$
$$= 0.7351 \text{ Å}$$
$$y_{CM} = \frac{\sum m_i y_i}{\sum m_i} = 0.0286 \text{ Å}$$
$$z_{CM} = \frac{\sum m_i z_i}{\sum m_i} = 0.$$

The coordinates of the atoms in the center of mass coordinate system are found by simple translations of the original coordinates and are summarized in the following table:

Atom	x	y	z
C	−0.7351	−0.0286	0
O	0.6949	−0.0286	0
H_1	−1.0932	1.0115	0
H_2	−1.0932	−0.5486	−0.9007
H_3	−1.0932	−0.5486	0.9007
H_4	1.0075	−0.8791	0

3. Determination of principal moments of inertia:

The moments of inertia and the products of inertia in the center of mass coordinate system are given by

$$I_{xx} = \sum_i m_i \left(y_i^2 + z_i^2\right) = m_C[(-0.0286)^2 + 0^2]$$
$$+ m_O[(-0.0286)^2 + 0^2]$$
$$+ m_{H_1}[(1.0115)^2 + 0^2] + m_{H_2}[(-0.5486)^2 + (-0.9007)^2]$$
$$+ m_{H_3}[(-0.5486)^2 + (0.9007)^2] + m_{H_4}[(0.8791)^2 + 0^2]$$
$$= 4.075 \text{ (g/mol) Å}^2$$
$$I_{yy} = \sum_i m_i \left(x_i^2 + z_i^2\right) = 20.489 \text{ (g/mol) Å}^2$$
$$I_{zz} = \sum_i m_i \left(x_i^2 + y_i^2\right) = 21.295 \text{ (g/mol) Å}^2$$
$$I_{xy} = \sum_i m_i x_i y_i = 0.922 \text{ (g/mol) Å}^2 = I_{yx}$$
$$I_{yz} = I_{xz} = 0.$$

If we are only interested in the product of the three principal moments of inertia, as, e.g., for the calculation of the molecular partition function of external rotation, cf. Section 3.6.3, we just have to evaluate the determinant

$$\det \mathbf{I} = \begin{vmatrix} I_{xx} & -I_{xy} & -I_{xz} \\ -I_{xy} & I_{yy} & -I_{yz} \\ -I_{xz} & -I_{yz} & I_{zz} \end{vmatrix} = I_A I_B I_C.$$

In the case of the methanol molecule this leads to

$$I_A I_B I_C = 1760 \text{ (g}^3/\text{mol}^3) \text{ Å}^6 = 8058 \cdot 10^{-120} \text{ g}^3\text{cm}^6,$$

in reasonable agreement with the data in [1].

The individual values of the principal moments and the principal axes of inertia are found from the eigenvalues of the matrix $\mathbf{I}$ of the moments and products of inertia. These eigenvalues are the principal moments. They thus follow from

$$\det(\mathbf{I} - \lambda \mathbf{E}) = 0,$$

or

$$\begin{vmatrix} I_{xx} - \lambda & -I_{xy} & -I_{xz} \\ -I_{xy} & I_{yy} - \lambda & -I_{yz} \\ -I_{xz} & -I_{yz} & I_{zz} - \lambda \end{vmatrix} = 0.$$

This leads to

$$\begin{vmatrix} 4.075 - \lambda & -0.922 & 0 \\ -0.922 & 20.489 - \lambda & 0 \\ 0 & 0 & 21.293 - \lambda \end{vmatrix} = 0$$

and, finally, to the characteristic equation

$$(21.293 - \lambda)[(4.075 - \lambda)(20.489 - \lambda) - (-0.922)^2] = 0.$$

The three principal moments of inertia thus are

$$\lambda_A = 4.024 \text{ (g/mol) Å}^2 = 6.682 \cdot 10^{-40} \text{ g cm}^2 = I_A$$

$$\lambda_B = 20.540 \text{ (g/mol) Å}^2 = 34.109 \cdot 10^{-40} \text{ g cm}^2 = I_B$$

$$\lambda_C = 21.293 \text{ (g/mol) Å}^2 = 35.358 \cdot 10^{-40} \text{ g cm}^2 = I_C.$$

Again there is reasonable agreement with the data of [1]. We further note that

$$\det I = \begin{vmatrix} I_A & 0 & 0 \\ 0 & I_B & 0 \\ 0 & 0 & I_C \end{vmatrix} = 8058 \cdot 10^{-120} \text{g}^3\text{cm}^6,$$

which is in agreement with the earlier result.

4. Determination of the principal axes:

The principal axes of inertia are a particular coordinate system in which

$$I_{xy} = I_{yz} = I_{xz} = 0$$

and

$$I_{xx} = I_A, \quad I_{yy} = I_B, \quad \text{and} \quad I_{zz} = I_C,$$

where I_A, I_B, and I_C are the principal moments of inertia. Clearly the z-axis is one of the principal axes, since it is perpendicular to a symmetry plane of the molecule. This is confirmed by the fact that $I_C = I_{zz}$. If we denote the three principal axes as r_A, r_B, and r_C, we thus have

$$r_C = \{0, 0, z\}.$$

The principal axis r_A is calculated from $I r_A - \lambda_A r_A = 0$. This gives

$$\begin{array}{ccc} (4.075 - 4.024)\, x_A & -0.922\, y_A + & 0\, z_A = 0 \\ -0.922\, x_A + & (20.489 - 4.024)\, y_A + & 0\, z_A = 0 \\ 0\, x_A + & 0\, y_A + & (21.293 - 4.024)\, z_A = 0. \end{array}$$

We thus find the absolute value of the angle between the principal axis r_A and the axis of internal rotation from

$$\tan \alpha = \frac{y_A}{x_A} = \frac{0.0516}{0.922} = \frac{0.922}{16.465},$$

which results in

$$\alpha = 3.2°.$$

Analogously, we find

$$\beta = 93.2°$$

as the angle between the principal axis r_B and the axis of internal rotation. Fig. E 2.3.2 shows the three principal axes. The moment of inertia of internal rotation resulting from these values is

$$I_{ir} = 5.432 \left[1 - 5.432 \left(\frac{\cos^2 3.2°}{6.682} + \frac{\cos^2 93.2°}{34.109} + \frac{\cos^2 90°}{35.358} \right) \right] = 1.027 \cdot 10^{-40} \text{g cm}^2.$$

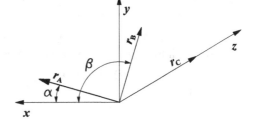

Fig. E 2.3.2. Principal axes of methanol.

The difference from the simple approximation of two symmetrical rotating groups is not more than 7% for the methanol molecule.

[1] Landolt–Börnstein. *Zahlenwerte und Funktionen aus Physik, Chemie, Geographie und Technik*, volume 2,4. Springer, Berlin, 1961.

2.4 Potential Energy of a Molecular System: Classical Electrostatics

We noted in the preceding section that the energy of a single molecule can be expressed in terms of its mechanical degrees of freedom. For each degree of freedom there is a classical mechanical Hamilton function of generalized position coordinates and momenta. It contains the molecular geometry, represented by various moments of inertia, the atomic masses, and properties related to the *intra*molecular force field, such as the vibration frequencies and the barriers to internal rotation. The energy of the single molecules constitutes one part of the total energy of a molecular system, referred to as the kinetic part. In addition, there is a potential part, originating from the *inter*molecular potential energy in the Hamilton function of a molecular system. In the general case, all degrees of freedom, i.e., the translation, rotation, and internal motions of the molecules, will be influenced by their neighbors. This reaction to the molecular environment can be expressed in terms of the forces acting between the molecules, which in turn are derived from the intermolecular potential energy. So, although the simple equations for the kinetic energies associated with the various degrees of freedom continue to be valid, they have to be supplemented by an expression for the intermolecular potential energy in terms of the molecular configuration of the system.

In order to acquire information about the potential energy contribution we must take notice of the fact that the molecules are not simply geometrical objects. They are rather charge clouds, due to the positive charges on the nuclei and the negative charges on the electrons. Although electrically neutral as a whole, the charge clouds of different molecules interact and lead to an intermolecular potential energy of the molecular system. Further, in many important applications, e.g., in electrolyte solutions and plasmas, we have to deal not only with neutral molecules, but also with positively or negatively charged ions, with severe consequences for the total energy of the molecular system.

The interactions between charges are described by classical electrostatics. It is therefore important to summarize the concepts and basic equations of this discipline [7]. Necessary additions based on quantum mechanics will be treated in Section 2.5.

2.4.1 Basic Equations of Classical Electrostatics

The basic law of classical electrostatics is Coulomb's law, according to which the scalar force F_{ab} in N between two charges q_a and q_b, separated by a distance r_{ab} in m in a vacuum, is obtained from

$$F_{ab}(r_{ab}) = \frac{q_a q_b}{r_{ab}^2}, \tag{2.110}$$

where the dimension of a charge is $kg^{1/2}\ m^{3/2}\ s^{-1}$ in electrostatic units, or equivalently

$$F_{ab}(r_{ab}) = \frac{1}{4\pi\varepsilon_0}\frac{q_a q_b}{r_{ab}^2}, \tag{2.111}$$

with a new dimension C (coulomb) for the charge and ε_0 as the so-called vacuum permittivity, a universal constant specified in App. 1. This defines the coulomb as that charge leading to a repulsive force of $8.9876 \cdot 10^9$ N on an equal charge at a distance of 1 m. The force is thus obtained in newtons when the charges are expressed in coulombs and the distance in meters. Because charges can be positive or negative, Coulomb's law leads to positive (repulsive) forces if the two charges have equal signs and to negative (attractive) forces for charges of unequal signs.

From the force we can derive the electrostatic potential energy $\phi_{ab}(r_{ab})$ associated with a pair of charges q_a and q_b at a distance r_{ab}. It is identical to the work involved in bringing up a charge q_a from infinity to a distance r_{ab} away from charge q_b and is given in a vacuum by

$$\phi_{ab}(r_{ab}) = -\int_{\infty}^{r_{ab}} F_{ab}(r_{ab})dr_{ab} = \frac{1}{4\pi\varepsilon_0}\frac{q_a q_b}{r_{ab}}. \tag{2.112}$$

This is consistent with our earlier statement that the force is represented by the negative gradient of the potential energy in conservative systems; cf. Section 2.3. It further turns out to be useful to express the electrostatic potential energy $\phi(r)$ in terms of the electrostatic potential $\psi(r)$. For a system of two charges, q_a and q_b, separated by r_{ab} in a vacuum, we have

$$\phi_{ab}(r_{ab}) = q_a\psi_b(r_{ab}), \tag{2.113}$$

where

$$\psi_b(r_{ab}) = \frac{1}{4\pi\varepsilon_0}\frac{q_b}{r_{ab}} \tag{2.114}$$

is the electrostatic potential due to a charge q_b at a distance r_{ab} away from it, i.e., at the location of charge q_a, again with a vacuum between the two charges. Finally, just as the energy of a system of two charges q_a and q_b can be calculated from the electrostatic potential ψ_b at the location of charge q_a due to charge q_b by (2.113), so the force on q_a can be written as

$$F_{ab} = q_a E_b(r_{ab}), \tag{2.115}$$

where E_b is the magnitude of the electric field strength arising from q_b. For the electric field strength due to charge q_b at a distance r_{ab} away from it, this gives

$$E_b(r_{ab}) = \frac{1}{4\pi\varepsilon_0}\frac{q_b}{r_{ab}^2}. \tag{2.116}$$

In a typical molecular system the charge distribution may be rather complex. Instead of a single point charge we rather have many charges, which frequently can be represented by a continuous charge density distribution $\varrho(r)$ in the unit C/m^3. The electric field strength, which, like the force, is actually a vector quantity, is then the negative gradient of the electrostatic potential; i.e.,

$$E = -\nabla\psi, \tag{2.117}$$

and the electrostatic potential arising from a given charge density distribution $\varrho(r)$ at some location r is the solution to Poisson's equation,

$$\nabla^2\psi = -\frac{\varrho(r)}{\varepsilon_0}, \tag{2.118}$$

where $\nabla^2 = \partial^2/\partial x^2 + \partial^2/\partial y^2 + \partial^2/\partial z^2$ in cartesian coordinates and analogous expressions hold in other coordinate systems.

EXERCISE 2.4

Calculate the electrostatic potential $\psi(r = R)$ caused by a point charge q located at $r = 0$ from the Poisson equation.

Solution

Using the spherical symmetry of the problem, the Poisson equation can be written as

$$\frac{1}{r^2}\frac{d}{dr}\left(r^2\frac{d\psi}{dr}\right) = -\frac{\varrho(r)}{\varepsilon_0}.$$

Fig. E 2.4.1 shows the configuration of the problem. The charge density associated with a point charge q at $r = 0$ is given by

$$\varrho(r) = \frac{q}{2\pi r^2}\delta(r),$$

where $\delta(r)$ is the Dirac delta function. We can verify this expression by making use of the properties of the delta function, i.e., $\delta(r \neq 0) = 0$ and $\int f(r)\delta(r)dr = f(0)$. Whereas the

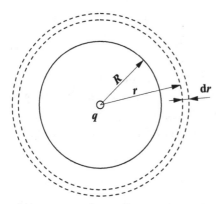

Fig. E 2.4.1. Charge density in a spherical shell of thickness dr at a distance r.

first property introduces the condition of a point charge at $r = 0$, the integral property shows that the volume integral over the charge density is

$$\int_0^\infty \varrho(r)4\pi r^2 dr = \int_0^\infty \frac{q\delta(r)}{2\pi r^2}4\pi r^2 dr = q\int_{-\infty}^{+\infty}\delta(r)dr = q,$$

as it must be. So the Poisson equation reads

$$\frac{1}{r^2}\frac{d}{dr}\left(r^2\frac{d\psi}{dr}\right) = -\frac{q\delta(r)}{2\pi\varepsilon_0 r^2}.$$

Integration between $r = 0$ and $r = r^*$ gives

$$r^{*2}\frac{d\psi}{dr}\bigg|_{r=r^*} = -\frac{q}{4\pi\varepsilon_0},$$

where we use $\int_0^{r^*}\delta(r)dr = \frac{1}{2}\int_{-r^*}^{r^*}\delta(r)dr = \frac{1}{2}$. Integration from $r^* = \infty$ to $r^* = R$ with $\psi(\infty) = 0$ gives

$$\psi(R) = \frac{1}{4\pi\varepsilon_0}\frac{q}{R},$$

in agreement with (2.114).

In a dense fluid, e.g., a liquid, the interacting particles are closely spaced. In principle, even then there is vacuum between them. However, considering the interaction of two particles, e.g., solute molecules, separated by a liquid of a different component, it is frequently useful to assume that the interacting particles are separated by a dielectric continuum, not by molecules in a vacuum. Thus, the electrostatic equations have to be modified. When two charges q_a and q_b at a distance r_{ab} are separated by a dielectric continuous medium other than a vacuum, the force is reduced to

$$F_{ab}(r_{ab}) = \frac{1}{4\pi\varepsilon}\frac{q_a q_b}{r_{ab}^2}, \tag{2.119}$$

where ε is the permittivity of the medium, normally written as $\varepsilon = \varepsilon_r\varepsilon_0$. The dimensionless quantity ε_r ($\varepsilon_r > 1$) is the relative permittivity or dielectric constant of the medium. The electrostatic effect of the dielectric continuous medium is

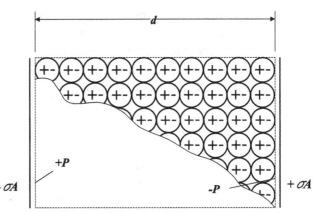

Fig. 2.9. Polarization of a di-electric.

thus that of a shielding of the charges q_a and q_b from each other. Clearly the same effect applies to the electrostatic potential energy, the potential, and the field strength. The continuum model thus replaces the effect of the complicated electrostatic interactions involving the molecules of the continuum by a macroscopic material property, the dielectric constant.

The shielding observed for two charges separated by a medium other than vacuum is due to an effect known as polarization. In order to visualize this effect and express it in terms of classical electrostatics, we consider in Fig. 2.9 a parallel-plate capacitor with the region between the plates initially evacuated, and let the charge per unit area on one plate be $+\sigma$ and $-\sigma$ on the other. The area of the plates is denoted by A. The associated electrical field strength is E_0. When the space between the plates is filled with a dielectric continuum, polarization takes place, and the electric field strength is reduced to $E = E_0(\varepsilon_0/\varepsilon_r)$. We can model the effect as arising from the presence of an opposing surface charge $\pm PA$ on either surface of the medium; cf. Fig. 2.9. This induced surface charge density P represents the polarization of the medium, and the electric field strength can thus also be expressed as $E = E_0(\sigma - P)/\sigma$. We thus see that the shielding of molecular charges by a dielectric, i.e., the volumetric effect of polarization, can be modeled in terms of an effect resulting from surface charges at the border of the continuum. We shall return to this interpretation below.

We now associate the macroscopic effect of polarization with molecular properties of the dielectric continuum. The effect of the induced surface charge density P can be interpreted as a macroscopic dipole moment density of the dielectric continuum, i.e., the macroscopic dipole moment divided by the volume of the continuum. Here the macroscopic dipole moment is defined as the sum of the surface charges of the sample multiplied by their distances from the center, i.e., $[PAd/2 + (-PA)(-d/2)]/Ad = P$. On a molecular scale the macroscopic dipole moment density of the dielectric continuum results from the dipole moments of the single molecules and can be expressed by the mean dipole moment of a molecule $< \mu >$ multiplied by the number density N/V of the molecules; i.e., $P = < \mu > N/V$. The molecular dipole moment is due to the fact that the electrons of the molecules constituting the dielectric continuum are displaced from their equilibrium positions under the influence of the electric field. The centers of the positive and negative charges thus no longer coincide and the associated molecular dipole moment is given (cf. Fig. 2.9) by

$$\mu = q_+ r_+ + q_- r_- \qquad (2.120)$$

for two charges q_+ and q_- separated from the center by r_+ and r_-, respectively. Because this dipole moment has been induced by the electric field, it is referred to as the induced dipole moment. All molecular induced dipoles being aligned, as in Fig. 2.9, produce an overall macroscopic dipole moment that reduces the field strength. The induced dipole moment in a molecule is proportional to the electric field and thus is given by

$$\mu = \alpha E, \qquad (2.121)$$

with α, i.e., the proportionality factor, as the polarizability of the molecule. Here, it was assumed that the dipole and the electrical field strength have the same direction and can thus be considered as scalars. Generally, both are vectors and the polarizability is thus represented by a tensor.

The molecular polarizability is a measure for the distortability of the electrical charges in the molecules of the continuum. So it should increase if the electrons are under less tight control of the nuclei. In line with this view, the polarizability can be expected to increase with the size of the molecule and the number of electrons it contains. Quantitatively, it can be obtained either from experiments or from quantum-chemical computer codes; cf. Section 2.5. From the equations above it is clear that it is related to the dielectric constant. The shielding effect of a dielectric continuum can be quite significant, depending on the value of the dielectric constant. For water we have $\varepsilon_r \approx 80$, which is large as compared to nonpolar solvents with typically $\varepsilon_r \approx 2$. The shielding effect becomes complete for an ideal electrical conductor, i.e., a substance containing freely moving electrons, with $\varepsilon_r = \infty$.

2.4.2 The Multipole Expansion

We consider a molecular system made up of many molecules. It may be a gas or a liquid. There is no other medium between the molecules, so the electrostatic equations for charges in a vacuum apply. There are many charges associated with the individual molecules. The electrostatic potential energy resulting from all of them is obtained by summing up the electrostatic energies between all pairs of charges in an otherwise empty space, i.e., in a vacuum. As noted before, two basically different kinds of contributions arise. In Fig. 2.10a we

(a)

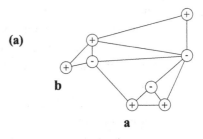

b

a

Fig. 2.10. (a) System of interacting charges (interactions only partly shown as tielines); (b) system of interacting molecules. --- intramolecular interactions; – intermolecular interactions.

(b)

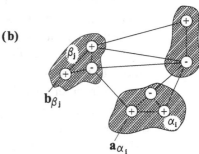

show a system of unconstrained interacting charges. In a molecular system the individual charges may be associated with particular molecules; cf. Fig. 2.10b. Those interactions that are effective between the charges within a molecule are responsible for the *intra*molecular force field, which determines the vibration frequencies and the barriers of internal rotation. On the other hand, there are interactions between the charges of different molecules. These interactions lead to the *inter*molecular potential energy and will be studied here.

We consider the electrostatic pair potential energy resulting from the charges of two molecules α_i and β_j in a vacuum, where α, β denote the component and i, j the particular molecule. It is obtained from Coulomb's law by summing up all individual charge pair contributions to give, cf. (2.112),

$$\phi_{\alpha_i \beta_j} = \sum_a^{K_\alpha} \sum_b^{K_\beta} \frac{1}{4\pi \varepsilon_0} \frac{q_{a_{\alpha_i}} q_{b_{\beta_j}}}{r_{a_{\alpha_i} b_{\beta_j}}}. \tag{2.122}$$

Here $q_{a_{\alpha_i}}$ is the charge a on molecule i of component α, with a similar notation for $q_{b_{\beta_j}}$, and $r_{a_{\alpha_i} b_{\beta_j}}$ is the distance between these two charges. We note that $q_{a_{\alpha_i}}$ and $q_{b_{\beta_j}}$ belong to different molecules α_i and β_j, so that $\phi_{\alpha_i \beta_j}$ is the potential energy describing the interaction of these two molecules due to electrostatic forces. To find the total electrostatic interaction energy of the system we would have to sum over all molecular pairs α_i, β_j. The reference state for calculating the potential energy on this basis is obviously the ideal gas state, with $r_{a_{\alpha_i} b_{\beta_j}} \to \infty$ and $\phi_{\alpha_i \beta_j} = 0$.

The electrostatic charge distribution in a molecule is very complicated. So it is generally not practical to work out the summation in (2.122). However, the molecular charge distribution can be represented by particular molecular properties, the multipole moments, which can either be measured or obtained from quantum-chemical calculations. In order to transform (2.122) into a form containing the multipole moments we perform a multipole expansion. We shall make use of a particular version of this expansion, the two-center expansion in terms of so-called spherical harmonics. We assume that the charge distribution of an isolated molecule is available in terms of fixed locations within a molecule fixed coordinate system, originating in the molecular center. This is adequate for small rigid molecules, such as the ones shown in Fig. 1.13. We further assume that the charge distributions of the two molecules do not overlap; i.e., we restrict consideration to sufficiently large intermolecular distances. In the two-center expansion the distance $r_{a_{\alpha_i} b_{\beta_j}}$ between two charges a and b in different molecules α_i and β_j is replaced by the coordinates of the charges $r_a, r'_b, \theta_a, \theta'_b, \phi_a, \phi'_b$ in the molecule-fixed coordinate systems of the individual molecules, the center-to-center distance $r_{\alpha_i \beta_j}$, and the orientational angles ω_{α_i} and ω_{β_j} of the two molecule-fixed coordinate systems in a space-fixed coordinate system. Here, ω_{α_i} and ω_{β_i} are the Euler angles as defined in Fig. 2.7. In particular, $\phi_{\alpha_i}, \theta_{\alpha_i}$ are the polar angles of the molecule-fixed z-axis, and χ_{α_i} is the twist angle about it. If we place both molecular centers on the z-axis of the space-fixed coordinate system and let the z-axes of both molecule-fixed coordinate systems coincide with it, cf. Fig. 2.11, this gives [8–10]

$$
\frac{1}{r_{a_{\alpha_i} b_{\beta_j}}} = \sum_{l_1=0}^{\infty} \sum_{l_2=0}^{\infty} \sum_{\substack{m_1 \\ m_2}} \sum_{\substack{n_1 \\ n_2}} \frac{r_a^{l_1} r_b'^{l_2}}{r_{\alpha_i \beta_j}^{l_1+l_2+1}} Y_{l_1}^{n_1}(\theta_a, \phi_a) Y_{l_2}^{n_2}(\theta'_b, \phi'_b)
$$

$$
\cdot \frac{4\pi(-1)^{l_2}}{2l_1 + 2l_2 + 1} \sqrt{\frac{4\pi(2l_1 + 2l_2 + 1)!}{((2l_1 + 1)!((2l_2 + 1)!}}
$$

$$
\cdot \sqrt{\frac{2l_1 + 2l_2 + 1}{4\pi}} C(l_1 l_2 l_1 + l_2; m_1 m_2 0)
$$

$$
\cdot D_{m_1 n_1}^{l_1}(\omega_{\alpha_i})^* D_{m_2 n_2}^{l_2}(\omega_{\beta_j})^* . \tag{2.123}
$$

Here, we have $-l \leq m, n \leq l$ and $m_1 = -m_2$. The $Y_l^n(\vartheta, \phi)$ are spherical harmonics, the $D_{mn}^l(\omega)^*$ rotation matrices, and the C Clebsch–Gordan coefficients, which are all well-defined functions; cf. App. 4. The asterisk, as usual, denotes the complex conjugate. As a measure for the charge distribution we now introduce with

$$
{}^{\gamma}Q_l^n = \sum_a^{K_\gamma} r_a^l q_a Y_l^n(\theta_a \phi_a) \tag{2.124}
$$

the multipole moments of a molecule of species γ, for a discrete charge distribution in the molecule-fixed coordinate system originating at the molecular

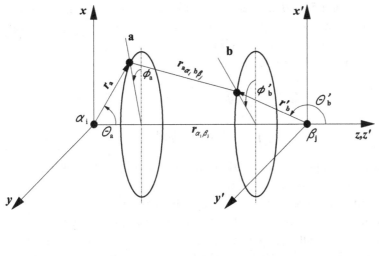

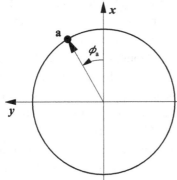

Fig. 2.11. Space-fixed coordinate system of the two-center expansion.

center. Here r_a is the scalar distance of charge a from the center. Combining (2.124) and (2.123) with (2.122), we find

$$\phi_{\alpha_i \beta_j}^{\text{mult}} = \sum_{l_1=0}^{\infty} \sum_{l_2=0}^{\infty} \sum_{\substack{m_1 \\ m_2}} \sum_{\substack{n_1 \\ n_2}} E_{\alpha_i \beta_j}^{\text{mult}} \left(l_1 l_2 l_1 + l_2; n_1 n_2; r_{\alpha_i \beta_j} \right) \sqrt{\frac{2l_1 + 2l_2 + 1}{4\pi}}$$

$$\cdot C \left(l_1 l_2 l_1 + l_2; m_1 m_2 0 \right) D_{m_1 n_1}^{l_1} \left(\omega_{\alpha_i} \right)^* D_{m_2 n_2}^{l_2} \left(\omega_{\beta_j} \right)^*, \tag{2.125}$$

with

$$E_{\alpha_i \beta_j}^{\text{mult}} \left(l_1 l_2 l_1 + l_2; n_1 n_2; r_{\alpha_i \beta_j} \right) = \frac{4\pi (-1)^{l_2}}{r_{\alpha_i \beta_j}^{l_1 + l_2 + 1}} \sqrt{\frac{4\pi (2l_1 + 2l_2 + 1)!}{(2l_1 + 1)!(2l_2 + 1)!}} \frac{{}^{\alpha} Q_{l_1}^{n_1} {}^{\beta} Q_{l_2}^{n_2}}{2l_1 + 2l_2 + 1} \tag{2.126}$$

for the expansion coefficients of the multipole forces. This is the formal spherical harmonic multipole expansion of the electrostatic potential energy between molecule i of species α and molecule j of species β. We note that this formulation is an approximation valid for large intermolecular distances $r_{\alpha_i \beta_j}$. At short distances the series will not converge. Practically, this restriction is not too serious because at short distances the repulsive overlap forces will dominate (cf. Chapter 5) and at intermediate distances a simple superposition of the short-range and long-range contributions has been shown to give good results in thermodynamic calculations [8].

The multipole moments are defined as

$$
{}^\gamma Q_l^n = \sum_a^{K_\gamma} r_a^l q_a Y_l^n(\theta_a, \phi_a)
$$
$$
= \int r^l \varrho_\gamma(r) Y_l^n(\theta, \phi) dr, \tag{2.127}
$$

where in the second equation summation over the individual charges has been replaced by integration over the continuous charge density $\varrho(r)$, with dr a volume element of the position space. They represent the permanent charge density distribution in an isolated molecule. In a neutral spherically symmetric molecule such as argon all multipole moments are zero and there is no electrostatic potential energy in such fluids. Multipoles thus reflect the asymmetry of a charge distribution. They are ordered by the index l. Permanent multipoles with $l = 0, 1, 2, 3, 4, \ldots$, are denoted as monopole, dipole, quadrupole, octopole, hexadecapole, ..., respectively. A monopole is the charge itself. Experimental data for dipole moments, i.e., the first-order multipole moments, are available for many molecules. For quadrupole moments, which are the second-order multipole moments, experimental data are relatively scarce, and even more so for the higher order moments [8]. Charge distributions of single molecules can be obtained from "ab initio" quantum mechanical calculations, and from such charge distributions the multipole moments can be computed; cf. Section 2.5.

Generally, multipole moments are tensors with $(2l + 1)$ elements because of $-l \le n \le l$. The number of tensor elements that differ from zero and are independent can be minimized by making use of the symmetry of a molecule, i.e., by defining a proper molecule-fixed coordinate system in cartesian coordinates. For this purpose it is useful to transform the multipole moments defined in spherical coordinates by (2.127) into cartesian coordinates of the molecule-fixed coordinate system. We start from the general relations between cartesian and spherical coordinates,

$$
x = r \sin \theta \cos \phi
$$
$$
y = r \sin \theta \sin \phi
$$
$$
z = r \cos \theta,
$$

where

$$
r^2 = x^2 + y^2 + z^2.
$$

With

$$
x \pm iy = r \sin \theta (\cos \phi \pm i \sin \phi) = r \sin \theta e^{\pm i\phi},
$$

we can express the term $r^n (\sin \theta)^n e^{in\phi}$ arising in the spherical harmonics of (2.127) by

$$
r^n (\sin \theta)^n e^{\pm in\phi} = (x \pm iy)^n.
$$

With this and the definition of the spherical harmonics (cf. App. 4) we find the following representations of the multipole moments in cartesian coordinates up to the quadrupole tensor:

$$Q_0^0 = \int \rho \frac{1}{\sqrt{4\pi}} d\mathbf{r} \tag{2.128}$$

$$Q_1^0 = \int \rho r Y_1^0 d\mathbf{r} = \int \rho \frac{1}{2} \sqrt{\frac{3}{\pi}} z d\mathbf{r} \tag{2.129}$$

$$Q_1^{\pm 1} = \int \rho r Y_1^{\pm 1} d\mathbf{r} = \int \rho \left(\mp \frac{1}{2} \sqrt{\frac{3}{2\pi}} \right) (x \pm iy) d\mathbf{r} \tag{2.130}$$

$$Q_2^0 = \int \rho r^2 Y_2^0 d\mathbf{r} = \int \rho \frac{1}{4} \sqrt{\frac{5}{\pi}} (3z^2 - r^2) d\mathbf{r} \tag{2.131}$$

$$Q_2^{\pm 1} = \int \rho r^2 Y_2^{\pm 1} d\mathbf{r} = \int \rho \left(\mp \frac{1}{2} \sqrt{\frac{15}{2\pi}} \right) z(x \pm iy) d\mathbf{r} \tag{2.132}$$

$$Q_2^{\pm 2} = \int \rho r^2 Y_2^{\pm 2} d\mathbf{r} = \int \rho \left(\frac{1}{4} \sqrt{\frac{15}{2\pi}} \right) (x \pm iy)^2 d\mathbf{r}. \tag{2.133}$$

The complex nature of some of these expressions need not disturb us, because generally products of these quantities arise in the final expressions of the inter-molecular forces, which will make all terms real.

The multipole moments represent well-defined integrals over the charge distribution of a molecule. We have one moment of zero order, which is the charge itself, i.e.,

$$Q_0^0 = \frac{q}{\sqrt{4\pi}}, \tag{2.134}$$

with the charge being defined as

$$q = \int \rho d\mathbf{r}. \tag{2.135}$$

There are three moments of first order, which constitute the components of the dipole vector

$$\boldsymbol{\mu} = \{\mu_x; \mu_y; \mu_z\}$$

with

$$\mu_\alpha = \int \rho \alpha d\mathbf{r}, \tag{2.136}$$

where α stands for x, y, or z. As an illustration, for a linear molecule with its molecular axis being the z-axis of the molecule-fixed coordinate system, only the z-component will survive because there are no charges in the x- or y-direction, and we find

$$Q_1^0 = \frac{1}{2}\sqrt{\frac{3}{\pi}}\mu_z \qquad (2.137)$$

with

$$\mu_z = \int \varrho z \mathrm{d}\boldsymbol{r}$$
$$= \sum_i q_i z_i, \qquad (2.138)$$

where the final formulation is valid for a discrete charge distribution; cf. (2.120).

EXERCISE 2.5

The linear molecule HCl has a dipole moment of $\mu_{HCl} = 3.67 \cdot 10^{-30}$ C m. The interatomic distance is $d_{HCl} = 0.127 \cdot 10^{-9}$m. Show that this dipole moment can be reproduced by two elementary charges ($e = 1.602 \cdot 10^{-19}$C) of different sign being placed at a distance of $z = \pm 0.01145 \cdot 10^{-9}$m away from the geometric center(GC) of the molecule.

Solution
Fig. E 2.5.1 shows the coordinates of the molecule. The dipole moment follows from (2.138) as

$$\mu_{HCl} = \sum_i q_i z_i = 2ez = 2 \cdot 1.602 \cdot 10^{-19}\text{C} \cdot 0.01145 \cdot 10^{-9}\text{m}$$
$$= 3.67 \cdot 10^{-30} \text{ C m}.$$

There are five elements in the quadrupole tensor. Again, for linear molecules, only the element associated with $n = 0$ will survive, giving

$$Q_2^0 = \frac{1}{2}\sqrt{\frac{5}{\pi}}\theta_{zz} \qquad (2.139)$$

with

$$\theta_{zz} = \frac{1}{2}\int \rho\left(3z^2 - r^2\right)\mathrm{d}\boldsymbol{r}$$
$$= \sum_i q_i z_i^2, \qquad (2.140)$$

where again the final formulation is valid for a discrete charge distribution.

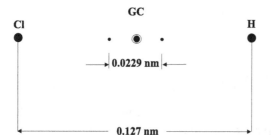

Fig. E 2.5.1. Dipole moment of HCl.

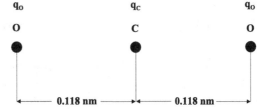

Fig. E 2.6.1. The quadrupole moment of CO_2.

EXERCISE 2.6

Set up a simple point charge model for the linear molecule CO_2 representing the scalar quadrupole moment $\theta = -1.44 \cdot 10^{-39} \, C \, m^2$ and the dipole moment $\mu = 0 \, C \, m$. Assume that the charges are placed in the locations of the atoms ($d_{CO} = 0.118 \cdot 10^{-9} m$).

Solution

Fig. E 2.6.1 shows the geometry of the molecule. From (2.140) the quadrupole moment of a linear molecule is given by

$$\theta_{zz} = \sum_i q_i z^2.$$

We use three partial charges, two on the O-atoms (q_O) and one on the C-atom (q_C). The electroneutrality of the molecule requires that $q_C = -2q_O$. By symmetry this also ensures a vanishing dipole moment. We thus have

$$\theta_{zz} = 2q_O d_{CO}^2 + q_C \cdot 0,$$

which gives

$$q_O = \frac{-1.44 \cdot 10^{-39} \, C \, m^2}{2 \cdot (0.118 \cdot 10^{-9} m)^2} = -0.517 \cdot 10^{-19} C$$

and

$$q_C = -2q_O = 1.03 \cdot 10^{-19} C.$$

For nonlinear molecules the definition of the multipole moments is more complicated because more tensor elements are required. Also, there is generally a dependence on the coordinate system, as is discussed in [8].

Once the multipole moments are defined, the multipolar pair potential can be evaluated from (2.125) and (2.126) by working the expansion coefficients and the rotation matrices out formally. The final results present the pair potential explicitly in terms of the center-to-center distance and the rotational orientations of the two molecules.

EXERCISE 2.7

Derive an explicit expression for the pair potential of the dipole-dipole interaction between two linear molecules (Stockmayer potential).

Solution

The general expression of the multipolar forces pair potential is, according to (2.125),

$$
\phi_{\alpha_i \beta_j}^{\text{mult}} \left(r_{\alpha_i \beta_j}, \omega_{\alpha_i} \omega_{\beta_j} \right) = \sum_{l_1=0}^{\infty} \sum_{l_2=0}^{\infty} \sum_{m_1, m_2} \sum_{n_1, n_2} E_{\alpha_i \beta_j}^{\text{mult}} \left(l_1 l_2 l_1 + l_2; n_1 n_2; r_{\alpha_i \beta_j} \right)
$$
$$
\cdot \sqrt{\frac{2l_1 + 2l_2 + 1}{4\pi}} C \left(l_1 l_2 l_1 + l_2; m_1 m_2 0 \right) \cdot D_{m_1 n_1}^{l_1} \left(\omega_{\alpha_i} \right)^*
$$
$$
\cdot D_{m_2 n_2}^{l_2} \left(\omega_{\beta_j} \right)^*.
$$

For the expansion coefficient of the multipole forces we have from (2.126)

$$
E_{\alpha_i \beta_j}^{\text{mult}} \left(l_1 l_2 l_1 + l_2; n_1 n_2; r_{\alpha_i \beta_j} \right) = \frac{4\pi (-1)^{l_2}}{r_{\alpha_i \beta_j}^{l_1 + l_2 + 1}} \cdot \sqrt{\frac{4\pi (2l_1 + 2l_2 + 1)!}{(2l_1 + 1)! (2l_2 + 1)!}} \cdot \frac{{}^{\alpha} Q_{l_1}^{n_1} \, {}^{\beta} Q_{l_2}^{n_2}}{2l_1 + 2l_2 + 1}.
$$

A dipole–dipole interaction is characterized by $l_1 = l_2 = 1$ and for linear molecules we find $n_1 = n_2 = 0$. The expansion coefficient for the Stockmayer potential thus reads

$$
E_{\alpha_i \beta_j}^{\mu\mu} \left(112; 00; r_{\alpha_i \beta_j} \right) = \frac{4\pi (-1)^1}{r_{\alpha_i \beta_j}^3} \cdot \sqrt{\frac{4\pi 5!}{3! 3!}} \cdot \frac{{}^{\alpha} Q_1^0 \, {}^{\beta} Q_1^0}{5}
$$

with, cf. (2.137), for an arbitrary molecule γ,

$$
{}^{\gamma} Q_1^0 = \frac{1}{2} \cdot \sqrt{\frac{3}{\pi}} \cdot {}^{\gamma} \mu_z,
$$

where $\mu_z = \int \rho z \, d\mathbf{r}$ is the z-component of the dipole moment, further denoted simply as μ. Introducing this into the equation for the expansion coefficient gives

$$
E_{\alpha_i \beta_j}^{\mu\mu} \left(112; 00; r_{\alpha_i \beta_j} \right) = -\frac{4\pi}{r_{\alpha_i \beta_j}^3} \cdot \sqrt{\frac{4\pi 10}{3}} \cdot \frac{1}{5} \cdot \left(\frac{1}{2} \cdot \sqrt{\frac{3}{\pi}} \cdot {}^{\alpha} \mu_z \right) \cdot \left(\frac{1}{2} \cdot \sqrt{\frac{3}{\pi}} \cdot {}^{\beta} \mu_z \right)
$$
$$
= -2 \cdot \sqrt{\frac{6\pi}{5}} \cdot \mu_\alpha \mu_\beta r_{\alpha_i \beta_j}^{-3}.
$$

The Clebsch–Gordan coefficients for the Stockmayer potential are, cf. App. 4,

$$
C(112; -110) = C(112; 1-10) = \sqrt{\frac{1}{6}} \quad \text{and} \quad C(112; 000) = \sqrt{\frac{2}{3}}.
$$

The general definition of the rotation matrices is, cf. App. 4,

$$
D_{mn}^l(\omega) = e^{-im\phi} \cdot d_{mn}^l(\theta) \cdot e^{-in\chi}
$$

with

$$
d_{mn}^l(\theta) = \sum_{\kappa} \frac{\sqrt{(l + m)!(l - m)!(l + n)!(l - n)!} \cdot (-1)^{\kappa}}{(l + m - \kappa)!(l - n - \kappa)!(\kappa - m + n)!\kappa!}
$$
$$
\cdot \left(\cos \frac{\theta}{2} \right)^{2l + m - n - 2\kappa} \cdot \left(\sin \frac{\theta}{2} \right)^{n - m + 2\kappa},
$$

where $(-x)! = \infty$. In addition, we have the identity

$$
D_{mn}^l(\omega)^* = (-1)^{m+n} \cdot D_{\underline{m}\underline{n}}^l(\omega),
$$

where $\underline{m} = -m$ and $\underline{n} = -n$.

As an example, we work out the result for $d^1_{-10}(\theta)$:

$$d^1_{-10}(\theta) = \sum_{\kappa} \frac{\sqrt{(1-1)!(1+1)!(1+0)!(1-0)!} \cdot (-1)^{\kappa}}{(1-1-\kappa)!(1-0-\kappa)!(\kappa+1+0)!\kappa!}$$
$$\cdot \left(\cos\frac{\theta}{2}\right)^{1-2\kappa} \cdot \left(\sin\frac{\theta}{2}\right)^{1+2\kappa}$$

$$= \sum_{\kappa} \frac{\sqrt{2}(-1)^{\kappa}}{(-\kappa)!(1-\kappa)!(\kappa+1)!\kappa!} \cdot \left(\cos\frac{\theta}{2}\right)^{1-2\kappa} \cdot \left(\sin\frac{\theta}{2}\right)^{1+2\kappa}$$

$$= \sqrt{2}\left(\cos\frac{\theta}{2}\right)\left(\sin\frac{\theta}{2}\right) = \sqrt{2}\sqrt{\frac{1+\cos\theta}{2}}\sqrt{\frac{1-\cos\theta}{2}}$$

$$= \frac{1}{\sqrt{2}}\sqrt{1-\cos^2\theta} = \frac{1}{\sqrt{2}}\sin\theta.$$

Analogous calculations give

$$d^1_{10}(\theta) = -\frac{1}{\sqrt{2}}\sin\theta$$

and

$$d^1_{00}(\theta) = \cos\theta.$$

Using all terms we finally arrive at

$$\phi^{\mu\mu}_{\alpha_i\beta_j}\left(r_{\alpha_i\beta_j}\omega_{\alpha_i}\omega_{\beta_j}\right) = -2 \cdot \sqrt{\frac{6\pi}{5}} \cdot \mu_\alpha\mu_\beta r^{-3}_{\alpha_i\beta_j}$$

$$\sqrt{\frac{5}{4\pi}} \cdot \left\{ \sqrt{\frac{1}{6}}\left[\left(e^{-i\phi_{\alpha_i}} \cdot \left(-\frac{\sin\theta_{\alpha_i}}{\sqrt{2}}\right) \cdot e^0\right) \cdot \left(e^{i\phi_{\beta_j}} \cdot \left(\frac{\sin\theta_{\beta_j}}{\sqrt{2}}\right) \cdot e^0\right)\right] \right.$$

$$+ \sqrt{\frac{2}{3}}\left[\left(e^0 \cdot \cos\theta_{\alpha_i} \cdot e^0\right) \cdot \left(e^0\cos\theta_{\beta_j} \cdot e^0\right)\right]$$

$$\left. + \sqrt{\frac{1}{6}}\left[\left(e^{i\phi_{\alpha_i}} \cdot \left(-\frac{\sin\theta_{\alpha_i}}{\sqrt{2}}\right) \cdot e^0\right) \cdot \left(e^{-i\phi_{\beta_j}} \cdot \left(\frac{\sin\theta_{\beta_j}}{\sqrt{2}}\right) \cdot e^0\right)\right] \right\}$$

$$= \frac{-\mu_\alpha\mu_\beta}{r^3_{\alpha_i\beta_j}}\left[-\frac{1}{2}e^{i(\phi_{\beta_j}-\phi_{\alpha_i})}\sin\theta_{\alpha_i}\sin\theta_{\beta_j} + 2\cos\theta_{\alpha_i}\cos\theta_{\beta_j}\right.$$

$$\left. -\frac{1}{2}e^{i(\phi_{\alpha_i}-\phi_{\beta_j})}\sin\theta_{\alpha_i}\sin\theta_{\beta_j}\right]$$

$$= -\frac{\mu_\alpha\mu_\beta}{r^3_{\alpha_i\beta_j}}\left[-\frac{1}{2}e^{i\phi}\sin\theta_{\alpha_i}\sin\theta_{\beta_j} + 2\cos\theta_{\alpha_i}\cos\theta_{\beta_j} - \frac{1}{2}e^{-i\phi}\sin\theta_{\alpha_i}\sin\theta_{\beta_j}\right]$$

$$= -\frac{\mu_\alpha\mu_\beta}{r^3_{\alpha_i\beta_j}}\left[-\frac{1}{2}\sin\theta_{\alpha_i}\sin\theta_{\beta_j}\left(e^{i\phi}+e^{-i\phi}\right) + 2\cos\theta_{\alpha_i}\cos\theta_{\beta_j}\right]$$

$$= -\frac{\mu_\alpha\mu_\beta}{r^3_{\alpha_i\beta_j}}\left[-\frac{1}{2}\sin\theta_{\alpha_i}\sin\theta_{\beta_j}\left(\cos\phi + i\sin\phi + \cos\phi - i\sin\phi\right)\right.$$

$$\left. + 2\cos\theta_{\alpha_i}\cos\theta_{\beta_j}\right]$$

$$= -\frac{\mu_\alpha\mu_\beta}{r^3_{\alpha_i\beta_j}}\left[2\cos\theta_{\alpha_i}\cos\theta_{\beta_j} - \sin\theta_{\alpha_i}\sin\theta_{\beta_j}\cos\phi\right],$$

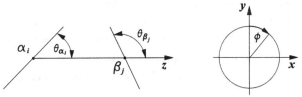

Fig. E 2.7.1. Coordinate system for the dipole–dipole interaction between two linear molecules.

with $\phi = \phi_{\beta_j} - \phi_{\alpha_i}$. The Stockmayer potential thus varies with r^{-3} and has a well-defined, simple angular dependence. Fig. E 2.7.1 shows the coordinate system to which this result applies.

In a manner analogous to that illustrated in Exercise 2.7, explicit equations for the higher multipolar contributions to the pair potential can be derived from (2.125) and (2.126). The number of terms of the multipole expansion to be included depends on the situation considered and must be chosen for each application specifically. The resulting expressions tend to become rather complicated, in particular for nonlinear molecules, where the third Euler angle χ has to be included. In a computer evaluation it is thus more economical to program the spherical harmonic expansion (2.125) directly. Also, for various types of fluid models, such as the perturbation models discussed in Sect. 5.5, integral relations for spherical harmonics and rules for the Clebsch–Gordan coefficients permit simple evaluations to be made using the general spherical harmonic expansion (2.125) instead of the explicit expressions. In some applications the Cartesian tensor representation of the multipole expansion is more readily evaluated on a computer. The final formulae for the Cartesian formulation, as well as the derivations, are documented in [8].

The two-center expansion with the multipole moments located in the molecular centers becomes inadequate for large molecules such as some of those shown in Fig. 1.14. Such molecules have no real centers and may also be so flexible that the assumption of a rigid charge distribution is not tenable. Their electrostatic interactions are more adequately represented by decentralized multipole moments in various locations of the molecules, along with the associated multicenter expansions. Such a formulation of the electrostatic interactions between complicated molecules is not common, however, and is too complicated to be treated by standard statistical mechanical models, although some studies with such force models have been performed. Alternatively, and more simply, the electrostatic properties of large molecules may be represented by a distribution of a few point charges at various strategic sites of the molecule, as discussed in Exercises 2.5 and 2.6. Then the electrostatic potential energy follows directly from (2.122).

Although (2.125) and (2.126) represent a complete formulation of the electrostatic pair potential energy associated with rigid charge distributions, classical electrostatics indicates that there are additional terms related to polarizability effects. In the equations above for the electrostatic interactions we considered rigid charge distributions of the isolated molecules. However, we must realize that the electrical field due to neighboring molecules will modify this distribution by inducing multipoles; cf. Section 2.4.2. So the permanent multipole moments of one molecule will produce a static electrical field that will induce multipoles in a second molecule, no matter if this second molecule has multipoles or not. This effect gives rise to a further type of electrostatic interaction, the induction energy. We here restrict consideration to induced dipole moments. Thus, when a molecule is placed in a uniform electric field E, there is a redistribution of charge and a dipole is induced in the molecule, which is proportional to the field strength; cf. (2.121). We have

$$\mu^{\text{ind}} = \alpha E, \tag{2.141}$$

with α as the polarizability of the molecule. In this equation the molecule is assumed to be isotropic, and the direction of the induced dipole vector coincides with that of the electric field vector. In general, for nonspherical molecules, the polarizability is a tensor quantity, and the directions of μ^{ind} and E may differ. We shall not present a formal derivation of the induction potential energy here, but rather we present the results later in the framework of quantum chemical perturbation theory in Section 2.5 and App. 5. However, we can give an indication here about the structure of the results to be expected. The interaction between two dipoles has been found to vary with $1/r^3$, cf. Exercise 2.7, together with a function of orientational angles. The potential at a distance away from a charge q is $\sim q/r$, cf. (2.114), and therefore that associated with the two charges of a dipole $\sim q_a/(|\boldsymbol{r} - \boldsymbol{r}_a|) + q_b/(|\boldsymbol{r} - \boldsymbol{r}_b|)$, where $\boldsymbol{r}_a, \boldsymbol{r}_b$ are the position vectors of the two charges with respect to the molecular center and $\boldsymbol{r}$ is the distance vector of the molecular center from the considered location. A Taylor expansion of $1/(|\boldsymbol{r} - \boldsymbol{r}_a|)$ in powers of $\boldsymbol{r}_a$ and of $1/(|\boldsymbol{r} - \boldsymbol{r}_b|)$ in powers of $\boldsymbol{r}_b$ and truncation after the dipole contribution reveals that the potential due to a dipole moment $\sim \mu/r^2$, giving the dependence of the electrical field strength arising from a dipole moment on r as $\sim \mu/r^3$; cf. (2.117). We therefore have an r^{-6} dependence on intermolecular distance for the induction pair potential and an additional dependence on orientation. When the electrical field strength results from the permanent dipole of species α and the induced dipole moment is that of species β, we thus expect an expression of the form

$$\phi_{\alpha_i \beta_j}^{\text{ind}} \sim -\frac{\mu_\alpha^2 \alpha_\beta}{r_{\alpha_i \beta_j}^6} f^{\text{ind}} \left(\omega_{\alpha_i} \omega_{\beta_j} \right). \tag{2.142}$$

Because the charge distributions of the molecules are not rigid and will generally be fluctuating, there is finally a further contribution to the pair potential energy, referred to as dispersion energy. Again, this contribution to the intermolecular potential energy will not be formally derived here, because it originates strictly in quantum-chemical effects. The results will be presented in Section 2.5 and App. 5. However, a simple interpretation within the framework of classical electrostatics can be given. When the charge distribution of molecule α_i fluctuates, it will develop a fluctuating dipole and produce a fluctuating electric field, which in turn will induce a fluctuating dipole moment in a neighboring molecule β_j. This induced dipole moment will interact with the fluctuating dipole moment of molecule α_i. So we have a fluctuating dipole-induced dipole interaction. The magnitude of the fluctuating dipole moment in molecule α_i is proportional to its polarizability α_α, because this molecular property controls the displacebility of the electrons. The fluctuating induced dipole moment is again given by (2.141). So we expect the same distance dependence of the pair potential as for the induction contribution. This gives

$$\phi_{\alpha_i \beta_j}^{\text{disp}} \sim -\frac{\alpha_\alpha \alpha_\beta}{r_{\alpha_i \beta_j}^6} f^{\text{disp}} \left(\omega_{\alpha_i} \omega_{\beta_j} \right), \qquad (2.143)$$

where we note that the polarizabilities here reflect reactions of a molecule to a fluctuating electric field and thus should be frequency-dependent. They are then referred to as dynamic polarizabilities, in contrast to the static polarizabilities adequate for the induction forces. The dispersion pair potential should thus result from an average over all frequencies of fluctuations.

We note in conclusion that higher order multipoles and polarizabilities contribute further additions to the potential energy, including multibody effects. An extensive derivation and summary of results is given in [9]. Summing up all contributions gives a rather sophisticated model for the long range and essentially attractive potential energy in a fluid system. At short range there are further interactions of electrostatic origin that are essentially repulsive. They arise when the atoms of two molecules approach each other so closely that the associated electrons overlap and repel each other. Obviously, short-range interactions reflect the shapes of the molecules. A simple electrostatic formulation of this contribution to the intermolecular potential energy, in contrast to the multipole and induction forces, is not possible. A discussion of comprehensive semiempirical potential models based on electrostatic effects will be given in Section 5.2.

2.4.3 Continuum Models

The formal representation of the electrostatic potential energy of a molecular system as presented in the preceding section is rather complicated. It is most commonly used when the distance and orientation dependence of the potential energy is important, e.g., in models for the equation of state. There are many situations, however, where such a detailed representation of the potential energy is not warranted. In such cases, notably in the liquid state, simpler formulations, although approximate, are frequently desirable. Such formulations can be based on a continuum model. As shown above, a continuum model introduces the dielectric constant as a measure for the polarization effect, i.e., the shielding effect that charges experience when instead of in a vacuum they are placed in a dielectric continuum. We shall see below that continuum models, contrary to the models discussed before, do not give a molecular potential energy that would be introduced into the partition function. Rather, they lead directly to macroscopic thermodynamic properties, notably changes of free energy, due to electrostatic (or other) interactions.

An important application of a continuum model is the estimation of the change in free energy due to electrostatic effects that is experienced by a molecule, the solute, being transferred from an ideal gas state into a liquid solvent, e.g., water. Then the dissolved molecule will experience interactions with the solvent molecules surrounding it but not with other solute molecules, and the situation is that of an ideal dilute solution. This is a frequently used reference state for calculating the chemical potential of a component in solution; cf. Section 2.1. This solution process is referred to as solvation, or, more specifically as hydration, when the solvent is water. The associated energy change is then the energy of solvation or hydration. From the corresponding ideal Gibbs free energy of solvation and the Gibbs free energy of formation of the solute molecule in the gas phase one easily finds its Gibbs free energy of formation in the ideal dilute solution state. This quantity is frequently required in thermodynamic calculations. In classical thermodynamics we rely on tabulated data for the enthalpy of formation, the standard entropy, and the Gibbs free energy of formation for components in this reference state. Such data may be obtained from solution experiments. The solvent is mostly water in such tabulations and the Henry coefficient of a component in water can easily be calculated from them; cf. Section 2.1. Here, we address the problem of setting up a simple continuum model for the ideal Gibbs free energy of solvation.

The basis of the continuum approach is to neglect the atomic structure of the solvent and replace its electrostatic properties by those of a dielectric continuum. The polarization of the medium and the back polarization of the solute are then the electrostatic reaction to placing the solute molecule into the solvent and thus a measure of the electrostatic interactions between solvent and solute. The macroscopic effect of this electrostatic interaction energy is calculated by a particular branch of continuum models, referred to as continuum solvation models, also denoted as dielectric continuum models or reaction field models. Since the first application of this approach [11], many different versions of the model have appeared in the literature [12, 13]. They all have in common that a cavity is constructed that represents the geometry of the solute molecule and the solvent is represented by an infinitely extended dielectric continuum outside this cavity. The electric field arising from the nuclei, as well as from the electrons of the solute molecule, is screened by the polarization of this continuum. The effect of this polarization can be represented by the surface charge density distribution it produces on the inner surface of the cavity, much like the polarization charges on the surface of a dielectric continuum placed between the plates of an electrical capacitor; cf. Fig. 2.9. Also, the screening charges back-polarize the solute and thus change its charge distribution. In a continuum solvation model the electrostatic interactions between solute and solvent molecules in a real solution, resulting from the interactions between the molecular charge densities, are thus modeled by a rather localized system, the charges of the solute in the solvent interacting with the associated charged surface of the cavity.

From the various available continuum solvation models there is one that has the merit of particular practical usefulness. In this model, referred to as the COSMO (conductor-like screening model), the continuum chosen to represent the solvent is an infinitely extended electrical conductor, i.e., a medium with $\varepsilon_r \to \infty$ [14]. This choice leads to a remarkably simple expression for the screening charges and the screening energy because use can be made of the condition that the resulting electrostatic potential is zero for every point on the surface of the cavity in the conductor. Because the electrical conductor screens the charges perfectly, it is intuitively clear that all electrostatic effects vanish behind the electrically conducting surface, including the electrostatic potential. More formally, this follows from the uniformity of the electrostatic potential in a conductor and the condition $\psi(\infty) = 0$; cf. (2.114). By

$$\boldsymbol{Q} = \{Q_1, Q_2 \cdots Q_n\}$$

we denote the vector of n point charges characterizing the electrostatic properties of the solute molecule and by

$$\boldsymbol{q} = \{q_\alpha, q_\beta \cdots q_\nu\}$$

the vector of ν screening charges on the inner surface of the cavity representing the solvent reaction in the conductor approximation. The Q's are localized at points $r_1, r_2, \ldots, r_n$ within the cavity and are assumed to be known from single-molecule quantum-chemical calculations; cf. Section 2.5. So we do not consider any change in the solute's charge distribution caused by the solution process; i.e., we neglect the back-polarization effect at this stage of the derivation. The results are generalized later; cf. Section 2.5. We assume that the shape of the cavity can be constructed from the geometrical structure of the solute molecule, i.e., in particular, from its atomic radii. The q's are localized at the various segments $\alpha, \beta, \ldots, \nu$ of its surface and are going to be computed. Fig. 2.12 shows the situation to which the electrostatic calculation has to be applied. Two charges, Q_i and Q_j, are shown, along with two surface segments, ν and μ. In a practical COSMO calculation the number of surface segments typically ranges from about one hundred for diatomic molecules to a few thousand for drug molecules.

We have three different contributions to the total electrostatic potential energy of the solute molecule in the conductor: the interactions between the Q's,

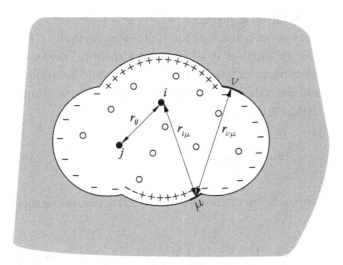

Fig. 2.12. Electrostatics in a continuum solvation model.

those between the Q's and the q's, and, finally, those between the q's. For the first contribution we have from Coulomb's law, using electrostatic units here and in what follows,

$$\phi_{QQ} = \sum_i \sum_{j>i} \frac{Q_i Q_j}{r_{ij}}, \qquad (2.144)$$

with $r_{ij} = |r_j - r_i|$ as the distance between charges Q_i and Q_j within the cavity representing the solute molecule. In vector notation this becomes

$$\phi_{QQ} = \frac{1}{2} \boldsymbol{Q} \mathbf{C} \boldsymbol{Q}, \qquad (2.145)$$

where the elements of the matrix $\mathbf{C}$ are given by $1/|r_j - r_i|$ with $1/r_{ii} = 0$ and the factor 1/2 avoids counting each interaction twice. The second contribution originates from interactions between the point charges Q_i within the cavity and the screening charges q_μ on its surface and is given by

$$\phi_{Qq} = \boldsymbol{Q}\mathbf{B}\boldsymbol{q}, \qquad (2.146)$$

with the elements of the matrix $\mathbf{B}$ being given by $1/|r_\mu - r_i|$. In this approximation we have assumed that all surface segments are chosen so small that the effects of curvature are negligible and that the surface of a segment μ, S_μ, can be considered to be flat and described by the vector r_μ. Finally, for the interaction between the surface segments, we have

$$\phi_{qq} = \frac{1}{2} \boldsymbol{q}\mathbf{A}\boldsymbol{q}, \qquad (2.147)$$

where the off-diagonal elements of the matrix $\mathbf{A}$ are given by $a_{\mu v} = 1/|r_v - r_\mu|$ for two surface segments, v and μ.

There is a conceptual difficulty in applying (2.147) for $\mu = \nu$, i.e., for $a_{\mu\mu} = a_{\nu\nu} = a_{\text{diag}}$. This term arises as an artifact of the finite segmentation of the surface. To derive a simple relation for it we consider a spherical cavity. A charge q in the center of the sphere results in a potential of q/R on the surface of the cavity; cf. (2.114) and Exercise 2.4. Charging the sphere to a total charge of Q then gives for the change in electrostatic potential energy

$$\Delta\phi = \int\limits_0^Q \frac{q}{R}\mathrm{d}q = \frac{Q^2}{2R}. \tag{2.148}$$

Representing this by a segmentation of the total spherical surface into n segments gives, according to (2.147),

$$\frac{Q^2}{2R} = \frac{n}{2}\sum_{\mu\neq\alpha}^n \frac{(Q/n)^2}{|r_\alpha - r_\mu|} + \frac{n}{2}\left(\frac{Q}{n}\right)^2 a_{\text{diag}}. \tag{2.149}$$

The sum represents the potential energy of a particular segment α interacting with all others. This term must be evaluated for all segments considered as fixed and therefore must be multiplied by the number of segments and divided by 2 to avoid counting each contribution twice. The second term represents the diagonal contribution. This gives an expression for a_{diag} for a spherical cavity,

$$a_{\text{diag}} = \frac{1}{R}\left(n - \sum_{\mu\neq\alpha}^n \frac{R}{|r_\alpha - r_\mu|}\right) = \sqrt{\frac{4\pi}{S_\mu}}\frac{1}{\sqrt{n}}\left(n - \sum_{\mu\neq\alpha}^n \frac{R}{|r_\alpha - r_\mu|}\right), \tag{2.150}$$

where S_μ is the surface area of one segment μ. Because r_μ scales linearly with R, the term in brackets becomes independent of R. Also, by evaluation of (2.150) for values of n from 4 to several thousands one finds a simple proportionality,

$$a_{\text{diag}} \approx 1.07\sqrt{\frac{4\pi}{S_\mu}}. \tag{2.151}$$

Because any cavity shape can be treated analogously if the segments are made sufficiently small, this result for the diagonal elements can be used for arbitrarily shaped cavities.

For given charge distributions Q and q the total electrostatic potential energy of the system solute plus conductor can be calculated as the sum of the three contributions evaluated above; i.e.,

$$\phi = \frac{1}{2}QCQ + QBq + \frac{1}{2}qAq. \tag{2.152}$$

To find the distribution of screening charges $q_\alpha^*, q_\beta^* \ldots$ along the inner surface of the cavity we use the condition of vanishing electrostatic potential at the surface of a conductor. The total potential there adds up from the potential of the solute point charges and from the potential of the screening charges. So we have, with (2.113) for the definition of the electrostatic potential,

$$\psi = \{\psi_\alpha, \psi_\beta \cdots \psi_\nu\} = \psi_Q + \psi_{q^*} = \mathbf{QB} + \mathbf{A}q^* = 0, \qquad (2.153)$$

with q^* as the actual vector of screening charges on the surface of the conductor, for which we thus obtain

$$q^* = -\mathbf{A}^{-1}\mathbf{B}Q. \qquad (2.154)$$

The energy change due to the dielectric continuum is obtained as

$$\Delta\phi_{\text{diel}} = \phi - \phi_{QQ} = -\frac{1}{2}\mathbf{QBA}^{-1}\mathbf{B}Q = \frac{1}{2}\psi_Q q^*. \qquad (2.155)$$

Here ϕ_{QQ} is the electrostatic energy of the solute molecule in the gas phase, which, under the assumption of negligible back polarization, is identical to the electrostatic energy of the solute molecule in the conductor. We note that (2.155) actually gives the work associated with the polarization of the solvent, i.e., of charging the cavity with q^*. This can be seen by realizing that the work ΔW_i associated with adding an increment of charge ΔQ_i into a system at a location where the electrostatic potential is ψ, i.e., with charging a particle there by this amount, is given by (2.113) as $\Delta W_i = \psi \Delta Q_i$. So we consider a process in which a point charge in the cavity at r is increased by infinitesimally small steps from zero to its actual value $Q(r)$ under the influence of the electrostatic potential generated by the charged cavity surface $\psi_q(r)$. Formulating the increasing charge as $\lambda Q(r)$, where λ increases from zero to one, we have for the work

$$W_Q = \int_{\lambda=0}^{1} \lambda Q(r)\psi_q(r)\mathrm{d}\lambda = \frac{1}{2}Q(r)\psi_q(r),$$

which is of course the same as $\frac{1}{2}\psi_Q q$ and thus identical to the energy change associated with one point charge in a dielectric. The same expression is obtained for the full charge distribution $Q(r)$ of the solute molecule. In view of this interpretation, we refer to the energy change (2.155) as the electrostatic Gibbs free energy of solvation, i.e., the change in Gibbs free energy associated with bringing a solute molecule from the gas phase into the solvent. If we wish to calculate this energy in a solvent with a finite dielectric constant we have to correct (2.155) by multiplying it by a suitable factor f, for which

$$f = \frac{\varepsilon_r - 1}{\varepsilon_r + 1/2} \qquad (2.156)$$

has been found to be of good accuracy [14]. For water with $\varepsilon_r \approx 80$, the finite dielectric correction is close to unity, whereas for small values of ε_r it becomes significantly smaller. Then, however, $\Delta\phi_{\text{diel}}$ is not very large and so the overall contribution of the error of (2.156) is again small.

As can be seen from (2.153), the screening charge vector q^* depends on the shape of the cavity. This dependence is rather strong, and so care must be exercised in finding the proper cavity shape. In COSMO this is done by modeling a solute molecule by a fused sphere geometry and choosing the radii of the spheres representing the atoms by fitting to experimental data. Also, the segmentation of the cavity surface must be done properly for reliable results. COSMO calculations have been incorporated into various quantum-chemical computer packages; cf. Section 2.5. So the results can be generated without bothering about every detail of their production. The thermodynamic energy of solvation obtained from COSMO will in general be no more than a crude approximation to the true experimental value, as shown in Section 2.5. This is to be expected, because the macroscopic dielectric constant cannot accurately reproduce the local electrostatic interactions between the solute and the solvent molecules. However, the COSMO energy can be used to define a more rational reference state for calculating the molecular potential energy of a solute in a solvent than the commonly used ideal gas state; cf. Section 2.4.2. Semiempirical potential models based on this reference state will be discussed in Section 4.2.

EXERCISE 2.8

Consider a spherical particle of charge Q. When it is embedded in a spherical cavity of radius R in an ideal electrical conductor, there will arise a screening charge distribution over the surface of the cavity and an associated change of free energy. Calculate these properties, exactly and by performing a segmentation into $1, 2, 4, \ldots$ segments.

Solution

Fig. E 2.8.1 shows the geometry to which the electrostatic calculation has to be applied. In view of the spherical symmetry of the problem, there will be a uniformly smeared screening charge distribution over the cavity surface defined exactly by $q^* = -Q$. The work required to charge the cavity surface from 0 to $-Q$ is obtained from (2.148) as

$$\Delta\phi_{\text{diel}} = -\int_0^Q \frac{q}{R}\mathrm{d}q = -\frac{Q^2}{2R}.$$

This is the change in electrostatic Gibbs free energy experienced by the particle as it dissolves in the electrical conductor. We now study how this exact result is reproduced by a progressively finer segmentation of the cavity surface upon applying the electrostatic interactions between the point charge at the center and the surface charges on the surface segments of the cavity. The general equation to be solved according to (2.154) is

$$q^* = -\mathbf{A}^{-1}\mathbf{B}Q.$$

Considering only one surface element, i.e., the surface of a spherical cavity, we get

$$\mathbf{A} = a_{\text{diag}} = 1.07\sqrt{\frac{4\pi}{S_\mu}} = \frac{1.07}{R}$$

and

$$\mathbf{B} = \frac{1}{R}$$

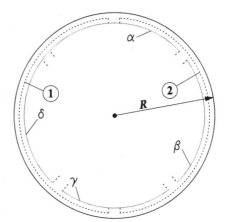

Fig. E 2.8.1. Segmentation of a sphere.

and thus

$$q^* = -\frac{R}{1.07} \cdot \frac{1}{R} \cdot Q = -0.93\, Q,$$

which is about 7% off the exact result. The Gibbs free energy change is accordingly

$$\Delta\phi_{\text{diel}} = -\frac{1}{2}\boldsymbol{Q}\mathbf{B}\mathbf{A}^{-1}\mathbf{B}\boldsymbol{Q} = -\frac{1}{2}\cdot\frac{1}{1.07}\cdot\frac{1}{R}Q^2 = -0.93\left(\frac{Q^2}{2R}\right).$$

If we proceed to a segmentation into two segments 1 and 2 as shown in Fig. E 2.8.1 we have again

$$\mathbf{B} = \frac{1}{R},$$

but for **A** we now get the matrix

$$\mathbf{A} = \begin{pmatrix} a_{11} & a_{12} \\ a_{21} & a_{22} \end{pmatrix},$$

where $a_{11} = a_{22}$ are the diagonal elements and $a_{12} = a_{21}$ the nondiagonal elements. The diagonal elements are given by

$$a_{11} = a_{22} = d = 1.07\sqrt{\frac{4\pi}{2\pi R^2}} = 1.07\frac{\sqrt{2}}{R}$$

and the nondiagonal elements by

$$a_{12} = a_{21} = n = \frac{1}{2R}.$$

The screening charge density is now calculated from

$$q^* = -\mathbf{A}^{-1}\mathbf{B}\boldsymbol{Q},$$

where, from the rules of inverting a matrix, we have

$$\mathbf{A}^{-1} = \begin{vmatrix} \frac{d}{d^2-n^2} & -\frac{n}{d^2-n^2} \\ -\frac{n}{d^2-n^2} & \frac{d}{d^2-n^2} \end{vmatrix} = \begin{vmatrix} A_{11} & A_{12} \\ A_{21} & A_{22} \end{vmatrix}.$$

So we find for the screening charges of the two surface segments 1 and 2

$$q^* = -A_{11}B_{1Q}Q - A_{12}B_{1Q}Q$$

$$q^* = -A_{21}B_{2Q}Q - A_{22}B_{2Q}Q.$$

With

$$A_{11} = \frac{1.07\frac{\sqrt{2}}{R}}{1.07^2\frac{2}{R^2} - \frac{1}{4R^2}} = 0.7418R = A_{22}$$

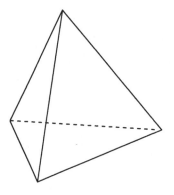

Fig. E 2.8.2. A regular tetrahedron.

and

$$A_{12} = \frac{-\frac{1}{2R}}{1.07^2 \frac{2}{R^2} - \frac{1}{4R^2}} = -0.2451\,R = A_{21},$$

the result for the screening charges is

$$q_1^* = -0.7418\,R\frac{1}{R}Q + 0.2451\,R\frac{1}{R}Q = -0.4967\,Q$$

$$q_2^* = 0.2451\,R\frac{1}{R}Q - 0.7418\,R\frac{1}{R}Q = -0.4967\,Q,$$

and so

$$q^* = q_1^* + q_2^* = -0.9934\,Q$$

and, accordingly,

$$\Delta\phi_{\text{diel}} = -\frac{1}{2}\mathbf{QBA}^{-1}\mathbf{BQ} = -0.9934\left(\frac{Q^2}{2R}\right).$$

Obviously, this is already rather close to the exact result.

We can go one step further and consider a segmentation into four segments α, β, γ, δ; cf. Fig. E 2.8.1. The surface of the sphere is now represented by four equal-sized surface segments, each of area πR^2. The centers of these surface segments are located on the four edges of a regular tetrahedron; cf. Fig. E 2.8.2. The geometrical functions of a regular tetrahedron are well known. The angle between the center of the tetrahedron, i.e., the location of the charge, and any two edges is the tetrahedral angle $\alpha = 109.47°$; cf. Fig. E 2.8.3. Then simple triogonometry shows that $a_{\nu\mu} = 1/1.633\,R$. The diagonal element is calculated as usual by $a_{\nu\nu} = 1.07\sqrt{4\pi/\pi}\,\bar{R^2} = 2.14/R$ and $\mathbf{B} = 1/R$. The calculation of the screening charges and the change in potential energy now involves inversion of a 6×6 matrix. Solving this on a computer gives $q^* = 1.004$ and $\Delta\phi = -1.004$, which

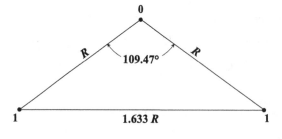

Fig. E 2.8.3. Trigonometry of a regular tetrahedron.

is off by no more than 0.4%. As can be seen, a progressively finer segmentation leads to a closer approximation to the exact result.

A second application for the continuum approach is to the interaction between different solute molecules in a solvent. This will lead to corrections to the ideal dilute solution properties. We again assume the solvent to be a dielectric continuum. In particular, we consider the interactions of ions in a solvent. Such ions exert strong and long-range electrostatic forces upon each other and thus do not interfere appreciably with the comparatively weak and short-range interactions of the solvent molecules. Thus, the continuum approximation of the solvent appears plausible here, at least at small ion concentrations. According to the molecular interpretation of the chemical potential, the additional effect of the ion–ion interactions on the chemical potential of an ion j in the solution, which we denote as $\Delta\mu_j^{el}$, is given by the change of Gibbs free energy, or alternatively, the work required on charging this ion from zero to its actual charge under the influence of all other ions in the solution. In thermodynamic terms this is related to the activity coefficient of the ion in the solution; cf. Section 2.1. Considering the ion j to be a point particle at the center ($r = 0$) of a coordinate system, we thus need to calculate the electrostatic potential $\psi(r = 0)$ arising from all ions in the solution at the location of ion j and then find the work, or electrostatic potential energy, from (2.113) as

$$\Delta\mu_j^{el} = RT \ln \gamma_j^{el} = \int_0^{Z_j e} \psi(r = 0) \mathrm{d}q_j. \qquad (2.157)$$

Here, $Z_j e$ is the charge of the ion j, with e as the elementary charge and Z_j as the charge number.

To calculate $\psi(r = 0)$ in the continuum approximation, we note that an ion j, considered as a central ion, will be surrounded by an ion cloud of many other ions with the dielectric continuum in between. If the ions were distributed completely at random, the chances of finding either a positive or a negative ion in the neighborhood of the central ion j would be identical. Such a random distribution would not produce an electrostatic potential at $r = 0$, because, on the average, negative contributions would be exactly balanced by positive ones. However, in the real situation there will be a local structure in the solution due to electrostatic interactions between the ions. Cations and anions are not uniformly distributed, but anions tend to be found in the vicinity of cations, and vice versa. Overall the solution is electrically neutral, but in the vicinity of any given ion there is a predominance of opposite charges, referred to as the ion atmosphere or ion cloud of a given ion; cf. Fig. 2.13. The electrostatic potential in the ion cloud at a location r is related to the charge density in the ion cloud at this location in the continuum approximation by Poisson's equation, cf. (2.118),

$$\nabla^2 \psi(\boldsymbol{r}) = -\frac{\varrho(\boldsymbol{r})}{\varepsilon}. \qquad (2.158)$$

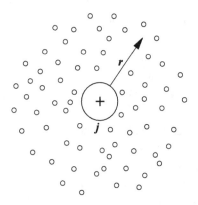

Fig. 2.13. Ion cloud around a central positive ion j having an average negative charge.

Here the charge density can generally be expressed as

$$\varrho(\mathbf{r}) = \sum_i Z_i e n_i(\mathbf{r}), \tag{2.159}$$

where $n_i(\mathbf{r})$ is the number density of ions of type i at $\mathbf{r}$ and $(Z_i e)$ is the charge on such an ion. The number density $n_i(\mathbf{r})$ is assumed to be related to the mean number density of ions of type i, n_i, by the Boltzmann factor, cf. (2.50),

$$\frac{n_i(\mathbf{r})}{n_i} = e^{-Z_i e \psi(\mathbf{r})/kT}. \tag{2.160}$$

Here the exponent of the Boltzmann factor represents the difference in electrostatic potential energy of an ion i in the ion cloud around ion j with respect to an energy zero associated with the mean number density. The local charge density of the ion cloud around ion j thus arises from a competition between the electrostatic action of the central ion and the distributing effects of thermal agitation. We note that (2.159) and (2.160) can only be realistic for dilute solutions, because any interactions between the ions of the ion cloud are neglected. In this limit of high dilution, the exponent may be expanded with $Z_i e \psi(\mathbf{r})/kT \ll 1$ to give the Poisson equation in the form

$$\nabla^2 \psi(\mathbf{r}) = -\frac{e}{\varepsilon} \sum_i Z_i n_i \left(1 - \frac{Z_i e \psi(\mathbf{r})}{kT}\right) = \frac{e^2}{\varepsilon kT} \sum_i Z_i^2 n_i \psi(\mathbf{r})$$

$$= K^2 \psi(\mathbf{r}), \tag{2.161}$$

where we used

$$\sum_i Z_i n_i = 0, \tag{2.162}$$

which is the condition of total electroneutrality of the solution, and

$$K^2 = \frac{e^2}{\varepsilon kT} \sum_i Z_i^2 n_i = \frac{N_A^2 \, 2 e^2 \varrho_0}{\varepsilon RT} I. \tag{2.163}$$

Here

$$I = \frac{1}{2} \sum Z_i^2 m_i \qquad (2.164)$$

is referred to as the ionic strength of the solution with $I \to 0$ for infinite dilution, and m_i, ϱ_0 are the molality of ion i and the density of the solvent, respectively. The introduction of m_i and ϱ_0 again makes use of the high dilution. In spherical coordinates the Poisson equation now reads

$$\frac{d^2[r\psi(r)]}{dr^2} - K^2[r\psi(r)] = 0.$$

It has the general solution

$$\psi(r) = A\frac{e^{-Kr}}{r} + B\frac{e^{+Kr}}{r}. \qquad (2.165)$$

Because $\psi(r \to \infty) = 0$, we have $B = 0$. Further, for high dilution $K \to 0$ and (2.165) degenerates to

$$\psi(r) = \frac{A}{r}e^{-Kr} = \frac{A}{r}(1 - Kr) = \frac{A}{r} - AK.$$

As $r \to 0$ we have $\psi(r) = A/r$ and the electrostatic potential will be determined by the charge on the central ion j, the effect of the ion cloud around it, as represented by K, being negligible. The associated electrostatic potential is $\psi(r) = A/r = Z_je/4\pi\varepsilon r$, in accordance with (2.114), giving $A = Z_je/4\pi\varepsilon$. We then find for the electrostatic potential

$$\psi(r) = \frac{Z_je}{4\pi\varepsilon r} - \frac{Z_je}{4\pi\varepsilon(1/K)}. \qquad (2.166)$$

Whereas the first term represents the contribution due to the central ion j, the second term adds the electrostatic potential at r due to the atmosphere of other ions around j. To calculate the chemical potential of ion j in high dilution, we need the electrostatic potential at the ion j arising from the ion atmosphere surrounding it. This is

$$\psi(0) = -\frac{Z_je}{4\pi\varepsilon(1/K)}. \qquad (2.167)$$

The work involved charging the central ion j situated at a location where the electrostatic potential is $\psi(0)$; i.e., the change in chemical potential associated with this process is given by

$$\Delta\mu_j^{el} = \int_0^{Z_je} \psi_j(0)dq = -\int_0^{Z_je} \frac{q}{4\pi\varepsilon(1/K)}dq = -Z_j^2e^2/8\pi\varepsilon(1/K). \qquad (2.168)$$

Transformation to a molar basis and expression in terms of the activity coefficient give

$$\ln\gamma_j^{el} = -\frac{N_A z_j^2 e^2 K}{8\pi\varepsilon RT}, \qquad (2.169)$$

which is the Debye–Hückel equation for the activity coefficient of an ion j in the limit of high dilution [15].

EXERCISE 2.9

Calculate the mean activity coefficient in water at 25°C for the electrolytes HCl, MgCl$_2$, and CuSO$_4$ at molalities between 0.001 and 1 mol/kg from the Debye-Hückel theory and compare to experimental data.

Solution

The result for the mean activity coefficient is obtained by introducing the Debye–Hückel equation for a single ion into the general equation for the mean activity coefficient, defined as

$$\gamma_\pm = \left(\gamma_+^{\nu_+}\gamma_-^{\nu_-}\right)^{1/\nu}.$$

This gives

$$\ln \gamma_\pm = \frac{-N_A}{\nu_+ + \nu_-} \frac{e^2 K}{8\pi\varepsilon RT}\left[\nu_+ Z_+^2 + \nu_- Z_-^2\right] = -\frac{N_A e^2 K}{8\pi\varepsilon RT}|Z_+ Z_-|,$$

where use has been made of $\nu = \nu_+ + \nu_-$ and of the electroneutrality condition $\nu_+ z_+ + \nu_- z_- = 0$. Usually, the logarithm of the mean activity coefficient is reported to base 10 and all natural constants are summarized into one, giving

$$\lg \gamma_\pm = -C|Z_+ Z_-|\sqrt{\frac{(I/\text{mol kg}^{-1})\varrho_0/(\text{g cm}^{-3})}{\varepsilon_r^3(T/K)^3}}$$

with

$$C = 1.825 \cdot 10^6.$$

For the solvent water at 25°C with $\varrho_0 = 0.997$ g/cm^3 and $\varepsilon_r = 75.5$, we have

$$\lg \gamma_\pm = -0.509|Z_+ Z_-|\sqrt{I/\text{mol kg}^{-1}}.$$

The ionic strength

$$I = \frac{1}{2}\sum Z_i^2 m_i$$

is related to the molality of the electrolyte by the general formula $I = km$, where k depends on the valencies of the ions in the solution. For HCl, which dissociates according to HCl $\rightarrow$ H$^+$ + Cl$^-$, we have $Z_+ = 1$ and $|Z_-| = 1$ and thus $I = \frac{1}{2}(m_+ + m_-) = 1m$. For MgCl$_2$, which dissociates according to MgCl$_2$ $\rightarrow$ Mg^{2+} + 2Cl$^-$, we find $I = \frac{1}{2}(4m_+ + m_-) = 3m$, because, $m_- = 2m$ and $m_+ = m$, and finally, for CuSO$_4$, dissociating according to CuSO$_4$ $\rightarrow$ Cu^{2+} + SO$_4^{2-}$, we have $I = \frac{1}{2}(4m_+ + 4m_-) = 4m$. So at $m = 0.01$ we find $I_{\text{HCl}} = 0.01$, $I_{\text{MgCl}_2} = 0.03$, and $I_{\text{CuSO}_4} = 0.04$. The corresponding mean activity coefficients are $\gamma_{\text{HCl}}^\pm = 0.889$, $\gamma_{\text{MgCl}}^\pm = 0.666$, $\gamma_{\text{CuSO}_4}^\pm = 0.392$. Fig. E 2.9.1 shows the comparison with experimental data. It is clear that the Debye–Hückel law is a good approximation at rather low molalities of $m < 0.01$.

As an illustration, we show in Fig. 2.14 a prediction of the solubility of SO$_2$ in water. The partial pressure of SO$_2$ in the gas phase is plotted against the total molality of dissolved SO$_2$. The inert component in the gas phase is N$_2$ and the

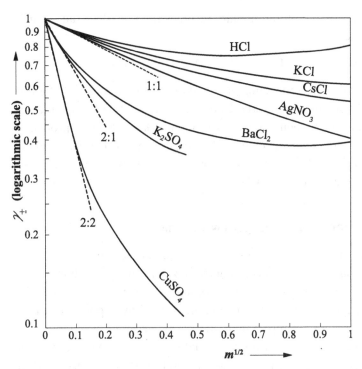

Fig. E 2.9.1. Mean activity coefficient for some electrolytes in water. (– data; --- Debye-Hückel limiting law; (2:2),(2:1),(1:1) = numbers of cations and anions produced on dissociation).

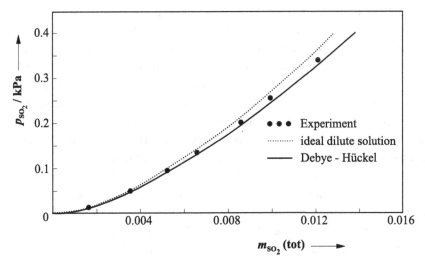

Fig. 2.14. Solubility of SO_2 in water at 298.15 K.

total pressure 105 kPa. The total molality includes the ion HSO_3^- (aq), which originates from the reaction

$$SO_2 \, (aq) + \, H_2O \rightleftharpoons H^+ \, (aq) + \, HSO_3^- \, (aq).$$

Whereas the dotted curve represents the ideal dilute solution results, the full curve shows the correction due to the Debye–Hückel limiting law for the activity coefficient. It can be seen that the experimental data [16] are reproduced very well by this model at these small molalities; cf. Fig. 1.7. For the ideal dilute solution results, tabulated values were used for the Gibbs free energies of formation in the associated standard states.

2.5 Molecular Properties: Quantum Mechanics

In Section 2.3 we have presented the laws of classical mechanics and introduced the Hamilton function in terms of generalized position coordinates and momenta. This is the adequate classical formulation of the total energy of a system of particles moving in a given force field. We have applied this theory to derive expressions for important models of single-molecule energy such as translation, rigid rotation, harmonic oscillation, and internal rotation. In Section 2.4 we have discussed the information about the interactions between molecules available from classical electrostatics. Although we shall make extensive use of all these results for setting up molecular models for fluids, they are incomplete and need additions and modifications.

Modifications of the classical mechanical results are required because this theory is not correct for molecules. Impressive experiments show that classical theory is unable to treat very small bodies, e.g., microparticles such as electrons, atoms, and molecules. A different theory is required on the atomic scale. The very assumption that the position and the velocity of a particle of atomic dimensions may be prescribed accurately at the same time is not tenable. So classical mechanical concepts are not valid for molecular systems. Rather, the properties of molecular systems must be described by quantum mechanics. Fortunately, the quantum theory adequate on the atomic scale converges to classical mechanical behavior for many practical applications. In particular, it turns out that the results in Section 2.3 can be used in setting up the correct quantum equations for the energy of a molecular system. In some cases, the final quantum results can even be reduced to the classical results by applying simple corrections. The correspondence between classical and quantum mechanical results under well-defined conditions is referred to as the correspondence principle. It is the basis of the semiclassical approximation; cf. Section 2.2.

Additions are needed because the classical mechanical and electrostatic results do not give us any information about molecular geometry, the vibration frequencies, the potential barrier of internal rotation, the enthalpy of formation, and further single-molecule properties such as multipole moments and polarizabilities. All these molecular properties are needed in the calculation of molecular energies and can be obtained from quantum mechanical calculations, as performed in practice by quantum-chemical computer codes. Finally, and most importantly, quantum mechanics reveals important contributions to the intermolecular potential energy beyond the classical electrostatic interactions discussed in the previous section. Thus, quantum mechanics makes fundamental contributions to molecular models for fluids [17,18].

2.5.1 Duality of Particle and Wave: The Wavefunction

Experimental observation requires that a dual nature be attributed to radiation, in the sense that some experiments show radiation to have wavelike characteristics, whereas in others particlelike characteristics (photons) become evident. An example of wave characteristics of radiation is the appearance of the well-known diffraction patterns. The particlelike characteristics of radiation became evident by Planck's derivation of the correct expression for the spectral energy distribution of blackbody radiation in 1900 and by Einstein's equation for the photoelectric effect in 1905. Along with the duality hypothesis of radiation, it became necessary to assume that its energy is not emitted or absorbed continuously, as implied by classical theory, but rather in small portions ε, so-called quanta, as $\Delta E = n\varepsilon$, where n is a discrete number. So although a standard electrical bulb sends out a continuous spectrum in a vacuum, dark lines appear when a cold gas is placed between the bulb and the prism. The dark lines represent those energies at which radiation is absorbed as it passes through the gas and reflect the discrete energy values allowed in the absorption process.

In 1924, L. de Broglie postulated on theoretical grounds an analogous dual nature for matter. So, in some experiments, matter was assumed to behave with well-known particlelike characteristics, whereas in others, wavelike characteristics were postulated. This duality has been formulated by associating a wavelength λ to a particle. The wavelength λ of a particle of momentum p is called its de Broglie wavelength. The actual verification of the wavelike characteristics of matter followed later from experiments in which electrons, protons, and whole molecules produced diffraction patterns under suitable conditions. Such phenomena are incompatible with the picture of particles moving along well-defined trajectories that are described by the classical mechanical equations, but can easily be accommodated in the wave picture.

A wave is characterized by a frequency ν and a wavelength λ, whereas particles are described in terms of energy ε or momentum p. The relationships between the two ways of characterizing radiation and matter are formulated in terms of a new universal constant h and are given by the equations

$$\varepsilon = h\nu \tag{2.170}$$

and

$$p = \frac{h\nu}{c} = \frac{h}{\lambda}. \tag{2.171}$$

Here c is the speed of light, p is the momentum of a particle (photon), and h is Planck's constant, of dimensions momentum times length. Both constants, c and h, are universal. Numerical values for them are given in App. 1. The relation between frequency and wavelength is given by

$$\nu = \frac{c}{\lambda}. \tag{2.172}$$

The duality of matter, as well as a lot of further experimental evidence, makes it clear that the classical concept of well-defined position and momentum coordinates for a particle, i.e., the concept of a trajectory, must be abandoned on the atomic scale. Instead, an element of probability and indeterminacy must be introduced into the description of matter on the molecular level. This is achieved by postulating that matter on the molecular scale is described in terms of a so-called wavefunction and the associated mechanics consists of setting up a scheme for calculating how the wavefunction is distributed in a molecular system.

The probability for a particular location of an atomic particle is related to its wavefunction $\psi(x, y, z)$, where x, y, z are the cartesian position coordinates. Such a wavefunction may typically be thought of as a superposition of sine functions or some related representation. The amplitude of a wave may be positive as well as negative. A probability, however, should be a positive number. Thus, the probability of a particular configuration of a molecular system in terms of the position coordinates of the wavefunction is characterized by the square of its amplitude. In other words, the probability of finding the particle at a particular location is largest where the squared amplitude of its wavefunction has its maximum and lower at other locations. An atomic particle can assume many different states. As stated before and unlike the classical concept, these states are not continuous but discrete. They are referred to as quantum states and indexed, e.g., by n. In every quantum state the particle has an individual energy value E_n and an individual wavefunction ψ_n. There may be multiple wavefunctions, i.e., quantum states, associated with a particular energy value, a phenomenon referred to as degeneracy. The wavefunction does not represent a real wave in a physical sense, but, instead, is an entirely conceptual function introduced to describe the experimentally observed behavior of atomic systems. In general, therefore, it may be complex. If we consider, as an example for an arbitrary molecular system, N microparticles in a cartesian coordinate system, we introduce the wavefunction $\psi(x_1, y_1, z_1, x_2, y_2, z_2, \ldots, x_N, y_N, z_N)$ to describe the molecular configuration of the system. The probability for a particular configuration, i.e., the probability of finding one particle at a location in the volume element $d\mathbf{r}_1$ at $\mathbf{r}_1$, while another is located in the volume element $d\mathbf{r}_2$ at $\mathbf{r}_2$, and so on, is then given by

$$dP = \psi^* \psi \, d\mathbf{r}_1 d\mathbf{r}_2 \ldots d\mathbf{r}_N, \tag{2.173}$$

where $r_i = (x_i, y_i, z_i)$ and the asterisk * denotes the complex conjugate. The product $\psi^*\psi$ is thus a probability density. Because the system must definitely be in some configuration we have the normalization condition

$$\int \psi^*\psi \, dr^N = 1, \tag{2.174}$$

where $dr^N = dr_1 \, dr_2 \cdots dr_N$ and the integration extends over the whole coordinate space of the system. Clearly, to produce physically meaningful values of the probability density, the wavefunction ψ must be single-valued, finite, and continuous and must also fulfil the normalization condition (2.174). In quantum mechanics, (2.173) for the probability of a particular configuration of N microparticles replaces the well-determined configuration of a molecular system in classical mechanics.

2.5.2 The Schrödinger Equation

We accept that the configuration of an atomic system in terms of a set of position coordinates $(x_1 \ldots z_N)$, where $(x_1 \ldots z_N)$ are the cartesian position coordinates of all N particles, can be described by its wavefunction $\psi(x_1 \ldots z_N)$. This wavefunction is determined from the Schrödinger equation, which replaces Newton's equation or rather Hamilton's equations, cf. Section 2.3, of classical mechanics, and reads

$$\hat{H}\psi = E\psi. \tag{2.175}$$

Here E is the total energy of the molecular system and $\hat{H}$ the so-called Hamilton operator, also referred to as the Hamiltonian. The Hamilton operator is the operator representing the total energy and is defined as

$$\hat{H} = \hat{T} + \hat{V}. \tag{2.176}$$

The first term $\hat{T}$ is the operator representing the kinetic energy and $\hat{V}$ is the operator representing the potential energy. For a single particle and cartesian coordinates we have

$$\hat{T} = -\frac{h^2}{8\pi^2 m}\nabla^2 = -\frac{h^2}{8\pi^2}\frac{1}{m}\left(\frac{\partial^2}{\partial x^2} + \frac{\partial^2}{\partial y^2} + \frac{\partial^2}{\partial z^2}\right), \quad \hat{V} = V(x, y, z). \tag{2.177}$$

Analogous representations apply for other coordinate systems. More complicated formulations for $\hat{T}$ and $\hat{V}$ in terms of more coordinates have to be introduced for a many-particle system, e.g., a molecule consisting of electrons and nuclei; cf. Section 2.5.4. Although ∇^2 is a differential operator, the operator $\hat{V}$ indicates just a multiplication; i.e., the wavefunction is multiplied by the function $V = V(x, y, z)$. We note that in quantum mechanics the classical Hamilton function, i.e., the total energy of a system, translates into an operator acting on the wavefunction.

To set up the Schrödinger equation for a particular atomic system we thus have to decide on adequate position coordinates, introduce the mass, and formulate the potential energy in terms of the charges. This information is all provided by classical mechanics and electrostatics and thus can be taken over from Sections 2.3 and 2.4. The resulting equation in the form of (2.175) is an eigenvalue equation, i.e., an equation in which an operator acting on a function, here the wavefunction, produces a multiple of the same function. Solutions of this equation are sought that fulfil particular boundary conditions that in turn are deduced from the physical problem under consideration. Together with these boundary conditions, the eigenvalue equation turns into an eigenvalue problem. The set of functions for which the equation, along with the boundary conditions, holds is its eigenfunctions and the associated values of the multiplicity factors are the eigenvalues, here the energy values. So, when solving the Schrödinger equation for a particular system, one usually finds that the system can exist in a large number of different states n, with the associated functions ψ_n and values E_n. Thus, the Schrödinger equation formalizes the experimental observation of quantization of atomic states, in particular of energy, to which we referred before. This quantization is a consequence of the boundary conditions. All values E_n that can be found from measurements of the total energy of a system are represented by the spectrum of eigenvalues of its Hamilton operator.

There are two important mathematical properties of the wavefunctions ψ_n. First, it can be shown that they are orthogonal; i.e.,

$$\int \psi_n^* \psi_m \mathrm{d}\boldsymbol{r} = 0 \tag{2.178}$$

for $n \neq m$, where $\boldsymbol{r}$ is shorthand for all position coordinates on which the wavefunction depends. Second, a general function can be expanded in terms of all the eigenfunctions of an operator, a so-called complete set of functions. This means that we can write for any function

$$f = \sum_n c_n \psi_n, \tag{2.179}$$

where c_n are constants and the sum is over a complete set of functions ψ_n.

Once a solution to the Schrödinger equation, i.e., the wavefunction ψ, has been found, one can calculate expectation values for any dynamical variable of the system, such as position or kinetic or potential energy. For example, according to (2.173), the expectation value of the location of a particle moving on the x-axis would be

$$\langle x \rangle = \int_{-\infty}^{+\infty} x \psi^* \psi \mathrm{d}x, \tag{2.180}$$

which in operator notation would become

$$\langle x \rangle = \int_{-\infty}^{+\infty} \psi^* \hat{x} \psi \mathrm{d}x, \tag{2.181}$$

where $\psi = \psi(x)$ is the wavefunction of the system and $\hat{x}$ the operator of multiplication associated with coordinate x. The physical significance of the expectation value of x is the average of x that would be found in a collection of many systems all described by a wavefunction ψ, or from a repetition of an experiment on a single system always prepared in the same state ψ. Analogous definitions apply to the expectation values of all other dynamical variables in terms of the associated operators; cf. Exercise 2.10.

Finally, given the fact that the classical position of a particle is represented in quantum mechanics by an associated multiplicative operator and the classical conjugate momentum of a particle is given by an associated differential operator, it can generally be shown that the position and the momentum of a particle in the same direction cannot have precise values at the same time. Rather, the combined uncertainty of position and momentum is characterized by a minimum value, given by

$$\sqrt{\langle \Delta x^2 \rangle}\sqrt{\langle \Delta p_x^2 \rangle} \geq \frac{h}{4\pi}. \tag{2.182}$$

Here $\sqrt{\langle \Delta x^2 \rangle} = \sqrt{\langle (x - \langle x \rangle)^2 \rangle} = \sqrt{\langle x^2 \rangle - \langle x \rangle^2}$ is the standard deviation from the mean value for x, with an analogous definition for $\sqrt{\langle \Delta p_x^2 \rangle}$. So, if we know the position of a particle to within a range Δx, then we can specify the linear momentum parallel to x to within a range Δp_x, with $\Delta x \Delta p_x \sim h$. The phase space formed by all values of the position and momentum coordinates is thus discretized, i.e., divided into subcells of order h. This constraint was used in formulating the semiclassical approximation in Section 2.2. Equation (2.182) is a special case of the Heisenberg uncertainty principle, which also applies to other conjugate pairs of variables. Although we have not derived it here mathematically we shall confirm it in a particularly simple case below; cf. Exercise 2.10.

Generally, solution of the Schrödinger equation for a molecular system is not possible in closed form, and the numerical calculation of high-quality approximate solutions is a very difficult task of computational chemistry. However, in exceptional but important cases analytical solutions can be worked out, e.g., for some one-particle problems. A well-known example is the hydrogen atom, which consists of one electron moving around the atomic nucleus, considered to be in a fixed position, so that only a Schrödinger equation of the electron remains. Thus, we arrive at a one-particle problem. Similarly, the three most important models of molecular motion of single molecules discussed before in the framework of classical mechanics, translation, vibration, and rotation, have been reduced to one-particle problems in the classical mechanical treatment. The associated Schrödinger equations can be solved analytically and we shall present the results of these solutions below.

2.5.3 Energy Levels of a Molecule

The energies of the relevant classical degrees of freedom for a single molecule were evaluated in Section 2.3. Here, we present the quantum mechanical treatment. Both types of results will be used in Chapter 3 and this will allow us to appreciate the conditions under which the classical mechanical results can be used and inform us about corrections that are necessary when we apply the semiclassical approximation.

We first consider the energy mode of translation. From classical mechanics we know that the kinetic energy of translation of a particle is most conveniently formulated in cartesian coordinates. If we first restrict ourselves to one-dimensional motion of a particle of mass m in the x direction without any potential energy the Schrödinger equation for this simple system reads

$$\frac{d^2\psi_x}{dx^2} + \frac{8\pi^2 m}{h^2}\varepsilon_x\psi_x = 0. \tag{2.183}$$

Here $\psi_x(x)$ is the wavefunction of the system, and ε_x its total energy. Specifically, we consider free movement within a container of dimension X. The coordinates of the particle are then restricted in that it is confined to within the boundaries of the system $0 < x < X$, and the potential energy is zero in this range, whereas it becomes infinite at the boundaries. This implies the boundary conditions

$$\psi_x(0) = 0 \tag{2.184}$$

and

$$\psi_x(X) = 0, \tag{2.185}$$

ensuring that the molecule cannot be found outside the container. Although we are primarily interested in the energy values of translational motion, we shall discuss the solution of the Schrödinger equation in some detail. This will allow us to derive some results in this particular case that we will generalize to others.

The general solution of the Schrödinger equation for this system is

$$\psi_x = C_1 \sin\left(x\sqrt{\frac{8\pi^2 m\varepsilon_x}{h^2}}\right) + C_2 \cos\left(x\sqrt{\frac{8\pi^2 m\varepsilon_x}{h^2}}\right). \tag{2.186}$$

The boundary condition at $x = 0$ gives

$$C_2 = 0, \tag{2.187}$$

whereas that at $x = X$ leads to

$$0 = C_1 \sin\left(X\sqrt{\frac{8\pi^2 m\varepsilon_x}{h^2}}\right). \tag{2.188}$$

Because $C_1 \neq 0$ in order to have the particle somewhere in the container, the boundary condition at $x = X$ requires that

$$\sin\left(X\sqrt{\frac{8\pi^2 m\varepsilon_x}{h^2}}\right) = 0, \tag{2.189}$$

or, alternatively,

$$X\sqrt{\frac{8\pi^2 m\varepsilon_x}{h^2}} = n_x\pi \tag{2.190}$$

with

$$n_x = 1, 2, 3, \cdots \tag{2.191}$$

as the translational quantum number in the x direction. The value $n_x = 0$ is excluded due to the normalization condition (2.173); cf. (2.194). Physically this would mean that there was no particle at all in the x-range. We thus are led to the result that the total energy of free translational motion of a single particle in a one-dimensional container can only have discrete values, according to

$$\varepsilon_{n_x} = \frac{h^2}{8mX^2}n_x^2. \tag{2.192}$$

The energy of translational motion of a particle that is confined to some range is quantized and the quantization results from the boundary conditions of the Schrödinger equation. In its lowest possible quantum state, $n_x = 1$, the particle has a specific energy and thus is never at rest. This is in accordance with the Heisenberg uncertainty principle: A particle confined within a finite region must have a corresponding uncertainty of momentum that excludes a precise energy value of zero; cf. Exercise 2.10.

The wavefunction ψ_{n_x} in the state n_x is given by

$$\psi_{n_x}(x) = C_1 \sin\left(n_x\pi\frac{x}{X}\right), \quad n_x = 1, 2, \cdots. \tag{2.193}$$

The still undetermined constant C_1 can be found from the normalization condition

$$\int_{x=0}^{X} \psi_{n_x}^2(x)\,dx = 1, \tag{2.194}$$

which simply states that the particle exists somewhere between $x = 0$ and $x = X$, so that

$$C_1 = \sqrt{\frac{2}{X}}. \tag{2.195}$$

One then has for the normalized wavefunction in state n_x

$$\psi_{n_x}(x) = \sqrt{\frac{2}{X}}\sin\left(n_x\pi\frac{x}{X}\right). \tag{2.196}$$

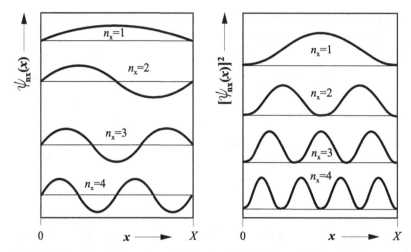

Fig. 2.15. Wavefunctions and probability densities for a particle in a box.

These wavefunctions and the associated probability densities are shown in Fig. 2.15 for the first four values of the translational quantum number. It can be seen that in the lowest translational state ($n_x = 1$) the molecule will most probably be found in the center of the x-range, whereas with increasing values of the quantum number the probability will be distributed more evenly over all x-values. The latter is expected for a macroscopic particle. This convergence of the quantum mechanical results to those of classical mechanics at large quantum numbers is a general phenomenon and has been referred to as the correspondence principle before. We reiterate that it is the basis of the semiclassical approximation introduced in Section 2.2. Fortunately, it turns out that this limit is actually attained in most practical cases.

EXERCISE 2.10

For a one-dimensional particle in a box of length X the wavefunction in quantum state n_x is given by

$$\psi_{n_x}(x) = \sqrt{\frac{2}{X}} \sin\left(n_x \pi \frac{x}{X}\right).$$

Calculate the expectation values of x, p_x, x^2, and p_x^2 as well as the combined uncertainty

$$\sqrt{\langle \Delta x^2 \rangle}\sqrt{\langle \Delta p_x^2 \rangle}$$

of position and momentum.

Solution

In quantum mechanics each classical variable is represented by a particular operator. So the kinetic energy operator is $-h^2/8\pi^2 m \nabla^2$ and the operators for the potential energy as well as for the position coordinate require multiplication by the respective functions $V(x)$ and x. From a generalization of (2.181) we find the expectation value of a dynamical quantity D in a particular quantum state by introducing the associated quantum mechanical operator $\hat{D}$ via

$$D = \int\limits_{-\infty}^{+\infty} \psi^* \hat{D} \psi \, dr^N.$$

We thus get

$$\langle x \rangle = \int\limits_0^X x \frac{2}{X} \sin^2 \left(n_x \pi \frac{x}{X} \right) dx = \frac{1}{X} \int\limits_0^X x \left[1 - \cos \left(\frac{2 n_x \pi x}{X} \right) \right] dx$$

$$= \frac{X}{2} - \frac{1}{X} \left\{ \left[\frac{\cos \frac{2 \pi n_x x}{X}}{\left(\frac{4 n_x^2 \pi^2}{X^2} \right)} \right]_0^X + \left[\frac{x \sin \frac{2 \pi n_x x}{X}}{\left(\frac{2 n_x \pi}{X} \right)} \right]_0^X \right\} = \frac{X}{2}$$

$$\langle p_x \rangle = \int\limits_0^X \sqrt{\frac{2}{X}} \sin \left(n_x \pi \frac{x}{X} \right) \frac{h}{2 \pi i} \frac{\partial}{\partial x} \sqrt{\frac{2}{X}} \sin \left(n_x \pi \frac{x}{X} \right) dx = 0,$$

where we used $p = \sqrt{2 \varepsilon m}$, and so, with (2.177), $\hat{p} = \frac{h}{2 \pi i} \frac{\partial}{\partial x}$,

$$\langle x^2 \rangle = \frac{2}{X} \int\limits_0^X x^2 \sin^2 \left(n_x \pi \frac{x}{X} \right) dx = \frac{1}{X} \int\limits_0^X x^2 \left[1 - \cos \left(\frac{2 n_x \pi x}{X} \right) \right] dx$$

$$= \frac{X^2}{3} - \frac{1}{X} \left\{ \left[\frac{x^2 \sin \frac{2 n_x \pi x}{X}}{\frac{2 n_x \pi x}{X}} \right]_0^X - \frac{2}{\frac{2 n_x \pi}{X}} \int\limits_0^X x \sin \frac{2 n_x \pi x}{X} dx \right\}$$

$$= \frac{X^2}{3} + \frac{2}{\frac{2 n_x \pi}{X}} \frac{1}{X} \left\{ \left[\frac{\sin \frac{2 n_x \pi}{X} x}{\left(\frac{2 n_x \pi}{X} \right)^2} \right]_0^X - \left[\frac{x \cos \frac{2 n_x \pi x}{X}}{\frac{2 n_x \pi}{X}} \right]_0^X \right\}$$

$$= \frac{X^2}{3} - \frac{1}{2} \left(\frac{X}{n_x \pi} \right)^2,$$

and finally,

$$\langle p_x^2 \rangle = -\frac{2}{X} \left(\frac{h}{2 \pi} \right)^2 \int\limits_0^X \sin \left(n_x \pi \frac{x}{X} \right) \frac{\partial^2}{\partial x^2} \sin \left(n_x \pi \frac{x}{X} \right) dx$$

$$= \frac{2}{X} \left(\frac{h}{2 \pi} \right)^2 \left(\frac{n_x \pi}{X} \right)^2 \int\limits_0^X \sin^2 \left(n_x \pi \frac{x}{X} \right) dx$$

$$= \frac{2}{X} \left(\frac{h}{2 \pi} \right)^2 \left(\frac{n_x \pi}{X} \right)^2 \left[-\frac{\sin \left(n_x \pi \frac{x}{X} \right) \cos \left(n_x \pi \frac{x}{X} \right)}{\frac{2 n_x \pi}{X}} + \frac{1}{2} x \right]_0^X$$

$$= \left(\frac{h}{2 \pi} \right)^2 \left(\frac{n_x \pi}{X} \right)^2.$$

We learn that the particle is expected to be found at the center of the one-dimensional container with zero momentum, independent of the particular quantum state. This is plausible for symmetry reasons, because the particle "sees" nothing except the repulsive walls at $x = 0$ and $x = X$ and is equally likely to be moving in either direction. Its expected kinetic energy corresponds to its total energy, which is given by (2.192). We further see that the uncertainty in position, i.e.,

$$\sqrt{\langle (\Delta x^2) \rangle} = \sqrt{\langle (x - \langle x \rangle)^2 \rangle} = \sqrt{\langle x^2 \rangle - \langle x \rangle^2},$$

is proportional to the length X, whereas that of momentum,

$$\sqrt{\langle \Delta p_x^2 \rangle} = \sqrt{\langle (p_x - \langle p_x \rangle)^2 \rangle} = \sqrt{\langle p_x^2 \rangle - \langle p_x \rangle^2},$$

is inversely proportional to X. Thus, in the limit $X \to \infty$, we have a free particle, with completely indefinite position, but there is no uncertainty of momentum. In the limit $X \to 0$, when the particle is precisely localized, we have an infinite uncertainty of momentum. For the combined uncertainty of position and momentum we find

$$\sqrt{\langle \Delta x^2 \rangle} \sqrt{\langle \Delta p_x^2 \rangle} = \frac{h}{2\pi} \sqrt{\left[\frac{X^2}{12} - \frac{1}{2} \left(\frac{X}{n_x \pi} \right)^2 \right] \left(\frac{n_x \pi}{X} \right)^2}$$

$$= \frac{h}{2\pi} \sqrt{\frac{(n_x \pi)^2}{12} - \frac{1}{2}}.$$

In the lowest quantum state, $n_x = 1$, we have

$$\sqrt{\langle \Delta x^2 \rangle} \sqrt{\langle \Delta p_x^2 \rangle} = \frac{h}{2\pi} \sqrt{\frac{\pi^2}{12} - \frac{1}{2}} = 1.136 \frac{h}{4\pi} > \frac{h}{4\pi}.$$

Thus, the Heisenberg uncertainty relation is verified for the particle in a box in its ground state and hence in all states with $n_x > 1$.

The results for the translational energy in one dimension can be generalized to three dimensions because of the independence of the three directions of motion. We thus arrive at

$$\varepsilon_{n_x n_y n_z} = \frac{h^2}{8m} \left(\frac{n_x^2}{X^2} + \frac{n_y^2}{Y^2} + \frac{n_z^2}{Z^2} \right), \tag{2.197}$$

or, if we assume a cubic container of volume V, without loss of generality,

$$\varepsilon_{n_x n_y n_z} = \frac{h^2}{8m V^{2/3}} \left(n_x^2 + n_y^2 + n_z^2 \right). \tag{2.198}$$

The energy eigenvalues of translational motion depend on volume. It can further be seen that different states may be associated with the same energy. As noted above, this phenomenon is called degeneracy. Fig. 2.16 shows the energy values $\varepsilon_{n_x n_y n_z}$, their degeneracies, g, and the associated quantum numbers for some states of the particle in a threedimensional box. Each individual triple of quantum numbers represents an individual quantum state, i.e., an individual wavefunction. Clearly, with increasing energy, the degeneracy of the levels also increases so that g becomes large for large values of n_x, n_y, n_z. For example, continuing the numbers in Fig. 2.16 on a computer leads to degeneracies between 1 and 264 for n_x, n_y, n_z up to 100 and to degeneracies between 1 and 1566 for n_x, n_y, n_z up to 500. In practical applications the translational quantum numbers are so high that they lead to a very large degeneracy of the translational energy levels of a molecule, which may be much larger than the number of molecules.

The solution of the Schrödinger equation for a particle in a box obeys the Heisenberg uncertainty relation, as shown in Exercise 2.10. It also allows us to verify the dimensional factor introduced in an ad hoc manner

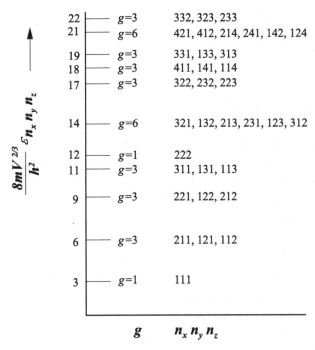

Fig. 2.16. Energy values, degeneracies, and quantum numbers for a particle in a box.

into the formulation of the semiclassical approximation to the partition function in Section 2.2.5 for the special case of translation, as demonstrated in Exercise 2.11.

EXERCISE 2.11

Derive the dimensional factor in the semiclassical approximation to the partition function for the particular case of the translational motion of one atom in a container of volume V.

Solution

The energy levels for translational motion of an atom in a volume V according to (2.198),

$$\varepsilon_{n_x n_y n_z} = \frac{h^2}{8mV^{2/3}} \left(n_x^2 + n_y^2 + n_z^2\right),$$

lead to the quantized momenta

$$p_{n_x} = (\pm)\sqrt{2m\varepsilon_{n_x}} = (\pm)\sqrt{\frac{h^2}{4V^{2/3}}n_x^2} = (\pm)\frac{h}{2V^{1/3}}n_x$$

with corresponding values for p_{n_y} and p_{n_z}. We therefore have

$$(\Delta p_{n_x})(\Delta p_{n_y})(\Delta p_{n_z}) = \left(\frac{h^3}{8V}\right)8,$$

where the factor 8 results from the fact that to each value of n_x, n_y, or n_z correspond a positive and negative momentum. The dimensional factor then follows from multiplication by $(\Delta x)(\Delta y)(\Delta z)$ as

$$(\Delta x)(\Delta y)(\Delta z)(\Delta p_{n_x})(\Delta p_{n_y})(\Delta p_{n_z}) = \Delta V\left(\frac{h^3}{8V}\right) 8,$$

which has the dimension of h^3 in agreement with (2.70).

We next consider the energy mode of vibration. The vibrational degree of freedom was modeled classically in Section 2.3.3 as a linear harmonic oscillator. It was shown to be energetically equivalent to a single particle of mass m_r undergoing a harmonic vibration in the x direction under the influence of an intramolecular potential energy

$$U = \frac{1}{2}\kappa(r - r_e)^2 = 2\pi^2 v_0^2 m_r x^2.$$

We thus have appropriate definitions for the mass, the coordinate, and the potential energies, and so the Schrödinger equation can be written as

$$\frac{d^2\psi}{dx^2} + \frac{8\pi^2 m_r}{h^2}\left(\varepsilon - 2\pi^2 v_0^2 m_r x^2\right)\psi = 0, \tag{2.199}$$

where ε is the total energy of the harmonic oscillator. When the notation is simplified by setting

$$\lambda = \frac{8\pi^2 m_r}{h^2}\varepsilon \quad \text{and} \quad \alpha = \frac{4\pi^2 m_r v_0}{h},$$

the Schrödinger equation for vibration takes the form

$$\frac{d^2\psi}{dx^2} + (\lambda - \alpha^2 x^2)\psi = 0. \tag{2.200}$$

Solution of this differential equation is more difficult than for the translation mode, but can be worked out analytically by suitable series expansions. We are here only interested in the energy levels. One then finds that physically acceptable solutions require the total energy of the oscillator to assume discrete values of the form

$$\varepsilon_v = \left(v + \frac{1}{2}\right)hv_0, \quad v = 0, 1, 2, 3\ldots. \tag{2.201}$$

Here, v is the vibrational quantum number. In the lowest quantum state, i.e., at $v = 0$, there is nonzero vibrational energy. This is in accordance with the uncertainty principle for an oscillating particle confined to a certain region by the potential energy. To evaluate the energy values quantitatively the frequency v_0 is required. It depends on the intramolecular potential energy of the molecule; cf. Section 2.3.3. The individual energy levels are nondegenerate, so that to each individual energy level there belongs only one wavefunction. Fig. 2.17 shows the intramolecular potential energy along with the quantized energy values of the harmonic oscillator. As the force constant decreases or the mass increases,

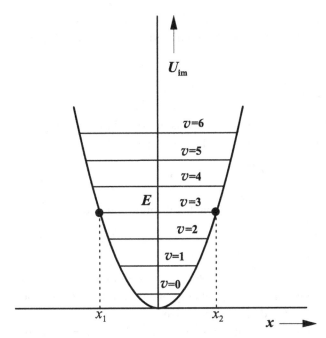

Fig. 2.17. Intramolecular potential energy of the harmonic oscillator.

we have a decreasing frequency and the separation between neighboring energy levels decreases too. In the limit of zero force constant the oscillating particle is unconfined and the energy quantization is lost. We note that a real diatomic molecule does not oscillate like a harmonic oscillator but rather will dissociate at sufficiently large energies. However, as will be shown in Chapter 3, the harmonic oscillator is a good model at normal temperatures for many purposes. For a real oscillator, the difference between successive quantum states becomes progressively smaller with increasing quantum number and quantization is lost in the limit of very large quantum numbers, as postulated by the correspondence principle.

The vibrational energy of polyatomic molecules, as discussed in Section 2.3.1, is formulated in terms of normal coordinates. The quantum mechanical expressions for the normal modes are then identical to those of a diatomic molecule. However, they may be degenerate; i.e., there may be more than one vibration with a particular frequency. An example for this is the bending mode for CO_2; cf. Fig. 2.3b. The quantum mechanical expression for a vibrational energy state ε_v of a polyatomic molecule can thus be written as

$$\varepsilon_v = \sum_j \varepsilon_{v_j} = \sum_j h\nu_{0j}\left(\upsilon_j + \frac{1}{2}\right), \quad \upsilon_j = 0, 1, 2\ldots, \tag{2.202}$$

where the summation goes over all normal modes.

We finally consider the rotational degree of freedom. Our treatment of the linear rigid rotator with moment of inertia I in the framework of classical mechanics has revealed that its energy can be described as the kinetic energy of a single particle of mass I rotating on a spherical surface of $r = 1$. It is thus adequately formulated in terms of polar coordinates with the constant value of $r = 1$. We then have the Schrödinger equation in the form

$$\frac{1}{\sin\theta}\frac{\partial}{\partial\theta}\left(\sin\theta\frac{\partial\psi}{\partial\theta}\right) + \frac{1}{\sin^2\theta}\frac{\partial^2\psi}{\partial\phi^2} + \frac{8\pi^2 I\varepsilon}{h^2} = 0.$$

In solving this equation by suitable series expansions one finds that the total energy of the linear rigid rotator is quantized according to

$$\varepsilon_l = l(l+1)\frac{h^2}{8\pi^2 I}, \quad l = 0, 1, 2, 3\ldots. \tag{2.203}$$

These energy values are degenerate. To one fixed rotational energy ε_l there belong

$$g = 2l + 1 \tag{2.204}$$

quantum states; i.e., each rotational energy level is $(2l + 1)$-fold degenerate. At $l = 0$, i.e., in the lowest quantum state, the energy of rotation is zero because the rotation is free, i.e., unconfined in rotation space.

The simple models considered above for the various degrees of freedom of a single molecule are surprisingly good approximations to the single-molecule energy. The associated quantum mechanical results for the energy states are those that have to be used in the rigorous quantum mechanical formulation of the partition function. By comparing them to the results obtained from classical mechanics in the framework of the semiclassical approximation, we will reassure ourselves that this approximation does not introduce substantial errors in many cases. However, we also will find exceptions to this rule, e.g., for the vibrational degree of freedom; cf. Chapter 3. Even in those applications where the semiclassical approximation is valid, the quantum mechanical treatment is necessary because it provides us with the proper correction factors due to symmetry effects and due to the Heisenberg uncertainty principle, anticipated in Section 2.2.5. We note that we have not presented a simple closed formula for the energy levels of the internal rotation. Although for the limiting cases of $U_{\text{ir,max}}/kT$ being very small or very large this energy mode can be transformed into those of external rotation and vibration, respectively, there is no simple analytical solution for the Schrödinger equation for this energy mode in the general case. Results from a numerical solution will be used in Chapter 3.

2.5.4 Electronic Structure of Molecules

The one-particle Schrödinger equation can frequently be solved analytically. Important results are obtained for the motional degrees of freedom associated with single molecules, such as translation, rotation, and vibration. Here, the particle to be considered is the atom or the rigid rotator or the harmonic oscillator, the latter both formulated in single-particle terms. These results, however, address only part of the problems associated with molecular energies. In the mechanical models discussed above, the geometries of the molecules, as well as their intramolecular force fields, have been assumed to be known. This information, however, must also be obtained from the Schrödinger equation in a fully predictive theory. This then leads to many-particle problems [19]. Because the particles to be considered are the electrons, we refer to the methods for solving the many-particle Schrödinger equation as electronic structure methods. Before we describe these methods we note that for many simple molecules spectroscopic data are available that have been analyzed within the framework of quantum mechanics to yield molecular properties such as moments of inertia and vibration frequencies. If such data are available in tables we can use them in our molecular models without bothering about electronic structure calculation. However, for complicated molecules such data are not normally available or are of questionable accuracy and thus the use of quantum-chemical computer codes is the only source of information for their properties.

In a molecule there is more than one nucleus and there are many electrons and the wavefunction depends on the positions of the electrons and the nuclei. The Hamilton operator associated with a molecule then reads

$$\hat{H} = -\frac{h^2}{8\pi^2} \sum_k \frac{1}{m_k} \left(\frac{\partial^2}{\partial x_k^2} + \frac{\partial^2}{\partial y_k^2} + \frac{\partial^2}{\partial z_k^2} \right) + \frac{1}{4\pi\varepsilon_0} \sum_j \sum_{k<j} \frac{q_j q_k}{\Delta r_{jk}}. \qquad (2.205)$$

The first term, the kinetic energy operator, contains a summation over all particles in the molecule, i.e., electrons and nuclei. The second term, the potential energy operator, is the Coulomb repulsion between all pairs of charged particles, with Δr_{jk} as the distance between and q_j, q_k as the charges on the particles j and k. For an electron, the charge is $-e$, the elementary charge, whereas for a nucleus, the charge is Ze with Z as the atomic number for the considered atom.

The kinetic energy operator is a sum of one part associated with the motion of the electrons and another part associated with the motion of the nuclei. The potential energy operator accounts for the attractions between the nuclei and the electrons, the repulsion between the electrons, and the repulsion between the nuclei. Because the mass of a typical nucleus is much greater (by more than a factor of 1000) than that of an electron, the nuclei are heavy enough to behave essentially like classical particles. This finally justifies modeling molecules as geometrical structures, which is the basis of all molecular models discussed in this book. As a consequence, the velocities of the nuclei are much smaller than those of the electrons. If we now restrict our attention to a region around a local minimum of a single electronic state, in particular the ground state, the Schrödinger equation can be approximately separated into one equation for the electronic wavefunction for a fixed nuclear geometry and another equation for the nuclear wavefunction, where the energy from the electronic wavefunction in the ground state is the potential energy. The nuclei thus move on the potential energy surfaces, which are solutions to the electronic Schrödinger equation. The total energy of the electrons, E^{el}, is calculated for fixed nuclei in a sufficiently large set of possible positions and its minimum defines (approximately) the geometrical structure of a molecule. This is the Born–Oppenheimer approximation briefly introduced in classical terms in Section 2.3.

On the basis of the Born–Oppenheimer approximation the electronic Schrödinger equation reads

$$\hat{H}^{el}\psi^{el}(r; R) = E^{el}(R)\psi^{el}(r; R) \tag{2.206}$$

with

$$\begin{aligned}
\hat{H}^{el} = & -\frac{h^2}{8\pi^2 m_e} \sum_i^n \left(\frac{\partial^2}{\partial x_i^2} + \frac{\partial^2}{\partial y_i^2} + \frac{\partial^2}{\partial z_i^2} \right) \\
& - \sum_i^n \sum_I^N \frac{Z_I e^2}{4\pi \varepsilon_0 |R_I - r_i|} + \sum_i^n \sum_{j<i}^n \frac{e^2}{4\pi \varepsilon_0 |r_i - r_j|} \\
& + \sum_I^N \sum_{J<I}^N \frac{Z_I Z_J e^2}{4\pi \varepsilon_0 |R_I - R_J|}.
\end{aligned} \tag{2.207}$$

Here, we adopt the notation of capital letters for the nuclei, of which the total number is N, and of small letters for the electrons, of which the total number is n. The notation $\psi(r; R)$ thus means a wavefunction depending on the positions r of the electrons, which also depends parametrically on the fixed positions R of the nuclei. The last term is constant for fixed positions of the nuclei. It thus simply shifts the energy and may be added afterward. We will therefore not consider it any further in (2.207). The resulting electronic energy $E^{el}(R)$ for a particular electronic state, such as the ground state, depends only on the fixed positions of the nuclei. It is used as the potential energy in the nuclear Hamiltonian, as, e.g., in the Schrödinger equation for harmonic vibration. In the following we drop the parametric dependence on R for notational simplicity.

There are essentially two complications with respect to simple one-particle solutions of the Schrödinger equation as adequate, e.g., for the hydrogen atom or also for the mechanical models discussed above. First, the fact that there is more than one particle makes the solutions more complex because of the higher dimensionality of the wavefunction. Second, the electrons interact, precluding a simple reduction to a one-particle solution. Analytical solutions are no longer possible for many-particle problems.

Let us consider, as an illustration, a four-electron problem, i.e., the solution of the Schrödinger equation for a system of four electrons. We then have a Hamilton operator of the form $\hat{H}(1, 2, 3, 4)$, where $1, \ldots, 4$ refer to the four particles. If we assume, as a first approximation, that we can neglect the energetic interactions between the electrons, the four-particle Hamilton operator can be separated to yield

$$\hat{H}(1, 2, 3, 4) = \hat{h}(1) + \hat{h}(2) + \hat{h}(3) + \hat{h}(4), \tag{2.208}$$

where $\hat{h}(1)$ is the one-particle Hamilton operator of the first electron, with an analogous notation for $\hat{h}(2)$, $\hat{h}(3)$, and $\hat{h}(4)$. These operators include the Coulomb interaction between the electrons and the nuclei, the latter assumed to be in fixed positions. The Schrödinger equation for the four-particle problem reads

$$\hat{H}(1, 2, 3, 4)\psi(r_1, r_2, r_3, r_4) = E\psi(r_1, r_2, r_3, r_4), \qquad (2.209)$$

where $r_i = \{x_i, y_i, z_i\}$ is the position vector of electron i. Due to the assumed independence of the electrons the Schrödinger equation can be solved by a product ansatz and we find for the wavefunction

$$\psi(r_1, r_2, r_3, r_4) = \phi_a(r_1)\phi_b(r_2)\phi_c(r_3)\phi_d(r_4), \qquad (2.210)$$

where $\phi_a(r_1)$ is the single-electron wavefunction of electron 1 as a function of its position in a quantum state a, i.e., one element of the set of eigenfunctions associated with the single-electron Hamilton operator $\hat{h}(1)$. The function $\phi_a(r_1)$ is referred to as a molecular orbital, with analogous notations for the other electrons. Although the solution (2.210) is correct from a purely mathematical point of view, it is physically unacceptable, even apart from the neglect of electron interactions. The reason is that it does not fulfill the Pauli exclusion principle. This is a physical principle, which postulates that electron wavefunctions must be antisymmetric under the exchange of any two electrons. Thus, any valid wavefunction must satisfy the condition

$$\psi(r_1, \ldots, r_i, \ldots r_j, \ldots r_n) = -\psi(r_1, \ldots, r_j, \ldots r_i, \ldots r_n). \qquad (2.211)$$

In correspondence with the Pauli principle, we must qualify the quantum state of an electron more precisely by noting that it is not defined by the orbital alone, but also has two different spins, i.e., intrinsic angular momenta. So we must multiply each molecular orbital by a spin function, which can have either of two values. The molecular orbital then is replaced by a spinorbital, a function of both the electron's position and its spin. A n-electron system is generally characterized by n occupied spinorbitals. If the $n/2$ lowest energy (spatial) orbitals (n being an even number) are each occupied with two electrons the system is referred to as a closed-shell system. For simplicity, we restrict further consideration here to closed shells. A systematic construction of antisymmetric and normalized wavefunctions from a product ansatz then leads to the introduction of so-called Slater determinants. For a four-electron problem the Slater determinant reads

$$\psi(r_1, r_2, r_3, r_4) = \frac{1}{\sqrt{4!}} \begin{vmatrix} \phi_\alpha(r_1) & \phi_\beta(r_1) & \phi_\gamma(r_1) & \phi_\delta(r_1) \\ \phi_\alpha(r_2) & \phi_\beta(r_2) & \phi_\gamma(r_2) & \phi_\delta(r_2) \\ \phi_\alpha(r_3) & \phi_\beta(r_3) & \phi_\gamma(r_3) & \phi_\delta(r_3) \\ \phi_\alpha(r_4) & \phi_\beta(r_4) & \phi_\gamma(r_4) & \phi_\delta(r_4) \end{vmatrix}. \qquad (2.212)$$

Here, the $\phi_\alpha \ldots \phi_\delta$ represent the single-electron wavefunctions, i.e., the spinor-bitals. Each of the 24 products contained in this Slater determinant and given by a different product of the spin orbitals represents a mathematically correct eigensolution of the four-electron Schrödinger equation, (2.209), with Hamiltonian (2.208), but not a physically acceptable wavefunction. The Slater determinant, however, as a linear combination of these, although also an eigenfunction, is in particular one that is a physically acceptable wavefunction obeying the Pauli exclusion principle. Each row is formed by representing all possible assignments of one particular electron r_i to all orbital-spin combinations. The prefactor ensures normalization. Interchanging two electrons means an interchange of two rows in the determinant, with the effect of changing its sign.

We now take electronic interactions into account. Then the Hamilton operator reads, cf. (2.207),

$$\hat{H} = \sum_i \hat{h}(i) + \sum_i \sum_{j<i} \frac{1}{4\pi\varepsilon_0} \frac{e^2}{|r_i - r_j|}, \tag{2.213}$$

where the second term is called the interelectronic repulsion term and introduces the Coulomb repulsion of any two electrons i and j, each with charge e. It is this term that is responsible for the fact that an exact solution to the Schrödinger equation in terms of a simple product ansatz with orbitals obtained from single-electron operators $\hat{h}$ is no longer possible. A many-particle solution is required, with the consequence that approximate methods must be used. One such method that is widely used is the Hartree–Fock (HF) method. In the Hartree–Fock method for closed-shell problems a wavefunction in the form of a single Slater determinant, such as (2.212) for a four-electron system, is sought. For an n-electron system the approximate wavefunction in the HF level would then be

$$\psi(1, 2, \ldots n) = \Phi_0, \tag{2.214}$$

where

$$\Phi_0 = \frac{1}{\sqrt{n!}} \begin{vmatrix} \phi_\alpha(1) & \phi_\beta(1) & \cdots & \phi_n(1) \\ \phi_\alpha(2) & \phi_\beta(2) & \cdots & \phi_n(2) \\ \vdots & & & \\ \phi_\alpha(n) & \phi_\beta(n) & \cdots & \phi_n(n) \end{vmatrix}, \tag{2.215}$$

with $\alpha, \beta, \ldots, n$ as the n occupied spinorbitals, is the Slater determinant in the HF level. The spinorbitals that give the best n-electron determinantal wavefunction are found from the variational principle. This general and easily proven principle states that the expectation value for the energy corresponding to any approximate wavefunction such as Φ_0 has a value above or equal to the exact energy, where equality holds only for the true wavefunction. In operational form the variational principle reads

$$\int \Phi_0^* \hat{H} \Phi_0 dr^n \rightarrow \text{Minimum}, \tag{2.216}$$

where $\hat{H}$ is given by (2.213). The application of this minimization procedure leads to the Hartree–Fock equations for the individual spinorbitals. When the HF method is analyzed more closely it becomes clear that it is an approximation, in the sense of a mean-field method. An electron is considered in the field of the fixed nuclei and the averaged field of all other electrons.

The variational principle ensures that the best orbitals consistent with the product ansatz, i.e., the Slater determinant, are obtained. The numerical solution of this minimization procedure is performed by expanding the molecular orbitals (more precisely, the spatial parts of the spinorbitals) in terms of so-called basis functions, as

$$\phi_a(\boldsymbol{r}) = \sum_{\mu=1}^{m} c_{\mu a} \chi_\mu(\boldsymbol{r}), \tag{2.217}$$

where the $c_{\mu a}$ are as yet unknown molecular orbital expansion coefficients and the $\chi_\mu(\boldsymbol{r})$ are the basis functions that are normalized. They are also called atomic orbitals, although they are generally not solutions to the atomic Hartree–Fock problem. The summation extends over the number of basis functions chosen. In principle, the basis functions can be chosen arbitrarily. In practice, either Gauss-type orbitals or Slater-type orbitals, which have a close relation to the wavefunctions of the one-electron atom, are used. Gauss functions have the general form

$$g_{i,j,k}(\alpha, \boldsymbol{r}) = c x^i y^j z^k e^{-\alpha r^2}, \tag{2.218}$$

where α is a constant determining the radial extent of the function. Here, $\boldsymbol{r} = \{x, y, z\}$ defines the location of an electron with respect to the center of the Gauss function and the Gauss functions are usually centered on the atomic nuclei. Usually, linear combinations of the primitive Gaussians with the form of (2.218) are used to build up the actual basis functions $\chi_\mu(\boldsymbol{r})$ and these are then called contracted Gaussians. Application of the variational principle implies that the molecular orbital expansion coefficients are determined by minimizing the energy. The set of equations obtained is referred to as the Roothaan–Hall equations. Their explicit appearance is of no interest to us here, because we do not wish to solve them but rather leave this to the established computer codes. It suffices to say that they are nonlinear and thus must be solved iteratively by a procedure that is known as the self-consistent field (SCF) method. So, at convergence, the energy is at a minimum, and the orbitals obtained generate a field that produces the same orbitals. Although we have as yet considered only the n actually occupied spinorbitals, the system of equations, like the standard Schrödinger equation, has in principle an infinite number of eigenfunctions. Thus, the solution produces two sets of orbitals, one set occupied ($\phi_a, \phi_b, \ldots$) and one set virtual, i.e., unoccupied ($\phi_{a'}, \phi_{b'}, \ldots$). The total number of spatial wavefunctions actually obtained in the numerical solution is equal to the number of basis functions used. If this number is m, with $m \geq n$, we have $2m$ different spinorbitals, i.e., n occupied spinorbitals and $(2m - n)$ virtual spinorbitals.

The Slater determinant Φ_0 composed of the n occupied spinorbitals is the HF ground-state wavefunction of the molecule.

The whole set of basis functions used in (2.217) is referred to as the basis set. The use of an infinite number of basis functions would result in a limiting energy called the Hartree–Fock limit. This HF limit is not the exact ground-state energy of the molecules, however, because it is based on the mean-field approximation for electronic interactions. Still, it is the best result that can be obtained on the Hartree–Fock level and serves to quantify the basis-set truncation error associated with the use of a finite basis set. One aspect of the art of computational chemistry is the right compromise between keeping the number of basis functions low, and nevertheless achieving a small basis-set truncation error by choosing them cleverly. The simplest type of basis set is the minimal basis set. It is constructed in such a way that one function is used to represent each of the orbitals of elementary valence theory. So a minimal basis set would include one function for H and five functions for O, which leads to seven functions as a minimal basis set for H_2O. Minimal basis sets result in wavefunctions and energies that are not close to the Hartree–Fock limit. Significant improvements are achieved by so-called double-zeta or triple-zeta basis sets, in which each basis function of the minimal basis set is replaced by two or three basis functions, respectively. When bonds form in molecules, atomic orbitals are distorted (polarized) by adjacent atoms. This effect is taken into account by extended basis sets, e.g., by adding particular polarization functions. Also, there are numerous ways to construct contracted basis sets.

While the Hartree–Fock approximation takes interactions between electrons into account in an average way it still neglects important aspects of these interactions, i.e., the correlation effects. The motions of electrons are correlated because they try to avoid each other due to Coulomb interactions. The associated correlation energy is defined as the difference between the true energy and the Hartree–Fock energy. Such correlation effects are covered by quantum mechanical methods beyond Hartree–Fock. Various approximate approaches are available.

One such method is referred to as configuration interaction (CI). Because, for closed shells, the HF method determines the best one-determinant wavefunction within a given basis set, it is plausible that, to improve on HF results, a trial wavefunction containing more than one Slater determinant should be used. Many different Slater determinants can be constructed by taking excited states into account. By replacing molecular orbitals, which are occupied in the HF determinant, with orbitals that are unoccupied, i.e., virtual, a whole series of determinants may be generated, referred to as configuration state functions. These can be classified according to the number of occupied HF molecular orbitals that have been replaced by virtual molecular orbitals. Thus Slater determinants that are singly, doubly, etc. excited relative to the HF determinant can be created. Thus, a singly excited determinant is one for which an orbital ϕ_a has been replaced by a virtual orbital ϕ_a'; in a doubly excited determinant two orbitals ϕ_a and ϕ_b are replaced by ϕ_a' and ϕ_b'; and so on. It can be proven that the exact ground-state wavefunction can be expressed as a linear combination of all possible n-electron Slater determinants arising from a complete set of spinorbitals in the form

$$\psi(1, 2 \ldots n) = a_0 \Phi_0 + \sum_j a_j \Phi_j, \qquad (2.219)$$

where Φ_0 is the Hartree–Fock limit and the Φ_j are excited determinants obtained by summing over all singly excited determinants, over all unique pairs of spinorbitals in doubly excited determinants, over all unique triples of spinorbitals in triply excited determinants, and so on. Clearly, it is practically impossible to evaluate an infinite set of n-electron Slater determinants with each determinant being composed from an infinite set of spinorbitals. Practically, all determinants are constructed from a finite basis set. We refer to a quantum chemical computation as "full CI" when all determinants are used for a given finite basis set. As the number of spinorbitals gets larger and larger "full CI" calculations eventually reach the exact solution to the Schrödinger equation. Practical CI methods augment the Hartree–Fock level by adding only a limited number of excited determinants. Thus, the CIS method adds single excitations, CID double excitations, CISD both single and double excitations, and so on. The expansion coefficients $a_0, a_1, \ldots$ in (2.219) and the corresponding expansion coefficients in limited configuration interaction are again determined from variation theory, i.e., by minimizing the expectation value of the energy.

There is an alternative approach to taking correlation effects into account, referred to as perturbation theory. In this method, the Hamiltonian is formulated as

$$\hat{H} = \hat{H}^{(0)} + g\hat{H}^{(1)}, \qquad (2.220)$$

where the zero-order Hamiltonian $\hat{H}^{(0)}$ is associated with the HF (ground-state) energy wavefunction and the HF (ground state) energy. Further, $\hat{H}^{(1)} = \hat{H} - \hat{H}^{(0)}$, where $\hat{H}$ is the electronic Hamiltonian, takes the perturbation due to electron correlation into account. Finally, g is a formal strength parameter between 0 and 1. This version of perturbation theory is referred to as Møller-Plesset (MP) perturbation theory. We do not specify $\hat{H}^{(0)}$ and $\hat{H}^{(1)}$ here explicitly, because we do not evaluate numerical results, but rather rely for that purpose on quantum-chemical computer codes.

To derive the formal equations for a basic understanding of the method we expand the wavefunctions and the energy values of the system with electron correlation formally around those of the unperturbed system. Restricting consideration to the electronic ground state, denoted by an index 0, we write

$$\psi_0 = \psi_0^{(0)} + g\psi_0^{(1)} + g^2\psi_0^{(2)} + \cdots \qquad (2.221)$$

$$E_0 = E_0^{(0)} + gE_0^{(1)} + g^2E_0^{(2)} + \cdots. \qquad (2.222)$$

Here, suffix (1) refers to a first-order correction to the unperturbed system, suffix (2) to a second-order, etc. The Schrödinger equation for the ground state reads

$$\hat{H}\psi_0 = E\psi_0. \tag{2.223}$$

Introducing the expansions (2.221) and (2.222) into the Schrödinger equation (2.223) and noting that the equation must be satisfied for any arbitrary value of the expansion parameter g leads, by comparing coefficients of the various powers of g, to a hierarchy of equations. The first three of them are

$$g^0 : \hat{H}^{(0)}\psi_0^{(0)} = E_0\psi_0^{(0)} \tag{2.224}$$

$$g^1 : \hat{H}^{(0)}\psi_0^{(1)} + \hat{H}^{(1)}\psi_0^{(0)} = E_0^{(0)}\psi_0^{(1)} + E_0^{(1)}\psi_0^{(0)} \tag{2.225}$$

$$g^2 : \hat{H}^{(0)}\psi_0^{(2)} + \hat{H}^{(1)}\psi_0^{(1)} = E_0^{(0)}\psi_0^{(2)} + E_0^{(1)}\psi_0^{(1)} + E_0^{(2)}\psi_0^{(0)}. \tag{2.226}$$

An expression for the perturbation energy, i.e., its expectation value, to first order is obtained by multiplying (2.225) from the left by $\psi_0^{(0)*}$ and integrating over all position coordinates; cf. (2.181). This gives

$$\int \psi_0^{(0)*}\hat{H}^{(0)}\psi_0^{(1)}d\mathbf{r} + \int \psi_0^{(0)*}\hat{H}^{(1)}\psi_0^{(0)}d\mathbf{r}$$
$$= \int \psi_0^{(0)*}E_0^{(0)}\psi_0^{(1)}d\mathbf{r} + \int \psi_0^{(0)*}E_0^{(1)}\psi_0^{(0)}d\mathbf{r}. \tag{2.227}$$

The first-order correction to the ground state wavefunction can be expanded in terms of the unperturbed wavefunctions of the system because the latter constitute a complete basis set of functions; i.e.,

$$\psi_0^{(1)} = \sum_{k\neq 0} a_k\psi_k^{(0)}, \tag{2.228}$$

where the sum is over all states and may include integration over a continuum, if this applies. Introducing (2.228) into (2.227) gives

$$\sum_{k\neq 0} a_k \int \psi_0^{(0)*}(\hat{H}^{(0)} - E_0^{(0)})\psi_k^{(0)}d\mathbf{r} = \sum_{k\neq 0} a_k \int \psi_0^{(0)*}(E_n^{(0)} - E_0^{(0)})\psi_k^{(0)}d\mathbf{r}$$
$$= E_0^{(1)} - \int \psi_0^{(0)*}\hat{H}^{(1)}\psi_0^{(0)}d\mathbf{r}.$$

Because $\psi_n^{(0)}$ are orthonormal functions, cf. (2.178), we find

$$E_0^{(1)} = \int \psi_0^{(0)*}\hat{H}^{(1)}\psi_0^{(0)}d\mathbf{r}. \tag{2.229}$$

An analogous procedure leads to the second-order correction to the energy. We multiply (2.226) from the left by $\psi_0^{(0)*}$ and integrate, giving

$$\int \psi_0^{(0)*}(\hat{H}^{(0)} - E_0^{(0)})\psi_0^{(2)}d\mathbf{r} = \int \psi_0^{(0)*}E_0^{(2)}\psi_0^{(0)}d\mathbf{r} + \int \psi_0^{(0)*}(E_0^{(1)} - \hat{H}^{(1)})\psi_0^{(1)}d\mathbf{r}. \tag{2.230}$$

The second-order correction to the ground state wavefunction is written, analogously to (2.228), as

$$\psi_0^{(2)} = \sum_{k \neq 0} b_k \psi_k^{(0)}. \tag{2.231}$$

Introducing this into (2.230) gives

$$E_0^{(2)} = \int \psi_0^{(0)*} \hat{H}^{(1)} \psi_0^{(1)} \mathrm{d}\boldsymbol{r} = \sum_{k \neq 0} a_k \int \psi_0^{(0)*} \hat{H}^{(1)} \psi_k^{(0)} \mathrm{d}\boldsymbol{r}. \tag{2.232}$$

We finally need the expansion coefficients a_k. Multiplying (2.225) from the left by $\psi_i^{(0)*}$ and substituting (2.228) for $\psi_0^{(1)}$ gives

$$\sum_{k \neq 0} a_k \int \psi_i^{(0)*} (\hat{H}^{(0)} - E_0^{(0)}) \psi_k^{(0)} \mathrm{d}\boldsymbol{r} = \int \psi_i^{(0)*} (E_0^{(1)} - \hat{H}^{(1)}) \psi_0^{(0)} \mathrm{d}\boldsymbol{r},$$

which can be solved, again making use of the orthonormality of wavefunctions, as

$$a_i = \frac{\int \psi_i^{(0)*} \hat{H}^{(1)} \psi_0^{(0)} \mathrm{d}\boldsymbol{r}}{E_0^{(0)} - E_i^{(0)}}, \tag{2.233}$$

where we assumed nondegenerate energy levels, which guarantee a nonzero denominator in (2.233). The second-order energy then finally reads

$$E_0^{(2)} = \sum_{k > 0} \frac{\int \psi_0^{(0)*} \hat{H}^{(1)} \psi_k^{(0)} \mathrm{d}\boldsymbol{r} \int \psi_k^{(0)*} \hat{H}^{(1)} \psi_0^{(0)} \mathrm{d}\boldsymbol{r}}{E_0^{(0)} - E_k^{(0)}}. \tag{2.234}$$

We note that, whereas the first-order correction $E_0^{(1)}$ can be evaluated from the unperturbed wavefunctions of the ground state, $E_0^{(2)}$, i.e., the second-order correction, depends on unperturbed wavefunctions in excited states. The sum $E_0^{(0)} + E_0^{(1)}$ gives the Hartree–Fock energy, whereas $E_0^{(2)}$ represents the first electron correlation correction to the Hartree–Fock level. Solution of (2.234) is demanding and can be executed using quantum-chemical computer codes. This level of perturbation theory is referred to as MP2. We note that MP2 is not variational; i.e., it does not in general give energies that are upper bounds to the exact energy.

There is a further method for accounting for electron correlation, referred to as the coupled-cluster method (CC). The CC method introduces the cluster operator, which relates the exact electronic wavefunction to the HF wavefunction. The effect of the cluster operator on Φ_0 is a linear combination of Slater determinants in which electrons from occupied spinorbitals have been excited to virtual spinorbitals, similarly to the CI method; cf. (2.219). The coupled cluster equations are obtained by substituting the linear combination of Slater determinants into the electronic Schrödinger equation.

All the above methods based on the HF level are widely used. Their limitations are due to the computational difficulty of executing accurate calculations for molecules containing many electrons and many atoms. So they are mostly applied today to small molecules. The final quality of such a quantum-chemical calculation is determined by the level of theory, i.e., how much electron correlation is included, and the size of the basis set. Different molecular properties such as molecular structure or vibration frequencies differ in their demands on computational level. Usually the molecular structure is much less sensitive to the theoretical level than relative energies. A major advantage of this route is that systematic improvement is possible by using progressively more sophisticated levels of calculation.

An entirely different quantum-chemical method is available that is not based on wave mechanics employing the exact Hamilton operator and making approximations to the wavefunction. This alternative approach is referred to as density functional theory (DFT) and is the preferred method for electronic structure calculations for nonsimple molecules. The basis of DFT is the fact that the ground state electronic energy E is completely determined by the electron density $\varrho(\mathbf{r})$. The electronic energy is thus a functional of the electron density, which is written as $E[\varrho]$. It has been proven that such a universal functional, i.e., one valid for all systems, exists. So, for a given function $\varrho(\mathbf{r})$, there is a single electronic energy E. The practical significance of this is that, whereas the wavefunction of an n-electron system depends on $3n$ spatial coordinates, the electron density depends only on three. This reduces the computational effort considerably. So, whereas the complexity of the wavefunction increases with the number of electrons, the electron density is determined by a number of variables that is independent of the system size. However, the universal functional connecting the electron density with the ground state energy is not known. Approximations to it must be found from DFT methods. The energy functional may be divided into three parts, the kinetic energy $T[\varrho]$, the attraction energy between nuclei and electrons $E_{ne}[\varrho]$, and the electron–electron repulsion energy $E_{ee}[\varrho]$. Further, the E_{ee} term may be divided into a Coulomb and an exchange-correlation part, $J[\varrho]$ and $E_{XC}[\varrho]$. Although the first three functionals are well defined, we do not know the exact analytical form of the exchange-correlation functional. The specification of a particular DFT method is thus associated with the selection of a suitable form for the exchange-correlation functional. The ground-state electron density is expressed in terms of one-electron functions, i.e., orbitals, such as

$$\varrho(\mathbf{r}) = \sum_{i=1}^{n} |\phi_i(\mathbf{r})|^2. \tag{2.235}$$

The $\phi_i(\mathbf{r})$ are referred to as the Kohn–Sham orbitals. They are found by the Kohn–Sham equations, which follow from applying the variational principle to the electronic energy $E[\varrho]$. The KS orbitals can be expressed in terms of a set of basis functions. Solving the KS equations then gives the coefficients in the basis set expansion. The computational problem is thus similar to that of the

HF method. Although DFT calculations are much less time-consuming than the methods based on wave mechanics, they cannot be improved systematically and a systematic approach to the exact result is impossible.

A number of computer codes are available that allow quantum-chemical calculations to be made for single molecules even for the nonspecialist [20, 21]. From such computer codes various important molecular properties related to the electronic structure of a molecule can be obtained. So it is possible to calculate the geometry of a molecule, i.e., the interatomic distances and angles determining the moments of inertia, by searching for the minimum of the energy hypersurface. It is further possible to calculate the multipole moments and polarizabilities as well as the vibrational frequencies and the barriers to internal rotation, if present. Also, formation and dissociation energies may be obtained. Whereas for small molecules such results are essentially exact when high-level quantum-chemical calculations are used, larger molecules would require excessive amounts of computer time, so lower computational levels are usually chosen. The reliability of the results must then be analyzed critically and with reference to the particular applications considered. A shorthand nomenclature for characterizing the quantum mechanical calculation procedure has established itself, indicating the level of electron correlation and the basis set taken into account. For example, MP stands for perturbation theory (Møller-Plesset); aug-cc-pVDZ defines a particular basis set, where aug stands for "augmented," (cc) means correlation consistent, and pVDZ polarized valence double zeta, i.e., a basis set of double zeta quality for the valence part plus a polarization function. The notation B3LYP defines the exchange-correlation functional of the DFT method and 6-31+ G(d,p) defines a particular contracted basis set of Gauss functions. The detailed information contained in this shorthand notation need not concern us here. It suffices to say that there are a great variety of methods among which to choose. So, although performing the calculation is relatively easy, considerable expertise is required to pick the level of calculation adequate for the particular information to be obtained and to properly estimate the accuracy of the results.

As an illustration, we consider some molecular properties of the ammonia molecule as obtained from a quantum-chemical computer code [21]. Table 2.1 summarizes the results for the moments of inertia, the harmonic vibration frequencies, the multipole moments, and the isotropic polarizability, as they are obtained directly from a computer listing. The moments of inertia I are communicated in 10^{-40} g cm^2, the vibration frequencies $\tilde{\nu} = \nu/c$ in cm^{-1}, where c is the speed of light, the dipole moment μ in D (debye), where 1 D = $3.336 \cdot 10^{-30}$ C m, the quadrupole moment θ in D Å, and the isotropic polarizability α in Å^3. The experimental sources are listed in the table. The agreement with the experimental data obtained from spectroscopy and other techniques is rather satisfactory, with the exception of the polarizability for the DFT level. We note that the calculated frequencies are harmonic while the experimental values

Table 2.1. Molecular parameters for NH_3 from the Gaussian 03 computer code

	B3LYP 6-31 + G(d,p)	MP2 aug-cc − pVDZ	MP2 aug-cc − pVTZ	exp. [8,22,23]
I_1	2.80	2.86	2.81	2.82; 2.78
I_2	2.80	2.86	2.81	2.82; 2.78
I_3	4.53	4.67	4.42	4.43; 4.31
$\tilde{v}_1$	999	1045	1037	932; 949
$\tilde{v}_2$	1674	1649	1669	1626; 1628
$\tilde{v}_3$	1674	1649	1669	1626; 1628
$\tilde{v}_4$	3485	3481	3502	3336; 3332
$\tilde{v}_5$	3628	3635	3649	3443; 3448
$\tilde{v}_6$	3628	3635	3649	3443; 3448
μ_z	1.765	1.641	1.593	1.47
θ_{zz}	−2.356	−2.043	−2.104	−2.32
α	1.621	2.081	2.103	2.22

represent true observed frequencies. The anharmonic frequencies are usually a few percent lower than the harmonic frequencies, as can also be seen from Table 2.1. Anharmonic corrections can also be obtained from computer codes. For the calculation of ideal gas heat capacities, the differences are not important at normal temperature; cf. Chapter 3.

2.5.5 Intermolecular Interactions

Although we have so far looked only at the properties of single molecules, quantum mechanics also determines the energy levels of whole molecular systems. Therefore, it can provide information about the intermolecular potential energy in a more general way than just considering classical electrostatics, as was done in Section 2.4. Generally, the intermolecular potential energy is given as

$$U = E - E(\infty), \tag{2.236}$$

i.e., by the difference between the total energy E of the molecular system of charges in the actual molecular configuration and the total energy in a state where all molecules are infinitely apart from each other, $E(\infty)$, i.e., in the state of single molecules. The total energy of a molecular system can be calculated, at least in principle, from its Schrödinger equation. The various contributions arise more clearly if we split the Hamilton operator of a system of charges into a kinetic and a potential term according to (cf. (2.176))

$$\hat{H} = \hat{H}^{\text{kin}} + \hat{H}^{\text{pot}} = -\frac{h^2}{8\pi^2} \sum_k \frac{1}{m_k} \nabla_k^2 + \frac{1}{2} \sum_a \sum_{b \neq a} \hat{V}_{ab}. \tag{2.237}$$

Here k denotes all particles in the system, i.e., electrons and nuclei. Further, a, b are the dummy indices of the charges. The interaction V_{ab} between the charges q_a and q_b is defined according to Coulomb's law, cf. (2.110), by

$$V_{ab} = \frac{1}{4\pi\varepsilon_0} \frac{q_a q_b}{r_{ab}}. \tag{2.238}$$

$\hat{V}_{ab}$ is a multiplicative operator, cf. Section 2.5.2, and thus represents the multiplication of $q_a q_b / 4\pi\varepsilon_0 r_{ab}$ by the wavefunction ψ, where q_a and q_b are the charges on sites a and b. Differentiating between charges within a molecule and those associated with different molecules, cf. Fig. 2.10, we finally get

$$\hat{H} = -\frac{h^2}{8\pi^2} \sum_k^K \frac{1}{m_k} \nabla_k^2 + \frac{1}{2} \sum_\alpha^\kappa \sum_i^{N_\alpha} \sum_a^{K_\alpha} \sum_{b\neq a}^{K_\alpha} \hat{V}_{a_{\alpha_i} b_{\alpha_i}}$$
$$+ \frac{1}{2} \sum_\alpha^\kappa \sum_\beta^\kappa \sum_i^{N_\alpha} \sum_j^{N_\beta} \sum_a^{K_\alpha} \sum_b^{K_\beta} \hat{V}_{a_{\alpha_i} b_{\beta_j}}, \tag{2.239}$$

where $i \neq j$ for $\alpha = \beta$ in the last summation. Here κ is the number of components, N_α the number of molecules of component α, and K_α the number of charges in a molecule of component α. An analogous notation applies to component β. Further, $V_{a_{\alpha_i} b_{\alpha_i}}$ denotes the interaction between charge q_a and charge q_b, both in molecule i of species α. This *intra*molecular interaction reflects the electronic structure of the single molecules. The resulting energy hypersurface yields, by suitable evaluations, the vibration frequencies and the potential of hindered internal rotation of single molecules and has been considered in Section 2.5.4. So the first two terms in (2.239) represent the operator for the energy of the single molecules. $V_{a_{\alpha_i} b_{\beta_j}}$ introduces the interaction between charge q_a of molecule α_i and charge q_b of molecule β_j. It is this contribution that is related to the *inter*molecular interactions.

A rigorous calculation of the intermolecular potential energy can be done for pairs of molecules in a large number of configurations. The potential energy surface for a molecular pair thus obtained can be represented by appropriate mathematical functions in terms of the configurational coordinates to yield the pair potential function. Although such calculations have been done in a few cases [24], they are currently still too computationally demanding and inaccurate for practical requirements. Yet valuable basic information can be obtained from them about the intermolecular potential energy and their practical utility is bound to increase as even more powerful computers become available in the future. However, for practically relevant molecules this is no more than a long time perspective. So, in the meantime, we must content ourselves with approximate approaches that make use of electronic structure calculations for single molecules as discussed in the preceding section.

One such approximation, designed in particular for liquid mixtures, is the continuum solvation model, discussed in the context of the electrostatic potential energy in Section 2.4. In this model the charge density distribution of a single molecule in the gas phase, as obtained from electronic structure calculations, is the starting point of a quantum-chemical iterative calculation procedure combined with classical electrostatics ending up with the total energy of the molecule, referred to as the solute, in a dielectric continuum. The dielectric responds to the charge distribution of the solute molecule by polarization. As discussed in Section 2.4, this effect can be modeled by a distribution of positive and negative screening charges on the inner surface of a cavity representing the solute molecule. In a self-consistent solution of the problem the effect of back polarization of the solute molecule by the screening charges is taken into account; i.e., the Hamilton operator of the solute in the dielectric is changed with respect to the gas phase. In a particular version of a continuum solvation model, referred to as COSMO, cf. Section 2.4, the solvent is modeled as a perfect conductor, i.e., with $\varepsilon_r = \infty$, and the results for the charges in each iteration are corrected for a finite dielectric constant of the solvent by an empirical factor $f_\varepsilon = (\varepsilon_r - 1)/(\varepsilon_r + 1/2)$.

The actual execution of the calculation is done with the help of quantum-chemical computer codes, in which COSMO has been implemented. Results of such a COSMO calculation are the charge density distribution over the surface of the cavity and the energy of the solute in the conductor, including the back polarization effect. The difference between the energy of the solute in the conductor and in the gas phase reflects the electrostatic intermolecular interactions. It is not a molecular potential energy, but rather has been shown in Section 2.4 to be identical to the electrostatic part of the thermodynamic free energy of solvation. It thus can be compared directly to experimental data. The ideal molar Gibbs free energy of solvation, $\Delta g_i^{s,0}$, is usually reported as the change of Gibbs free energy associated with the transfer of one mole of component i from the ideal gas state at $T^0 = 298.15$ K and $p^0 = 1$ bar to the state of an ideal dilute solution in the solvent S at $m^0 = 1$ and at the same temperature and pressure. For the purpose of assessing the potential model it is more straightforward to report the comparison between COSMO and experiment in terms of equal molar concentrations in the gas phase and in the solvent phase; i.e., $c_i^{ig} = 1$ mol/l and $c_i^{ids} = 1$ mol/l. We thus eliminate entropy changes due to compression effects in the comparison. The ideal molar Gibbs free energy of solvation in this definition, which we denote as $\Delta g_i^{s,1}$, is related to $\Delta g_i^{s,0}$ by

$$\Delta g_i^{s,1} = \Delta g_i^{s,0} - RT\ln(p^1/p^0) + RT\ln(x_i^1/x_i^0),$$

with $p^1 = c_i^{ig} RT$, $x_i^1 = 1/55.343$, $x_i^0 = 1/(1 + 1000/M_S)$, if we neglect any compression effects in the liquid. This gives for the solvent water

$$\Delta g_i^{s,1} = \Delta g_i^{s,0} - 7902 \text{ J/mol}.$$

In Table 2.2 we present a comparison of COSMO results as obtained from [20] for the free energy of solvation with experimental results for some solutes

Table 2.2. Gibbs free energy of solvation: Comparison of the COSMO model to experimental data [25] ($T = 298.15$ K; in kJ/mol)

Compound	$\Delta g_i^{s,0}$ exp	$\Delta g_i^{s,1}$ exp	$\Delta g_i^{s,1}$ COSMO	$\dfrac{\Delta g_{COSMO}^{s,1} - \Delta g_{exp}^{s,1}}{\Delta g_{exp}^{s,1}} \times 100$
Ammonia	−9.89	−17.57	−22.19	+26.3
Methanol	−13.44	−21.34	−18.98	−11.1
Methylamine	−11.35	−19.25	−17.78	−7.6
Methane	+16.27	+8.34	−1.02	−112
Oxygen	+16.40	+8.50	−0.78	−109
Butane	+16.09	+8.79	−1.34	−115

in water. The quantum-chemical calculations are based on the DFT method (BP86) with a TVZP basis. The cavity radii associated with the atoms were chosen as 1.2 times the corresponding van der Waals radii as obtained from [20]. A value of $\Delta g_i^{s,1} = 0$ would correspond to the neglect of any intermolecular potential energy, as if the solute in the solvent behaved like an ideal gas. We realize that for the small polar molecules (the first three listed) the COSMO model, although not in perfect agreement with the data, gives a significant improvement over the ideal gas model. Such performance is expected for these solutes because their intermolecular potential energy should indeed be dominated by electrostatic effects. This is not so for the three nonpolar solutes. The COSMO results here even show a slight trend opposite to experiment. However, the magnitude of the electrostatic contribution to the free energy of solvation for these solutes, as calculated by COSMO, is small, which is as it should be for such compounds and thus confirms the COSMO model. Clearly, the results for the nonpolar solutes must be interpreted in such a way that different kinds of intermolecular forces, not taken into account in COSMO, are responsible for the solvation behavior of these compounds. Such types of forces are notably of the dispersion type. Analogous results are found for other solvents. All in all, COSMO presents a first realistic approximation to electrostatic solvation effects. Although not generally reliable or sufficiently accurate, it will be shown to be a good starting point for a realistic semiempirical model of the molecular potential energy in liquid mixtures beyond the continuum solvation approximation; cf. Section 4.2 and App. 9.

A second more general approximate approach is based on quantum mechanical perturbation theory; cf. Section 2.5.4. In the spirit of perturbation theory we note that at large intermolecular distances the interaction energy of the charges belonging to different molecules may be considered as a perturbation of a reference system without such interactions. This reference system consists of independent, single molecules in the gas phase, the properties of which are known from the electronic structure calculations of Section 2.5.4. The intermolecular potential energy can then be calculated from quantum mechanics by a perturbation expansion in which the reference system (0) is defined by infinite distances between the molecules. For the Hamilton operator of the molecular system we thus set

$$\hat{H} = \hat{H}^{(0)} + g\hat{V},\tag{2.240}$$

where g is a perturbation parameter, having values from $g = 0$ for the unperturbed system and $g = 1$ for the real system, and

$$\hat{V} = \frac{1}{2}\sum_{\alpha}^{\kappa}\sum_{\beta}^{\kappa}\sum_{i}^{N_\alpha}\sum_{j}^{N_\beta}\sum_{a}^{K_\alpha}\sum_{b}^{K_\beta}V_{a_{\alpha_i}b_{\beta_j}},\tag{2.241}$$

where $i \neq j$ for $\alpha = \beta$. $\hat{H}^{(0)}$ is the Hamilton operator of the unperturbed system and $\hat{V}$ the interaction operator. Because g can take any value between zero and unity, the Hamilton operator of (2.240) can correspondingly reflect rather different strengths of the perturbation. The solution of the Schrödinger equation for the unperturbed system, i.e., a system of individual molecules without interactions, is assumed to be known; cf. Section 2.5.4. Its energy values are

$$E_0^{(0)}, E_1^{(0)}, E_2^{(0)}, \ldots, E_n^{(0)}, \ldots$$

and its wavefunctions

$$\psi_0^{(0)}, \psi_1^{(0)}, \psi_2^{(0)}, \ldots, \psi_n^{(0)}, \ldots.$$

We now expand the wavefunctions and the energy values of the actual system formally around those of the unperturbed system. As usual, we limit consideration to the lowest quantum state, i.e., the ground state of the molecules, denoted by an index 0. The development is entirely analogous to that in Section 2.5.4, so that the final equations can be written down without further derivation.

For the first-order correction to the energy we find, cf. (2.229),

$$E_0^{(1)} = \int \psi_0^{(0)*}\hat{V}\psi_0^{(0)}\mathrm{d}r.\tag{2.242}$$

The first-order perturbation term of the intermolecular interaction energy thus is the expectation value of the interaction operator in the unperturbed state. It can be expressed as

$$E_0^{(1)} = \int \psi_0^{(0)*}\left[\frac{1}{2}\sum_{\alpha}^{k}\sum_{\beta}^{k}\sum_{i}^{N_\alpha}\sum_{j}^{N_\beta}V_{\alpha_i\beta_j}\right]\psi_0^{(0)}\mathrm{d}r.$$

$$\alpha = \beta : i \neq j\tag{2.243}$$

The wavefunction of the unperturbed state can be reduced simply to the wavefunctions of the individual molecules due to their independence,

$$\psi_0^{(0)} = \prod_{\alpha}^{k}\prod_{i}^{N_\alpha}\psi_{0\alpha i}^{(0)}.\tag{2.244}$$

The first-order term thus reflects the charge distribution of the single molecules in their ground states. As noted in Section 2.4, this can be expressed in terms of permanent multipoles, which are available from single-molecule electronic structure calculations; cf. Table 2.1. An operator $\hat{V}_{\alpha_i \beta_j}$ does not depend on the coordinates of a wavefunction $\psi_{0\gamma k}$. Therefore, if we now introduce the two-center spherical harmonic multipole expansion of the interaction operator (2.238), i.e., if we assume, as in Section 2.4.2, that the molecules are rigid structures, we return to (2.125) and (2.126). To obtain explicit agreement we define the multipole moments of a molecule γ as expectation values of the multipole operators in the unperturbed ground state, through

$$^{\gamma}Q_l^n = \int \psi_{0\gamma}^{(0)*} \, ^{\gamma}\hat{M}_l^n \psi_{0\gamma}^{(0)} \, \mathrm{d}\boldsymbol{r}_\gamma,$$ (2.245)

where, cf. (2.148),

$$^{\gamma}\hat{M}_l^n = \sum_a^{K_\gamma} r_a^l q_a Y_l^n(\theta_a, \phi_\alpha)$$ (2.246)

is the multipole operator, which is a multiplicative operator; cf. (2.238). In the unperturbed ground state the molecular charge distribution of small molecules may be modeled as rigid and so it is plausible that quantum mechanical perturbation theory reproduces the electrostatic results obtained in Section 2.4.2.

For the second-order perturbation term of the molecular potential energy we find, cf. (2.234),

$$E_0^{(2)} = \sum_{k>0} \frac{\int \psi_0^{(0)*} \hat{V} \psi_k^{(0)} \mathrm{d}\boldsymbol{r} \int \psi_k^{(0)*} \hat{V} \psi_0^{(0)} \mathrm{d}\boldsymbol{r}}{E_0^{(0)} - E_k^{(0)}}$$ (2.247)

and a similar equation for $E_0^{(3)}$[9]. Thus (2.242) to (2.247) reveal that perturbation theory has transformed the problem of solving the Schrödinger equation of a many-molecule system, which is not yet possible in the general case with sufficient accuracy, to that of solving the Schrödinger equation for a single-molecule system. This is much easier and can be worked out to obtain many important properties related to the molecular charge density distribution; cf. Section 2.5.4. We note that, whereas the first-order term contains only ground state wavefunctions, the second-order perturbation term (and also the higher order terms) involves summations over single-molecule wavefunctions in excited states. So, in order to calculate these higher order contributions to the intermolecular potential energy, wavefunctions in all excited states are required. On this level of perturbation expansion we then definitely abandon the assumption of a fixed charge distribution. Instead of permanent multipole moments, which in quantum mechanical terms are charge operators averaged over ground state single-molecule wavefunctions, we now find terms in which the multipole operators are averaged over all the excited states (indexed by k) of the molecules. The associated molecular properties are the polarizabilities, which are introduced in simple electrostatic terms in Section 2.4. Their general definition for a molecule of type α in quantum mechanical terms is

$$\alpha \prod_{ll'}^{nn'} = -\sum_{k_\alpha > 0_\alpha} \frac{\int \psi_{0\alpha}^{(0)*} \,^\alpha \hat{M}_l^n \psi_{k\alpha}^{(0)} \mathrm{d}\boldsymbol{r} \int \psi_{k\alpha}^{(0)*} \,^\alpha \hat{M}_{l'}^{n'} \psi_{0\alpha}^{(0)} \mathrm{d}\boldsymbol{r}}{E_{0\alpha}^{(0)} - E_{k\alpha}^{(0)}}. \tag{2.248}$$

Clearly, the polarizability tensor will contain contributions from the dipole moment operator for $l, l' = 1$ and from higher multipole operators for $l, l' > 1$. The dipole polarizability tensor contains nine elements originating from $l, l' = 1$ and $n, n' = 0, \pm 1$. All these elements, i.e., the polarizabilities, are properties of single molecules, which are available from quantum mechanical calculations; cf. Section 2.5.4. In cartesian form the polarizabilities read

$$\,^\alpha \alpha_{\alpha\beta} = 2 \sum_{k_\alpha > 0_\alpha} \frac{\int \psi_{0\alpha}^{(0)*\alpha} \hat{\mu}_\alpha \psi_{k\alpha}^{(0)} \mathrm{d}\boldsymbol{r} \int \psi_{k\alpha}^{(0)*\alpha} \hat{\mu}_\beta \psi_{0\alpha}^{(0)} \mathrm{d}\boldsymbol{r}}{E_{k\alpha}^{(0)} - E_{0\alpha}^{(0)}}, \tag{2.249}$$

where $\,^\alpha \hat{\mu}_\beta$ is the dipole moment operator of component α in β-direction. For molecules with symmetry the number of elements in the polarizability tensor decreases [9]. In the simple case of a monatomic molecule, such as argon, the dipole polarizability tensor reduces to a scalar with $\alpha_{xx} = \alpha_{yy} = \alpha_{zz} = \alpha$ and $\alpha_{xy} = \alpha_{xz} = \alpha_{yz} = 0$. For a nonspherical molecule the isotropic polarizability is one third of the trace of the polarizability tensor, i.e., $\frac{1}{3}(\alpha_{xx} + \alpha_{yy} + \alpha_{zz})$. Concerning units, we note that division by the factor $4\pi\varepsilon_0$ gives the polarizability in terms of the unit of a volume; cf. Table 2.1. Also, it can be shown that its magnitude is approximately equal to the volume of the molecule.

When the interaction operator $\hat{V}$ in (2.247) is written out in terms of binary interactions, cf. (2.241), it is seen that the second-order perturbation term can be split into two different contributions. In the first, one molecule is excited whereas all others are in their ground states. This type of interaction is referred to as induction. Because here the permanent multipoles associated with the ground state induce static multipole moments in the excited molecule, the associated effect is simply given by the static, zero-frequency polarizability above. In fact, the potential energy of induction could be derived entirely by classical electrostatics, cf. Section 2.4, without going through quantum mechanical perturbation theory. When the binary interactions are again represented by their two-center spherical harmonic multipole expansion we arrive at a formulation analogous to (2.125) with (2.126) for the intermolecular pair potential induction forces in terms of static polarizabilities and multipole moments. These molecular properties are accessible by standard quantum mechanical calculations; cf. Section 2.5.4. We do not give the detailed derivation of the expansion terms [9], but rather put the results in App. 5; cf. (A 5.1) and (A 5.2).

The second contribution represents interactions between two excited molecules, i.e., dispersion interactions. Here, excited states and related multipoles are not induced by static electrical fields, but rather are genuine quantum mechanical phenomena. Charge fluctuations in one molecule, being represented by fluctuating multipole moments, induce fluctuating multipole moments in the other one and both interact. It can be shown that dispersion interactions between two molecules can be formulated in terms of a product of frequency-dependent dynamic polarizabilities, with integration over all frequencies. Dynamic polarizabilities are also available from quantum-chemical computer codes, although with much more effort than their static counterparts. So we also have explicit expressions in terms of quantum chemically defined molecular properties for the dispersion forces.

We will not present these results here, because it has become common practice to make use of a simplified version, suitable for introducing experimental information. This simplified version uses a semiclassical approach based on an electrostatic oscillator model, with the effect of replacing the dynamic polarizabilities by their static counterparts and the spectrum of excitation energies by the ionization energies of the molecules. The dispersion potential between two spherical molecules due to dipole polarizabilities is then found to be given by the London formula,

$$\phi_{\alpha_i \beta_j}^{\text{disp}} = -\frac{3}{2} \frac{\alpha_\alpha \alpha_\beta}{r_{\alpha_i \beta_j}^6} \frac{I_\alpha I_\beta}{I_\alpha + I_\beta}, \tag{2.250}$$

where $\alpha_\alpha, \alpha_\beta$ are the usual static polarizabilities of species α and β, I_α, I_β are the ionization energies of species α and β, and $r_{\alpha_i \beta_j}$ is the distance between the molecular centers of the two molecules. The general spherical harmonic expansion for the dispersion forces on the London level is put in App. 5; cf. (A 5.3) and (A 5.8). In practical calculations the ionization energies and the polarizabilities in the London formula are eliminated in favor of an empirical parameter to be fitted to experimental data. This makes it possible to introduce data for a system under consideration and has been found profitable to get good agreement with data when applying molecular models. With reference to a popular semiempirical potential function, i.e., the Lennard–Jones potential, cf. (5.15), we set

$$\phi_{\alpha_i \beta_j}^{\text{disp,LJ}} = -4\varepsilon_{\alpha\beta} \left(\frac{\sigma_{\alpha\beta}}{r_{\alpha_i \beta_j}} \right)^6, \tag{2.251}$$

where $\varepsilon_{\alpha\beta}$ is an energy parameter and $\sigma_{\alpha\beta}$ a distance parameter, with the physical interpretation to be discussed in Section 5.2.3. The quantities I_α and I_β are thus eliminated by

$$\frac{I_\alpha I_\beta}{I_\alpha + I_\beta} = \frac{2}{3} \frac{4\varepsilon_{\alpha\beta}\sigma_{\alpha\beta}^6}{\alpha_\alpha \beta_\beta}. \tag{2.252}$$

It should be noted that specific empirical parameters $\varepsilon_{\alpha\beta}$ and $\sigma_{\alpha\beta}$ for the unlike interaction now arise, in contrast to the multipolar and induction forces, where no empirical parameters had to be introduced. So, when we want to make predictions of mixture behavior from the properties of the pure components, we need combining rules to determine these parameters from those of the like interactions, which are assumed to be known from fitting to pure fluid data. The combination rule following from the London formula is found from (2.252), which gives

$$I_\alpha = \frac{4}{3} \frac{4\varepsilon_{\alpha\alpha}\sigma_{\alpha\alpha}^6}{\alpha_\alpha^2}$$

for the ionization energy of component α, and an analogous equation for I_β, as

$$\varepsilon_{\alpha\beta}\sigma_{\alpha\beta}^6 = \frac{2(\varepsilon_{\alpha\alpha}\sigma_{\alpha\alpha}^6)(\varepsilon_{\beta\beta}\sigma_{\beta\beta}^6)}{\varepsilon_{\alpha\alpha}\sigma_{\alpha\alpha}^6\alpha_\beta^2 + \varepsilon_{\beta\beta}\sigma_{\beta\beta}^6\alpha_\alpha^2}\alpha_\alpha\alpha_\beta. \tag{2.253}$$

To make this complete we need a combining rule for the distance parameter $\sigma_{\alpha\beta}$, for which we assume, by analogy to hard spheres,

$$\sigma_{\alpha\beta} = \frac{1}{2}\left(\sigma_{\alpha\alpha} + \sigma_{\beta\beta}\right). \tag{2.254}$$

It must be noted that the set of combining rules (2.253) and (2.254) is semi-empirical and limited to values of ε and σ that describe spherical interactions. When the potential model for a nonspherical system is formulated in terms of a spherical part and additional nonspherical contributions, as is usually the case, ε and σ refer to the spherical part and the polarizabilities in (2.253) are the isotropic polarizabilities. However, in situations where ε, σ are just empirical parameters incorporating nonspherical interactions in an effective way, they lose their physical basis. The London formula is then better replaced by the simple geometrical mean

$$\varepsilon_{\alpha\beta} = \sqrt{\varepsilon_{\alpha\alpha}\varepsilon_{\beta\beta}}, \tag{2.255}$$

which follows by setting $I_\alpha = I_\beta$ and also $\sigma_{\alpha\alpha} = \sigma_{\beta\beta}$. The combining rules (2.255) and (2.254) have become particularly popular in many applications, including empirical equations of state, and are referred to as Lorentz–Berthelot rules. Frequently, however, the more fundamental rule (2.253) has been found to be superior to the simple geometrical mean rule (2.255). As will be seen in the applications of Chapter 5, the London formula, when properly applied, usually performs quite well. Still, it is semiempirical and does not strictly represent the quantum mechanical solution for the dispersion forces. Unsafe models of dispersion are an important source of uncertainty in modeling the intermolecular potential energy and in prediction of mixture behavior from that of the pure components.

Both the induction and the dispersion forces have nonadditive three-body contributions. In particular, the nonadditive three-body dispersion forces of spherical molecules, arising from third-order perturbation theory, are given by

$$\phi_{\alpha_i \beta_j \gamma_k}^{\text{disp}} = \frac{\nu_{\alpha\beta\gamma}}{r_{\alpha_i \beta_j} r_{\alpha_i \gamma_k} r_{\beta_j \gamma_k}} \left(3 \cos \theta_{\alpha_i} \cos \theta_{\beta_j} \cos \theta_{\gamma_k} - 1\right), \tag{2.256}$$

where $\theta_{\alpha_i}, \theta_{\beta_j}, \theta_{\gamma_k}$ are the angles of the triangle formed by the lines connecting the centers of the three molecules and thus together with the distances between the molecular centers define the configuration. The dispersion coefficient is given by [9]

$$\nu_{\alpha\beta\gamma} = \frac{2\left(R_\alpha + R_\beta + R_\gamma\right) R_\alpha R_\beta R_\gamma}{\left(R_\alpha + R_\beta\right)\left(R_\alpha + R_\gamma\right)\left(R_\beta + R_\gamma\right)} \tag{2.257}$$

and

$$1/R_\alpha = \frac{1}{4}\left[\frac{1}{\alpha_\gamma \varepsilon_{\alpha\beta} \sigma_{\alpha\beta}^6} + \frac{1}{\alpha_\beta \varepsilon_{\alpha\gamma} \sigma_{\alpha\gamma}^6} + \frac{1}{\alpha_\alpha \varepsilon_{\beta\gamma} \sigma_{\beta\gamma}^6}\right], \tag{2.258}$$

when here again the mean excitation energies are eliminated in favor of the parameters of the Lennard–Jones pair potential and analogous expressions hold for R_β and R_γ. The nonadditive three-body dispersion forces of spherical molecules are referred to as Axilrod–Teller forces [26]. They frequently represent the major contribution of nonadditivity.

2.6 Experiments in Silico: Computer Simulation

Statistical mechanics provides basic relationships between the molecular model of a system, in particular its energy, and its macroscopic properties. It does not, however, indicate how these relationships may in practice be evaluated. For the kinetic part of the molecular energy, i.e., that of the single molecules, simple closed formulae will be derived in Chapter 3. Thus, the ideal gas properties can easily be obtained from the standard equations of Sections 2.2, 2.3, and 2.5. This is not the case for the potential part, which becomes relevant when molecules interact. So, for dense gases and liquids, we are left with the problem of obtaining solutions for the configurational part of the thermodynamic functions from the general statistical mechanical equations. Evaluating the partition function in the semiclassical approximation is a formidable task due to the many integration variables involved, which are proportional to the number of molecules. Although there are many specific approaches for particular applications, to be treated in the later chapters, there is one general route from molecular to macroscopic properties, referred to as computer simulation [27, 28]. In this method the goal is not an analytical equation for the fluid phase behavior, but rather providing numerical values for certain properties in particular thermodynamic states, such as experiments in a laboratory. Computer simulations are therefore frequently referred to as computer experiments, or experiments in silico. They replace laboratory experiments and not the smoothing, interpolation, and extrapolation of data by an analytical model, such as an equation of state or an excess function equation.

In Section 2.2.1 we realized that there are two fundamentally different but equivalent statistical approaches to the thermodynamic functions for molecular models. The first one is based on the fact that the dynamical variables pressure and energy fluctuate as the system runs through its different microstates. A macroscopically measurable quantity is thus an average value obtained as a time average, e.g., for the pressure in (2.46). The second one results from ensemble theory, which shows that any dynamical variable can equivalently be calculated as an ensemble average, cf. (2.48), from the relative probabilities of the microstates. Corresponding to these two different routes from the microscopic to the macroscopic world, two different methods of computer evaluation have established themselves, both referred to as computer simulations. The first one is the method of molecular dynamics, in which the equations of motion are solved numerically for a number of molecules, typically between 100 and 1000 for properties of homogeneous fluid phases. Significantly larger molecule numbers are required in applications with heterogeneous phases. The thermodynamic quantities are found by averaging over a large number of time steps. The second one is the Monte Carlo method, which numerically evaluates the ensemble average integrals in the statistical analogs of the thermodynamic functions, again for a similarly low number of molecules. The inaccuracies of the results due to the limited number of molecules in both methods are taken into account by ad hoc corrections. For a defined model of the molecules and their interactions, the two methods of computer simulation lead to the same results, as they should on the basis of the first postulate of statistical mechanics. If proper care is devoted to the execution of the calculations, essentially exact results can be obtained.

There are two different ways that computer simulations can be used to develop molecular models for fluids. One is testing potential energy models. So, if we compare computer-simulated data for a particular potential model with real data for a fluid, we get an indication of the quality of the potential model. Second, we can test statistical mechanical approximations. For this purpose, we define a potential model and generate essentially exact thermodynamic data for it by computer simulation. We then compare these data with those obtained from the statistical mechanical theory to be tested for the same potential. Thus, computer simulations are a very useful tool for developing molecular models and will be used frequently in this text. In addition to being indispensable for evaluating potential energy and statistical mechanical models, computer simulations are also frequently used to generate data of fluid phase behavior for direct use in applications. Many interesting simulations of this type have recently appeared in the literature. However, because of the considerable amount of computer time required, this approach is unlikely to replace the practical use of the simpler analytical or semianalytical models to be presented in Chapters 4 and 5 in any foreseeable future.

2.6.1 The Monte Carlo Method

To understand how a computer simulation by the Monte Carlo method works it is essential to cast the integrals to be solved into a particular form. We remember from (2.55) that the configurational internal energy can be calculated from the configuration integral via

$$U^C = kT^2 \left(\frac{\partial \ln Q^C}{\partial T} \right)_{V,N}, \tag{2.259}$$

from which we immediately find, for a system of N monatomic molecules, in the semiclassical approximation,

$$U^C = \int_{r^N} \left(U\left(r^N \right) e^{-U(r^N)/kT} / Q^C \right) dr^N. \tag{2.260}$$

Here, r^N is a shorthand for all the coordinates of the molecular centers. We note that the integral in (2.260) is a product of a function $U(r^N)$, the intermolecular potential energy, and the normalized Boltzmann factor

$$e^{-U(r^N)/kT} / Q^C.$$

This latter factor acts as a weight function for the molecular configurations over which the integration extends. An analogous expression can be derived for the pressure. We remember from (2.61) that

$$p = kT \left(\frac{\partial \ln Q^C}{\partial V} \right)_{T,N}, \tag{2.261}$$

because the nonconfigurational contributions of monatomic molecules to the partition function do not depend on volume. Without loss of generality the system volume V may be considered to be that of a cube of length $V^{1/3}$, because the pressure does not depend on the shape of the container. We thus have, again for N monatomic molecules,

$$Q^C = \int_0^{V^{1/3}} \cdots \int e^{-U(r^N)/kT} dr^N.$$

To find the derivative with respect to volume it is convenient to introduce dimensionless variables that are independent of volume by setting

$$x_i = V^{1/3} x_i', \qquad y_i = V^{1/3} y_i', \qquad z_i = V^{1/3} z_i'.$$

The configuration integral then reads

$$Q^C = V^N \int_0^1 \cdots \int e^{-U(r^N)/kT} dx_1' dy_1' dz_1' \cdots dx_N' dy_N' dz_N'.$$

For the pressure we get from (2.261)

$$p = \frac{kT}{Q^C} \left[NV^{N-1} \int_0^1 \cdots \int e^{-U(r^N)/kT} dx_1' \cdots dz_N' \right.$$

$$\left. - \frac{1}{kT} V^N \int_0^1 \cdots \int \left(\frac{\partial U(r^N)}{\partial V} \right)_{T,N} e^{-U(r^N)/kT} dx_1' \cdots dz_N' \right]. \qquad (2.262)$$

The volume derivative of the intermolecular potential energy may be written as

$$\frac{\partial U(r^N)}{\partial V} = \sum_i^N \frac{dr_i}{dV} \cdot \nabla_{r_i} U.$$

With

$$\frac{dx}{dV} = \frac{x}{3V},$$

we thus get for the residual pressure, with $p^{res} = p - p^{ig}$, and noting that the first summand in (2.262) is the ideal gas pressure,

$$p^{res} = -\frac{1}{3V} \int \left(\sum_i r_i \cdot \nabla_{r_i} U \right) \left(e^{-U(r^N)/kT} / Q^C \right) dr^N. \qquad (2.263)$$

The integrand for the residual pressure appears as a product of a function

$$\sum_i r_i \cdot \nabla_{r_i} U(r^N),$$

referred to as the virial, and the normalized Boltzmann factor

$$e^{-U(r^N)/kT} / Q^C,$$

which again acts as a weight function for the molecular configurations r^N over which the integration extends. Both integrals, that for the configurational internal energy and that for the residual pressure, or, generally, that for the configurational part of any dynamical quantity D, i.e., D^C, can then be written in a general form as

$$\langle D^C \rangle = \int D^C (\Gamma^{N,\text{pot}}) \varrho^C (\Gamma^{N,\text{pot}}) d\Gamma^{N,\text{pot}}, \qquad (2.264)$$

where we generalize to arbitrarily complex molecules by replacing all coordinates of the molecular centers r^N by all coordinates in the molecular phase space

$\Gamma^{N,\text{pot}}$ on which the intermolecular potential energy depends; cf. Section 2.2. We denote by $\langle D^C \rangle$ the canonical ensemble average of D^C and by

$$\varrho^C = e^{-U(\Gamma^{N,\text{pot}})/kT}/Q^C \qquad (2.265)$$

the normalized Boltzmann factor, i.e., the probability density of a particular configuration in the canonical ensemble. In fact, (2.264), along with (2.265), has the general form of an ensemble average of a dynamical quantity in the semiclassical approximation.

To perform the integration numerically the configurational phase space must be discretized, i.e., divided into as many cells as configurations are to be considered. Because their number would be on the order of the number of molecules, far too many volume elements or configurations would have to be included to render such a procedure feasible by standard numerical integration techniques. As a first step to a reduction of the problem let us assume, therefore, that systems of "only" a few hundred molecules are to be studied. We will later discuss measures that can be taken to enable the transfer of the results obtained on systems with such small numbers of molecules to real thermodynamic systems. This still leaves us with integrals of formidable dimensionality to be solved, and thus particular techniques different from standard numerical integration schemes must be used. An algorithm particularly suited for multidimensional integrals does not consider all possible volume elements of the discretized configuration space but instead only a sample of S elements chosen at random from a uniform distribution. An estimate of the canonical ensemble average of D^C is then

$$\langle D^C \rangle = \sum_{k=1}^{S} D^C \left[(\Gamma^{\text{pot}}) \right]_k \varrho^C \left[(\Gamma^{\text{pot}}) \right]_k, \qquad (2.266)$$

where k refers to a particular configuration of the molecules for which ϱ^C and D^C have well-defined values. Because of the random choice of the elements of the configuration space, this approach is referred to as the Monte Carlo method. Although this sample mean integration from a uniform distribution may in favorable cases be more effective than a straightforward integration method for multidimensional integrals, it is still not practical for the thermodynamic properties of fluids. This can be understood by noting that most of the configurations selected from a uniform distribution will contribute almost nothing to the value of the integral (2.264), because the associated normalized Boltzmann factors will be very small. The reason is the extremely high intermolecular energies $U \left[(\Gamma^{\text{pot}}) \right]_k$ associated with any of the numerous configurations where two molecules overlap significantly. Thus, the rate of convergence of the Monte Carlo estimate to the true value of $\langle D^C \rangle$ in the course of subsequent evaluations of configurations would be very small. A uniform distribution Monte Carlo procedure, therefore, would still require far too many configurations to be evaluated.

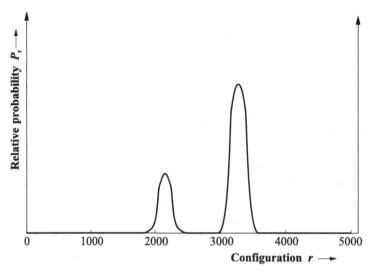

Fig. 2.18. A particular probability distribution.

However, if the sample mean integration is performed in such a way that the random configurations of the molecules are chosen from the nonuniform distribution $\varrho^C \left(\Gamma^{N,\text{pot}} \right)$ of (2.265), i.e., that they appear with statistical weights according to $\varrho^C \left(\Gamma^{N,\text{pot}} \right)$ as the computer experiment proceeds, the sample and thus the number of configurations to be evaluated will be dramatically reduced. Their statistical weights, i.e., their number within the sample of S states, will correspond to ϱ^C and the ensemble average of D^C is then simply calculated from

$$\langle D^C \rangle = \frac{1}{S} \sum_{k=1}^{S} D^C \left[\left(\Gamma^{\text{pot}} \right) \right]_k . \tag{2.267}$$

We note in particular that the normalized Boltzmann factor is eliminated in this equation.

The choice of random configurations from a distribution ϱ^C that restricts evaluation of a function to those regions of space that make important contributions to the integral is referred to as importance sampling. The idea behind importance sampling may be visualized in Fig. 2.18. Here we plot on the abscissa an ordering of configurations from $r = 0$ to $r = 5000$. The sample thus consists of 5000 configurations. To each configuration there belongs a certain relative probability, which is plotted on the ordinate in arbitrary units. If we now consider a systematic numerical integration routine, we will start with configuration 1, evaluate the integrand, proceed to configuration 2, evaluate the integrand, add it to the first step, and so forth. In the first 1900 steps, we would find practically no contribution to the integral, because the relative probabilities, i.e., the statistical weights of these configurations are essentially zero. We would then proceed and find important contributions between configurations 1900 and 2400. From configuration 2400 to 3000 we would again find no contribution, although, proceeding, we would find contributions in the region from configuration 3000 to 3500. Clearly all 5000 configurations would have to be fully evaluated. Standard Monte Carlo methods would again be useless because choosing a limited number of configurations from a uniform distribution will not normally yield the correct integral value in the situation considered. If we want to restrict the number of configurations to be evaluated effectively and still find a good approximation to the integral we have to select these from the important regions. Importance sampling ensures us that during the computer experiment the configurations to be evaluated are selected from the sample with a statistical weight, i.e., a frequency that corresponds to the prescribed probability density function of Fig. 2.18. In the example of Fig. 2.18 only about 1000 different and well-defined configurations should be evaluated in the whole sample. To evaluate the integral we simply add the integrand values for these 1000 configurations, taking into account their proper statistical weights. The rest of the 4000 configurations do not have to be evaluated. In a die-tossing experiment with importance sampling the different sides of the die would have different adhesive properties in relation to the playing board. The 1000 interesting configurations would occur with their proper frequencies during the 5000 trials.

How can we construct such a particular die capable of importance sampling on a computer? A random number generator will not suffice. To obtain configurations with a defined statistical weight from a random choice, we define transition probabilities. Two configurations, i and j, are linked by a transition probability P_{ij} that is the probability of going from state i to state j in one step. For S different configurations in the sample the transition probabilities P_{ij} form an $S \times S$ matrix,

$$
\mathbf{P} = \begin{pmatrix} P_{11} & P_{12} & \cdots & P_{1S} \\ P_{21} & P_{22} & \cdots & P_{2S} \\ \vdots & \vdots & & \vdots \\ P_{S1} & P_{S2} & \cdots & P_{SS} \end{pmatrix} , \tag{2.268}
$$

with

$$
\sum_{j=1}^{S} P_{ij} = 1 \quad \text{for } i = 1, 2, \ldots, S.
$$

The probability of moving from configuration i to configuration j in two steps is given by

$$
P_{ij}^{(2)} = \sum_{l=1}^{S} P_{il} P_{lj}, \tag{2.269}
$$

which becomes clear when we realize that in a sample of three configurations the probability of going from 1 to 2 in two steps is

$$
P_{12}^{(2)} = P_{11} P_{12} + P_{12} P_{22} + P_{13} P_{32}.
$$

Because (2.269) is the procedure for squaring a matrix, we have

$$\mathbf{P}^{(2)} = \mathbf{P}^2,$$

or, more generally, for n steps between 1 and all S other configurations

$$\mathbf{P}^{(n)} = \mathbf{P}^n. \tag{2.270}$$

For an arbitrary matrix of transition probabilities it can be shown that

$$\mathbf{P}^n_{n \to \infty} = \begin{pmatrix} \varrho_1 & \varrho_2 & \cdots & \varrho_S \\ \varrho_1 & \varrho_2 & \cdots & \varrho_S \\ \vdots & \vdots & & \vdots \\ \varrho_1 & \varrho_2 & \cdots & \varrho_S \end{pmatrix} = \mathbf{P}_{\lim}, \tag{2.271}$$

with

$$\mathbf{P}^2_{\lim} = \mathbf{P}^3_{\lim} = \ldots = \mathbf{P}_{\lim}. \tag{2.272}$$

Due to the particular form of $\mathbf{P}_{\lim}$, i.e., the identical values of ϱ in each column, we find

$$\varrho_{\lim} = (\varrho_1, \varrho_2, \ldots, \varrho_S) = (\varrho_1^0, \varrho_2^0, \ldots, \varrho_S^0)\mathbf{P}_{\lim}, \tag{2.273}$$

where $(\varrho_1^0, \varrho_2^0, \ldots, \varrho_S^0)$ is an arbitrary initial probability distribution of configurations and $\varrho_{\lim}$ is the required probability distribution for which the ensemble average is to be performed. In the canonical ensemble this would be ϱ^C of (2.265). We thus have the result that by choosing a suitable matrix $\mathbf{P}$ of transition probabilities, we can transform an arbitrary initial distribution into the desired distribution of configurations. This is the theoretical basis of importance sampling.

In a practical computer experiment for the canonical average of some dynamical state quantity we start from an arbitrary initial configuration of the sample. If this configuration is denoted by 1 the initial probability distribution is $(\varrho_1^0 = 1, \varrho_2^0 = \ldots = \varrho_S^0 = 0)$. By random modification and subsequent acceptance or rejection of a new configuration according to a prescribed transition probability, (2.273) assures us that the configurations, during the course of a simulation, will eventually appear with the stable probability distribution associated with the prescribed transition probability. If this probability distribution corresponds to the normalized Boltzmann distribution of (2.265), the D_k^C for all configurations of the simulation will have the desired statistical weights, at least after some initial period of stabilization, and simple summation and division by the number of configurations will yield the desired ensemble average.

EXERCISE 2.12

We study a statistical experiment for a system capable of three configurations. To visualize we consider an ensemble of these systems, here made up of 1000 systems, cf. Fig. E 2.12.1, from which only 10 systems are shown. Each system consists of a box

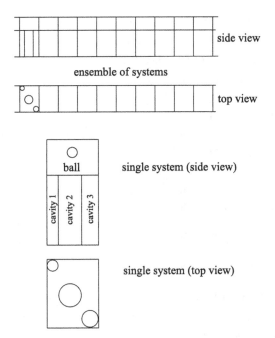

Fig. E 2.12.1. A statistical experiment: tossing a ball into one of three cavities.

into which three cylindrical cavities have been drilled. The areas of the openings are different, and in particular, from left to right, are 0.4, 1, and 0.6, all in arbitrary units. To each system belongs a small ball of cross-section area 0.3. The three configurations of the system are those in which the ball is located in any of the three cavities. Now, assume that we have an initial distribution of the balls in the cavities given by $\varrho_0 = (0.3, 0.4, 0.3)$, which means that in 300 systems the ball is in the left cavity, in 400 systems the ball is in the middle cavity, and in another 300 systems the ball is in the right cavity. If we mutate the configurations in the systems (e.g., by turning the ensemble upside down, so that the balls roll out of the cavities, and then putting it back in the original position, so that the balls roll again into one of the cavities), we get a new distribution of configurations. According to the areas of the openings this distribution will be $\varrho = (0.2, 0.5, 0.3)$, which will be found again and again in any trial, provided that the ensemble is large enough. Alternatively, if the ensemble degenerates into a single system, we will find this distribution as an average over many trials. So, starting from an arbitrary initial distribution, we find a stable distribution either in one step for the ensemble or in many steps for one system.

We wish to show that this experiment can be carried out on a computer by repeated execution for a single system with a matrix of transition probabilities defined by

$$\mathbf{P} = \begin{pmatrix} 0.33\bar{3} & 0.33\bar{3} & 0.33\bar{3} \\ 0.13\bar{3} & 0.66\bar{7} & 0.200 \\ 0.22\bar{2} & 0.33\bar{3} & 0.44\bar{4} \end{pmatrix}.$$

Solution

From the statements made above, cf. (2.271) and (2.272), we know that the matrix of transition probabilities will, on repeated multiplication with itself, degenerate to a matrix with identical rows, the elements of the rows being the elements of a stable probability distribution. As an example we calculate the element

$$P_{12}^{(2)} = P_{11}P_{12} + P_{12}P_{22} + P_{13}P_{32}$$
$$= 0.33\bar{3} \cdot 0.33\bar{3} + 0.33\bar{3} \cdot 0.66\bar{7} + 0.33\bar{3} \cdot 0.33\bar{3}$$
$$= 0.44\bar{4}.$$

We thus find

$$\mathbf{P}^2 = \begin{pmatrix} 0.230 & 0.444 & 0.326 \\ 0.178 & 0.556 & 0.267 \\ 0.217 & 0.444 & 0.338 \end{pmatrix}$$

$$\mathbf{P}^4 = \begin{pmatrix} 0.203 & 0.495 & 0.303 \\ 0.198 & 0.506 & 0.296 \\ 0.202 & 0.494 & 0.304 \end{pmatrix}$$

$$\mathbf{P}^8 = \begin{pmatrix} 0.200 & 0.500 & 0.300 \\ 0.200 & 0.500 & 0.300 \\ 0.200 & 0.500 & 0.300 \end{pmatrix}$$

$$\mathbf{P}^9 = \mathbf{P}^8 = \mathbf{P}_{\text{lim}}.$$

So we have

$$\varrho^{(0)}\mathbf{P}_{\text{lim}} = (0.3, 0.4, 0.3) \begin{pmatrix} 0.200 & 0.500 & 0.300 \\ 0.200 & 0.500 & 0.300 \\ 0.200 & 0.500 & 0.300 \end{pmatrix} = (0.200, \quad 0.500, \quad 0.300)$$

as required. The limiting probability density distribution is reached after eight steps. On a computer we would start with the configuration ϱ_0 and transform it in a first step into a configuration ϱ_1 by applying the matrix of transition probabilities, according to

$$(0.3, 0.4, 0.3) \begin{pmatrix} 0.33\bar{3} & 0.33\bar{3} & 0.33\bar{3} \\ 0.13\bar{3} & 0.66\bar{7} & 0.20\bar{0} \\ 0.22\bar{2} & 0.33\bar{3} & 0.44\bar{4} \end{pmatrix} = (0.220, 0.467, 0.313).$$

By repeating the same procedure with ϱ_1 we find $\varrho_2 = (0.205, 0.489, 0.305)$, and proceeding we eventually arrive at $\varrho_{\text{lim}} = \varrho = (0.200, 0.500, 0.300)$.

Equations (2.271) and (2.272) tell us that any matrix of transition probabilities has an associated limiting probability distribution, which is stable in the further course of the computer experiment. When using (2.267) to compute the configurational properties of a liquid, we know that the limiting probability density distribution, from which we choose at random the molecular configurations of our system, must be $\varrho^C(\Gamma^{N,\text{pot}})$ as given by (2.265). To any of the defined molecular configurations $1 \cdots S$ a normalized Boltzmann factor $\varrho^C(1) \cdots \varrho^C(S)$ is associated. Our problem, then, is to find a matrix of transition probabilities by which randomly mutated configurations are accepted or rejected in such a way that the configurations of the final sample have the correct statistical weight. Because we cannot evaluate the normalized Boltzmann factor, such a matrix of transition probabilities is not obvious. The problem has been solved by a scheme referred to as the asymmetrical solution [29]. Let N_δ be the number of different configurations that can be generated in one step of a computer simulation. This number depends on the extent of mutation allowed with respect to the previous configuration. In the preceding Exercise 2.12 we had $N_\delta = 3$. The matrix of transition probabilities is then defined in the asymmetrical solution as

$$P_{ij} = \frac{1}{N_\delta} \quad \text{for} \quad \frac{\varrho^C(j)}{\varrho^C(i)} \geq 1, \qquad (2.274)$$
$$\scriptstyle i \neq j$$

$$P_{ij} = \frac{1}{N_\delta} \frac{\varrho^C(j)}{\varrho^C(i)} \quad \text{for} \quad \frac{\varrho^C(j)}{\varrho^C(i)} < 1, \qquad (2.275)$$
$$\scriptstyle i \neq j$$

and

$$P_{ii} = 1 - \sum_{j \neq i} P_{ij}. \qquad (2.276)$$

The appropriate element of the matrix of transition probabilities for the transition from state i to state j depends on the ratio of the Boltzmann factors of the two states. This ratio can easily be evaluated, in contrast to the normalized Boltzmann factor itself, because the configuration integral cancels. If configuration j has a lower intermolecular energy than configuration i, the probability of configuration j is greater than that of configuration i and the new configuration is accepted. This is ensured by letting the transition probability be $1/N_\delta$, which reduces to 1 if only one configuration can be generated in each step. However, if configuration j has a higher intermolecular energy than configuration i, the probability of configuration j is smaller than that of configuration i. We will not reject all such steps but instead accept them in proportion to their relative probability in the desired distribution. So according to (2.275), such a configuration is accepted with a smaller probability, determined by the ratio $\varrho^C(j)/\varrho^C(i) < 1$.

EXERCISE 2.13

Consider a system capable of three states. The energies associated with these three states are such that the associated probability densities are $\varrho_1 = 0.2$, $\varrho_2 = 0.5$, and $\varrho_3 = 0.3$. Show that the asymmetric solution leads to a matrix of transition probabilities which is consistent with the required probability density distribution.

Solution

According to the asymmetric solution we arrive at the following matrix of transition probabilities:

$$\mathbf{P} = \begin{pmatrix} P_{11} = 1 - P_{12} - P_{13} & P_{12} = 1/N_\delta & P_{13} = 1/N_\delta \\ P_{21} = (1/N_\delta)\varrho_1/\varrho_2 & P_{22} = 1 - P_{21} - P_{23} & P_{23} = (1/N_\delta)\varrho_3/\varrho_2 \\ P_{31} = (1/N_\delta)\varrho_1/\varrho_3 & P_{32} = 1/N_\delta & P_{33} = 1 - P_{31} - P_{32} \end{pmatrix}$$

$$= \begin{pmatrix} 1 - \frac{1}{3} - \frac{1}{3} & \frac{1}{3} & \frac{1}{3} \\ \frac{1}{3} \cdot \frac{0.2}{0.5} & 1 - \frac{1}{3} \cdot \frac{0.2}{0.5} - \frac{1}{3} \cdot \frac{0.3}{0.5} & \frac{1}{3} \cdot \frac{0.3}{0.5} \\ \frac{1}{3} \cdot \frac{0.2}{0.3} & \frac{1}{3} & 1 - \frac{1}{3} \cdot \frac{0.2}{0.3} - \frac{1}{3} \end{pmatrix}.$$

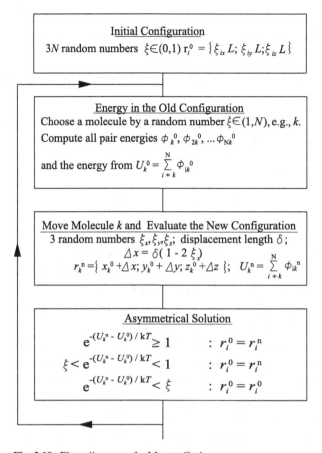

Fig. 2.19. Flow diagram of a Monte Carlo step.

Clearly, this is identical to the matrix studied in Exercise 2.12, from which we remember that

$$\mathbf{P}_{\text{lim}} = \begin{pmatrix} 0.200 & 0.500 & 0.300 \\ 0.200 & 0.500 & 0.300 \\ 0.200 & 0.500 & 0.300 \end{pmatrix}.$$

This confirms the asymmetric solution.

A flow diagram of a Monte Carlo step is shown in Fig. 2.19. Because we work in the canonical ensemble we specify the temperature T, the volume V, and the number of molecules N. This determines the number density $n = N/V$. For a cubic box we now know the length L of one side. By generating $3N$ random numbers ξ_{1x} to ξ_{Nz} between 0 and 1 we are able to specify a random configuration to which we refer as the old configuration, as defined by the position vectors of the molecules r_i^0 for $i = 1, \ldots, N$. To evaluate the intermolecular energy in the old configuration we choose at random one molecule, e.g., molecule k, compute all pair energies between this molecule k and all others, and sum the pair energies to obtain a characteristic part of the intermolecular energy function in the old configuration U_k^0. We now move molecule k at random in all three directions of space within a displacement cube of side length δ and compute the analogous characteristic part of the intermolecular energy function in the new configuration as defined by $\mathbf{r}_k^n$ and $\mathbf{r}_i^0$ for $i = 1, \ldots, N \neq k$. The length δ limits the moves of the selected molecule. Due to the finite number of digits on a computer the phase space is automatically discretized and there are a finite number of states N_δ within the displacement cube. The displacement length is an adjustable parameter. If it is too small, a large fraction of moves are accepted, but the phase space is explored very slowly. If the displacement length is too large, most trial moves are rejected, with again very little movement through phase space. Practically, δ is adjusted so that the ratio of accepted to rejected moves is about 0.5. To ensure Boltzmann sampling we finally use the asymmetrical solution. If the probability density of the new configuration is greater than or equal to that of the old one, the new configuration is accepted and the new position vectors become the old position vectors for the following step. The same is true when the ratio of the probability densities of the new to the old configuration is smaller than 1 but greater than a random number ξ. So, if $\varrho^C(j)/\varrho^C(i) = 0.25$, we have to accept the new configuration with a probability of 0.25, which can be carried out by producing a random number and accepting when this is smaller or equal 0.25. If, however, the ratio of Boltzmann factors is smaller than ξ, the new configuration is rejected and the configuration for the next step will be identical to the old configuration of the previous step. A new step begins with evaluating the energy in the old configuration. Note that we only have to evaluate U_k and not the full intermolecular energy function in order to apply the asymmetrical solution. The complete evaluation of a configuration is performed only on a subset of all configurations generated.

The starting configuration will not normally be consistent with the temperature and density considered. It is, therefore, necessary to run the Monte Carlo simulation for some initial period so that the system can come to equilibrium. This process is called equilibration. During the process of equilibration the configurations of the system are not considered for the calculation of the ensemble averages. The equilibration period is extended until the energy and the pressure cease to show a systematic drift and start to oscillate about steady mean values. No general rules can be given for the number of configurations needed for the equilibration period, but the order of magnitude in typical cases is $10^4 N$, with N being the number of molecules. Because neighboring configurations are highly correlated it is not appropriate to include every configuration in the computation of the ensemble average. Typically, at intervals of 10^3 configurations we store the configuration of the system for evaluation of the energy and the pressure. Only these configurations have to be fully evaluated in order to get the ensemble average of the configurational internal energy and the pressure. About $10^5 N$ configurations are created for a complete simulation of the ensemble averages.

When there are more configurational coordinates than just the locations of the centers of mass the phase space becomes more complicated. For rigid molecules we need translational as well as rotational moves. The translational part of the move is quite analogous to that for monatomic molecules. If the orientation of the molecule is described in terms of Euler angles, a change in orientation can be achieved by taking small random displacements of all three angles θ, ϕ, χ in terms of appropriate maximum displacements $\delta_\theta, \delta_\phi$, and δ_χ. Alternatively, one can displace the atoms of the molecule randomly with the constraint that the geometry of the molecule remains fixed. For more complicated molecules more coordinates are required.

2.6.2 Molecular Dynamics

In the method of molecular dynamics we follow the time evolution of the molecular system [30]. In a monatomic liquid we specify the momenta and positions of all atoms at an initial time t_0 and solve Newton's equations of motion in order to find the momenta and positions of the atoms at some later time $t_0 + \Delta t$. We remember from Section 2.5 that the atoms may be considered as classical particles moving in a force field produced by the charges on the electrons and on the nuclei, which in turn must basically be obtained from quantum mechanics. The calculations are continued for a time t considered long enough to perform the averages. Clearly, as for the Monte Carlo method, this can practically be performed only for a limited number of molecules.

For a molecule i the equation of motion reads, cf. (2.76),

$$\frac{d^2}{dt^2}r_i = \frac{F_i}{m_i}. \tag{2.277}$$

Here r_i is the position vector of molecule i, m_i its mass, and F_i the force on atom i resulting from its interaction with all other atoms. The calculation of the force is responsible for most of the computer time required in a MD calculation. As for MC, it scales as N^2 unless tricks are used to reduce it to scaling as N. Obviously, we need initial conditions for the locations and the momenta of the molecules. Various methods have been suggested to solve the equations of motion for the molecules. They all use stepwise numerical integration algorithms based on the Taylor expansion of the position and velocity vectors in time. Thus we can write

$$r(t + \Delta t) = r\left(t + \frac{1}{2}\Delta t\right) + \left.\frac{dr}{dt}\right|_{r(t+\frac{1}{2}\Delta t)} \frac{\Delta t}{2} + \frac{1}{2}\left.\frac{d^2r}{dt^2}\right|_{r(t+\frac{1}{2}\Delta t)} \left(\frac{\Delta t}{2}\right)^2 + O\left(\Delta t^3\right) \tag{2.278}$$

and

$$r(t) = r\left(t + \frac{1}{2}\Delta t\right) + \left.\frac{dr}{dt}\right|_{r(t+\frac{1}{2}\Delta t)} \left(-\frac{\Delta t}{2}\right) + \frac{1}{2}\left.\frac{d^2r}{dt^2}\right|_{r(t+\frac{1}{2}\Delta t)} \left(\frac{\Delta t}{2}\right)^2 + O\left(\Delta t^3\right). \tag{2.279}$$

Subtracting and noting that the time derivative of the position can be eliminated in terms of the momentum via $p/m = \mathrm{d}r/\mathrm{d}t$ gives

$$r(t + \Delta t) = r(t) + \frac{1}{m} p \left(t + \frac{\Delta t}{2} \right) \Delta t + O\left(\Delta t^3\right). \qquad (2.280)$$

Analogously, one can find for the momentum at time $\left(t + \frac{\Delta t}{2} \right)$

$$\frac{1}{m} p \left(t + \frac{\Delta t}{2} \right) = \frac{1}{m} p \left(t - \frac{\Delta t}{2} \right) + \frac{F(t)}{m} \Delta t + O\left(\Delta t^3\right), \qquad (2.281)$$

from which

$$\frac{1}{m} p(t) = \frac{\frac{1}{m} p\left(t + \frac{\Delta t}{2} \right) + \frac{1}{m} p\left(t - \frac{\Delta t}{2} \right)}{2}. \qquad (2.282)$$

The algorithm above is referred to as the leapfrog algorithm. Various other algorithms have been proposed. We note that the term in Δt^2 has been eliminated by this formulation, which allows larger time steps to be made. Similarly to the displacement length in the Monte Carlo method, here the size of a time step is a strategic parameter, which must be chosen properly to ensure rapid progress without increasing the fluctuations beyond a desired tolerance. So, if the time step is chosen small, we find an accurate solution and good energy conservation, but the computational progress is small. The opposite effects, i.e., rapid computational progress but unsatisfactory energy conservation, appear when the time step is chosen large.

Because we have to specify the positions and momenta of all atoms at $t = t_0$, we also fix the total momentum, which clearly must be zero in a nonmoving system, and the total energy. We do not fix the temperature, however. Instead, the temperature will fluctuate, as if it were a dynamical quantity like the pressure or the energy. The instantaneous temperature can be calculated from the kinetic energy of the atoms according to, cf. (1.3),

$$T(t) = \frac{2}{3} \frac{E_{\text{kin}}(t)}{Nk}, \qquad (2.283)$$

where

$$E_{\text{kin}}(t) = \sum_{i=1}^{N} \frac{p_i^2}{2m_i} \qquad (2.284)$$

and (2.282) is used for $p(t)$. So the temperature cannot be specified exactly but may deviate somewhat from the desired value. Methods have been proposed to keep this fluctuation small.

Thermodynamic properties such as the configurational contributions to the pressure and the internal energy are calculated as time averages over a sufficiently long observation time. So, for an arbitrary dynamical configurational quantity, D^C, we have, cf. (2.47),

$$D^C = \frac{1}{\tau} \sum_{k=1}^{\tau} D_k^C(t_k) = \frac{1}{\tau} \sum_{k=1}^{\tau} D^C(k\Delta t), \qquad (2.285)$$

where τ is a sufficiently large finite number of time steps of length Δt, so that the observation time is

$$t = \tau \Delta t. \tag{2.286}$$

As in a Monte Carlo simulation, we need to provide an equilibration period in which the system loses its memory of the starting configuration. Typically this may take several hundred time steps and is extended until the energy and pressure cease to show a systematic drift and start to oscillate about steady mean values. Because successive time steps are highly correlated, we do not use every step for computing averages. Rather, at intervals of typically 10 time steps we store molecular positions, velocities, etc. to compute quantities such as energy and pressure. These are then averaged over the total number of time steps, which is typically about 3000 for a liquidlike configuration.

In a system of polyatomic molecules we must also consider the equations of rotational motion. The rotational motion of molecule i is governed by the torque τ_i about the center of mass, which enters the rotational equations of motion in the same way as the force enters the translational equations. The numerical solution techniques are similar but the equations for the rotational motion are more complicated. We note that the evaluation of torques is not necessary in the Monte Carlo method.

2.6.3 Effects Due to Small Numbers of Molecules

In any computer simulation the number of molecules is limited by the speed of execution of the program. The time taken to evaluate the intermolecular energy and the forces in a basic program is proportional to N^2, where N is the number of molecules. There are, however, methods, such as the linked cell algorithm, that effectively lead to proportionality to the molecule number. Still, practical simulations in homogeneous phases are normally performed with no more than several hundred molecules. This is an extremely small number compared to the typical number of molecules in a thermodynamic system, which is on the order of 10^{23} for one mole. For a given fluid density this implies an extremely small dimension of the system, i.e., defines a very small length L of the associated cubic box. A major problem associated with such small systems is the fact that a substantial fraction of the molecules would experience interactions with the boundaries of the system and the simulation would not be representative for a small volume element in the bulk of a liquid. This problem of surface effects can be overcome by replicating the basic system throughout space to form an infinite lattice. This is shown in Fig. 2.20 for the two-dimensional case. The central box contains molecules i, j, k, l, and m. It is surrounded by eight identical boxes that contain images of these molecules, which can be labelled by the number of the box. As the simulation proceeds, not only does the molecule i in the central box move, but also, in exactly the same way, all its periodic images i_1, i_2, i_3, etc. move as well. There are no walls around the central box, which

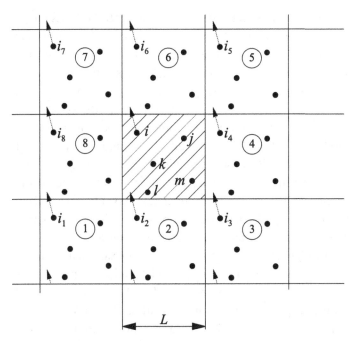

Fig. 2.20. A two-dimensional periodic system (central box shaded).

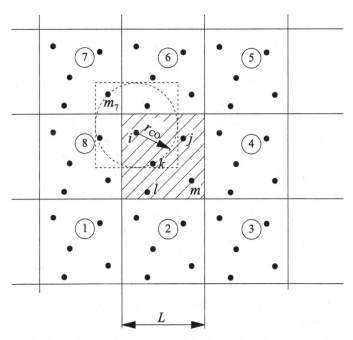

Fig. 2.21. Minimum image convention and cutoff radius.

eliminates the surface effects. As a molecule i leaves the central box, its image i_2 will enter through the opposite face. The number density in the central box is thus conserved and any density fluctuations with a wavelength greater than L will be suppressed. This approach to eliminating surface effects in a small system is referred to as the method of periodic boundary conditions. Clearly, it is important to investigate whether a small, infinitely periodic system can be used to simulate the properties of a macroscopic thermodynamic system. Because of the periodicity, each molecule actually "sees" multiple images of itself and each of its neighbors. Because interactions of any molecule with a neighbor should not be counted more than once, they are restricted to the nearest images of neighbors. This is the minimum image convention, visualized in Fig. 2.21. Thus, the artificial periodicity will not be noticed if the interaction between any two molecules is of shorter range than the dimension of the boxes. Molecule i can interact with m_7, but not with m or any other image of m. Usually, we further introduce a spherical cutoff radius r_{CO} which limits the interactions to be considered. In Fig. 2.21 we have chosen $r_{CO} = L/2$. The error introduced into the thermodynamic properties due to this truncation of the potential can be corrected by assuming a random distribution of the molecules for $r > r_{CO}$. For interactions of longer range more complicated corrections must be applied.

2.7 Summary

The development and the application of molecular models for fluids rests on interdisciplinary foundations. Application is performed on the basis of thermodynamic functions and their relations to macroscopic phenomena, as formulated by classical thermodynamics. The basic relationship between a molecular model of a fluid and its thermodynamic functions is provided by statistical mechanics. The thermodynamic functions are related to the partition function, which in turn can be evaluated as a sum (or double sum) over the Boltzmann factors of the system. Practical applications are in most cases based on the semiclassical approximation. To evaluate the partition function, the energy of a molecular system is needed. The energy of a single molecule can be found to a first approximation from classical mechanics. Classical mechanics makes it possible to formulate the Hamilton function of a single molecule in terms of its generalized coordinates and momenta. Although classical mechanics strictly does not apply to molecules, its results for the energy modes are very useful when suitable corrections are added in the final equations. Because molecules can be considered to be charge clouds formed by the positively charged atomic nuclei and the negatively charged electrons, there will be electrostatic interactions between them. These are expressed by the laws of classical electrostatics. For molecules in a vacuum these interactions can be quantitatively formulated in terms of multipoles and the configuration of the molecular system. When molecules are considered in solution the solvent can frequently be approximated by a dielectric continuum screening the charges of the solute molecule in proportion to the dielectric constant. Quantum mechanics is the adequate theory on the atomic scale. It provides proper information about the energy levels of single molecules and, in particular, their geometry, vibration frequencies, and charge distributions. Although analytical formulae for the degrees of freedom of single molecules can be derived, all more detailed information must be obtained from quantum-chemical computer codes, which are now readily available even to the nonspecialist. Analytical formulae for intermolecular interactions containing some molecular properties such as multipole moments and polarizabilities can be derived by quantum mechanical perturbation theory at large intermolecular distances. As will be seen in the remainder of this book, the progress that has been made in applying quantum mechanics allows very useful and predictive molecular models to be developed even for rather complex fluid systems. The statistical mechanical equations contain integrations over many variables, and generally cannot be solved analytically. A general approach to obtaining numerical results is provided by computer simulation. Methods summarized under this heading are the Monte Carlo method and the method of molecular dynamics. Both consider ensembles of only relatively few molecules and provide essentially exact numerical data for their macroscopic properties when the model for the molecular potential energy is specified, albeit with considerable effort in computer time.

2.8 References to Chapter 2

[1] S. I. Sandler. *Chemical, Biochemical and Engineering Thermodynamics*, 4th edition. Wiley, New York, 2006.

[2] J. O'Connell and J. Haile. *Thermodynamics. Fundamentals for Applications*. Cambridge University Press, Cambridge, 2005.

[3] R. C. Tolman. *The Principles of Statistical Mechanics*. Dover, New York, 1938.

[4] L. Meirovitch. *Methods of Analytical Dynamics*. McGraw-Hill, New York, 1970.

[5] B. J. McClelland. *Statistical Thermodynamics*. Chapman and Hall, London, 1973.

[6] J. E. Mayer and M. G. Mayer. *Statistical Mechanics*. Wiley, New York, 1940.

[7] J. D. Jackson. *Classical Electrodynamics*. Wiley, New York, 1988.

[8] C. G. Gray and K. E. Gubbins. *Theory of Molecular Fluids I*. Clarendon, Oxford, 1984.

[9] K. Lucas. *Applied Statistical Thermodynamics*. Springer, Berlin, 1991.

[10] M. E. Rose. *Elementary Theory of Angular Momentum*. Wiley, New York, 1957.

[11] M. Born. *Z. Phys.*, 1:45, 1920.

[12] M. Amovilli and B. Mennucci. *J. Phys. Chem. B*, 101:1051, 1997.

[13] J. Tomasi and M. Persico. *Chem. Rev.*, 94:2027, 1994.

[14] A. Klamt and G. Schüürmann. *J. Chem. Soc. Perkin Trans.*, 2:799, 1993.

[15] P. Debye and E. Hückel. *Phys. Z.*, 24:105, 1923.

[16] J. Krissmann, M.A. Siddiqi, and K. Lucas. *Fluid Phase Equilibria*, 141:221, 1997.

[17] D. A McQuarrie. *Quantum Chemistry*. University Science Books, Mill Valley, CA, 1983.

[18] P. W. Atkins and R. Friedman. *Molecular Quantum Mechanics*. Oxford University Press, New York, 2005.

[19] F. Jensen. *Introduction to Computational Chemistry*. Wiley, New York, 1999.

[20] TURBOMOLE. Available at http//:www.turbomole.com., vs-7-1, 2004.

[21] Gaussian 03 Revision B.05. *Gaussian Inc.* Pittsburgh, PA, 2003.

[22] G. Herzberg. *Molecular Spectra and Molecular Structure. III. Electronic Spectra and Electronic Structure of Polyatomic Molecules*. Van Nostrand Reinhold, New York, 1966.

[23] Landolt-Börnstein. *Zahlenwerte und Funktionen aus Physik, Chemie, Geographie und Technik*, volume 2,4. Springer, Berlin, 1961.

[24] A. K. Sum and S. I. Sandler. *Fluid Phase Equilibria*, 158–160:375, 1999.

[25] C. J. Cramer and D. G. Truhlar. *Rev. Comp. Chem. Vol. VI*, VCH, Weinheim, 1995.

[26] B. M. Axilrod and E. J. Teller. *J. Chem. Phys.*, 11:299, 1943.

[27] D. Frenkel and B. Smit. *Understanding Molecular Simulation*. Computational Science Series. Academic Press, New York, 2002.

[28] F. Vesely. *Computational Physics: An Introduction*. Kluwer Academic, New York, 2001.

[29] N. Metropolis, A. W. Rosenbluth, M. N. Rosenbluth, A. H. Teller, and E. Teller. *J. Chem. Phys.*, 21:1087, 1953.

[30] B. J. Alder and T. E. Wainwright. *J. Chem. Phys.*, 27:1208, 1957.

3 The Ideal Gas

3.1 Definition and Significance

An ideal gas is a particular molecular model system consisting of independent molecules without volume and without intermolecular forces. The molecules of an ideal gas are thus point particles, which are, when belonging to the same chemical compound, indistinguishable. Such a model system cannot exist in reality. However, in the limit of very low density, the molecules of a real gas are, on the average, so far apart from each other that the forces between them and their volumes have no influence on most of the thermodynamic properties of the system. We may therefore expect that important information about the thermodynamic behavior of real gases in the limit of low density can be obtained from the ideal gas model.

It is known from experimental observation that all real gases in the limit of zero density have the equation of state

$$pV = NRT$$

and that their internal energy becomes independent of volume in that limit; i.e.,

$$\left(\frac{\partial U}{\partial V}\right)_{T,N} = 0.$$

In practice, this universal limiting behavior is found at normal pressures and temperatures and thus has wide technical application. We show below that it is reproduced correctly by the ideal gas model. Also, some of the more specific thermodynamic properties of real gases at low density can be calculated quite accurately from the ideal gas model, e.g., the heat capacity. Further, the ideal gas serves as a reference state for which basic data for thermodynamic calculations such as the enthalpy of formation or the absolute entropy are defined and tabulated. Generally, all thermodynamic quantities may be computed from the equation of state of a system in combination with its heat capacities in the ideal gas state. Thus, the concept of the ideal gas state has great technical significance.

3.2 The Canonical Partition Function

The canonical partition function reads, in quantum mechanical terms, cf. (2.51),

$$Q = \sum_i e^{-E_i/kT}. \tag{3.1}$$

In an ideal gas the molecules are independent of each other. Hence, the energy of the system E_i in each system state i is the sum of the "private" energies $\{\varepsilon_j\}$ of the molecules distributed over their various molecular states in a particular distribution denoted by i. This simple subdivision of the total system energy is not possible when intermolecular forces are present and thus is limited to the model of the ideal gas.

To make this more explicit, we write for the first few energy states of an ideal gas system consisting of the distinguishable molecules $a, b, \ldots, N$, each one in a molecular state $1, 2, \ldots,$

$$E_1^{ig} = \varepsilon_{a1} + \varepsilon_{b1} + \cdots + \varepsilon_{N1}$$
$$E_2^{ig} = \varepsilon_{a2} + \varepsilon_{b1} + \cdots + \varepsilon_{N1}$$
$$E_3^{ig} = \varepsilon_{a1} + \varepsilon_{b2} + \cdots + \varepsilon_{N1}$$
$$\cdots$$

This allows us to write

$$\sum_i e^{-E_i^{ig}/kT} = e^{-(\varepsilon_{a1}+\varepsilon_{b1}+\cdots+\varepsilon_{N1})/kT}$$
$$+ e^{-(\varepsilon_{a2}+\varepsilon_{b1}+\cdots+\varepsilon_{N1})/kT}$$
$$+ e^{-(\varepsilon_{a1}+\varepsilon_{b2}+\cdots+\varepsilon_{N1})/kT}$$
$$+ \cdots$$
$$= \left[e^{-\varepsilon_{a1}/kT} + e^{-\varepsilon_{a2}/kT} + \cdots \right] \left[e^{-\varepsilon_{b1}/kT} + e^{-\varepsilon_{b2}/kT} + \cdots \right]$$
$$\cdots \left[e^{-\varepsilon_{N1}/kT} + e^{-\varepsilon_{N2}/kT} + \cdots \right] = \prod_{s=1}^{N} \left(\sum_i e^{-\varepsilon_{si}/kT} \right).$$

If we define a molecular partition function q_s for molecule s by

$$q_s = \sum_i e^{-\varepsilon_{si}/kT}, \tag{3.2}$$

we obtain

$$\sum_i e^{-E_i^{ig}/kT} = \prod_{s=1}^{N} q_s.$$

The molecules of a pure ideal gas are indistinguishable. There is thus no difference between $a, b, c, \ldots$ and we get for the canonical partition function of an ideal gas

$$Q^{ig} = \frac{1}{N!} q^N. \tag{3.3}$$

To take the indistinguishability of the molecules into account we again have divided by $N!$; cf. Section 2.2.5. Correspondingly, due to the indistinguishability of the individual molecules of each component $\alpha, \beta, \ldots$, the canonical partition function of an ideal gas mixture can be written as

$$Q^{\mathrm{ig}} = \frac{1}{N_\alpha! N_\beta! \cdots N_K!} q_\alpha^{N_\alpha} q_\beta^{N_\beta} \cdots q_K^{N_K}. \tag{3.4}$$

Here $q_\alpha, q_\beta, \ldots$ are the molecular partition functions of components $\alpha, \beta, \ldots$ in the mixture of volume V and $N_\alpha, N_\beta, \ldots$ the associated molecule numbers. Taking care of the indistinguishability by a factor $N!$, referred to as corrected Boltzmann statistics, is not exact under all circumstances. It may have to be replaced by more complicated statistics for particular gases at very low temperatures. Such applications are not considered in this book.

Frequently, the energy levels of a molecule are degenerate; i.e., the same energy value is associated with different quantum states, and (3.2) for the molecular partition function contains a number of identical terms. If we denote by g_j the degeneracy of energy level ε_j, we can write for the molecular partition function

$$q = \sum_j g_j e^{-\varepsilon_j / kT}, \tag{3.5}$$

where the summation now extends over all energy levels.

The energy values in the molecular partition function cannot be determined absolutely. As in classical thermodynamics, we here also introduce a zero of energy, i.e., a reference state in relation to which the energy is measured. In the thermodynamic functions we denote the energy of this reference state as U^0. We choose as the zero molecular energy ε^0 that of a free molecule in the nuclear and electronic ground state with the lowest possible quantum numbers of all other energy modes. Transferred into macroscopic terms, U^0 thus refers to the state of an ideal gas at $T = 0$ K.

3.3 Factorization of the Molecular Partition Function

To evaluate the molecular partition function and thus the thermodynamic functions of the ideal gas, we need the molecular energy states or the molecular energy levels with the associated degeneracies. These must be obtained from quantum mechanics. Alternatively, we can use classical mechanics for the Hamilton function of a single molecule and work out the molecular partition function in the semiclassical approximation. Simplified models for the energy of single molecules have been discussed in Sections 2.3 and 2.5 in the frameworks of classical and quantum mechanics. The molecular parameters contained in the final equations are again accessible using quantum-chemical computer codes. We shall now use these results. With the introduction of these molecular models, the molecular partition function can be evaluated.

The first simplification in calculating the energy of a molecule is achieved by making use of the independence of the translational motion of the center of mass from the other degrees of freedom. Due to the large average distance between the molecules, this is exact for an ideal gas. Denoting all nontranslational degrees of freedom associated with the motions of the atomic nuclei as ε_{ntr}, we thus have

$$\varepsilon = \varepsilon_{tr} + \varepsilon_{ntr}. \tag{3.6}$$

Introducing this into the equation for the canonical partition function of the ideal gas yields

$$
\begin{aligned}
Q^{ig} &= \frac{1}{N!} \left[\sum_{tr} \sum_{ntr} e^{-(\varepsilon_{i,tr} + \varepsilon_{i,ntr})/kT} \right]^N \\
&= \frac{1}{N!} \left[\left(\sum_{tr} e^{-\varepsilon_{i,tr}/kT} \right) \left(\sum_{ntr} e^{-\varepsilon_{i,ntr}/kT} \right) \right]^N \\
&= \frac{1}{N!} q_{tr}^N q_{ntr}^N.
\end{aligned} \tag{3.7}
$$

The molecular partition functions for the translational and nontranslational degrees of freedom in an ideal gas are thus independent of each other. According to (2.198), the single-molecule energy values of translational motion depend on volume, and thus the molecular partition function of translation depends on temperature and volume. By contrast, the energy values of the nontranslational degrees of freedom for a single molecule are independent of volume, and thus the associated contributions to the molecular partition function are just functions of temperature.

We thus find for the thermodynamic functions of an ideal gas

$$
\begin{aligned}
A^{ig} - U^0 &= -kT \ln Q^{ig} = -kT \left[-\ln N! + N \ln q_{tr} + N \ln q_{ntr} \right] \\
&= kT \ln N! - NkT \ln q_{tr} - NkT \ln q_{ntr} \\
&= A_{tr}^{ig} + A_{ntr}^{ig} - U^0
\end{aligned} \tag{3.8}
$$

$$
\begin{aligned}
U^{ig} - U^0 &= kT^2 \left(\frac{\partial \ln Q^{ig}}{\partial T} \right)_{V,N} \\
&= kT^2 \left[N \left(\frac{\partial \ln q_{tr}}{\partial T} \right)_{N,V} + N \left(\frac{\partial \ln q_{ntr}}{\partial T} \right)_{V,N} \right] \\
&= U_{tr}^{ig} + U_{ntr}^{ig} - U^0
\end{aligned} \tag{3.9}
$$

$$S^{ig} = -\frac{A^{ig}}{T} + \frac{U^{ig}}{T}$$

$$= Nk \left[-\frac{\ln N!}{N} + \ln q_{tr} + T \left(\frac{\partial \ln q_{tr}}{\partial T} \right)_{V,N} \right]$$

$$+ Nk \left[\ln q_{ntr} + T \left(\frac{\partial \ln q_{ntr}}{\partial T} \right)_{V,N} \right] = S_{tr}^{ig} + S_{ntr}^{ig} \qquad (3.10)$$

$$p^{ig} = kT \left(\frac{\partial \ln Q^{ig}}{\partial V} \right)_{N,T} = NkT \left(\frac{\partial \ln q_{tr}}{\partial V} \right)_{N,T} = p_{tr}^{ig}. \qquad (3.11)$$

The entropy does not contain a contribution of the zero point of energy, in agreement with the absolute character of this thermodynamic state quantity, due to the third law. The pressure is the only thermodynamic function of the ideal gas that depends exclusively on the translational partition function and not on the nontranslational motions of the molecules, because their energy states do not depend on volume. Because the individuality of a molecule is contained in the nontranslational degrees of freedom, we may expect a simple and universal expression for the pressure; not, however, for any of the caloric state quantities of the ideal gas.

The nontranslational degrees of freedom of a molecule are composed of rotations around axes through the center of mass and of vibrations and internal rotations of single molecular groups. A further simplifying assumption is that all these contributions to the nontranslational energy of a molecule are independent of each other. The contributions of external rotation and internal motions can thus be evaluated by the relatively simple classical and/or quantum mechanical models discussed in Sections 2.3 and 2.5. Although fulfilled in many practical cases to a high degree of accuracy, this is strictly an approximation. For very accurate calculations, coupling effects between rotation and vibration have to be considered; cf. Section 3.6.6. If we finally add the electronic energy as an independent contribution in the sense of the Born–Oppenheimer approximation we arrive at a simple additive expression for the total energy of a molecule,

$$\varepsilon = \varepsilon_{tr} + \varepsilon_r + \varepsilon_v + \varepsilon_{ir} + \varepsilon_{el}. \qquad (3.12)$$

Correspondingly, the molecular partition function factors as

$$q = q_{tr} q_r q_v q_{ir} q_{el}. \qquad (3.13)$$

This factorization permits us to investigate each degree of freedom separately, which is considerably simpler than considering the complex dynamical behavior of a molecule as a whole. Due to the logarithmic relationship between the thermodynamic functions and the molecular partition function, the individual elements of the molecular energy contribute additively to the thermodynamic functions. We can thus evaluate each contribution separately and finally sum them all up to the total function.

3.4 The Equation of State

From (3.11) we learn that only the translational degrees of freedom contribute to the pressure in an ideal gas, because only these depend on volume. To derive the equation of state for the ideal gas we thus need only the molecular partition function of translation.

In ideal gases the energy of the translational motion of the center of mass is entirely kinetic, for there are no forces between the molecules. The quantum mechanical energy states of kinetic translational motion have been derived in Section 2.5 from the Schrödinger equation; cf. (2.198). We thus have for the translational molecular partition function

$$q_{tr} = \sum_{n_x} \sum_{n_y} \sum_{n_z} e^{-\frac{h^2}{8mkT} V^{-2/3}(n_x^2 + n_y^2 + n_z^2)}. \tag{3.14}$$

The individual energy states of translation are very closely spaced. The summation over discrete terms can, therefore, be replaced by an integration, cf. App. 6:

$$q_{tr} = \int_{n_x=0}^{\infty} \int_{n_y=0}^{\infty} \int_{n_z=0}^{\infty} e^{-\frac{h^2}{8mkT} V^{-2/3}(n_x^2 + n_y^2 + n_z^2)} \, dn_x dn_y dn_z$$

$$= \left[\int_{n_x=0}^{\infty} e^{-\frac{h^2}{8mkT} V^{-2/3} n_x^2} \, dn_x \right]^3 = \left(\frac{2\pi mkT}{h^2} \right)^{3/2} V. \tag{3.15}$$

Here we have integrated over all positive quantum numbers from 0 to ∞, although a value of zero for translational quantum numbers does not exist. This approximation introduces an error that is entirely negligible. Also, due to our definition of energy zero, the translational energy should be evaluated strictly with respect to that of the lowest quantum state, i.e., $n_x, n_y, n_z = 1$. Again, the associated inaccuracy in the evaluation of the translational partition function by Eq. (3.15) is without practical consequences. The correct term in the exponent of (3.15) would be

$$\varepsilon_{tr} - \varepsilon_{tr}^0 = \frac{h^2}{8mV^{2/3}} (n_x^2 + n_y^2 + n_z^2 - 3),$$

where the contribution due to ε_{tr}^0 to the molecular partition function is entirely negligible because of the smallness of the term $h^2/8mV^{2/3}$.

For the equation of state of an ideal gas we find

$$p^{ig} = p_{tr}^{ig} = NkT \left(\frac{\partial \ln q_{tr}}{\partial V} \right)_{N,T} = \frac{NkT}{V}. \tag{3.16}$$

For $N = N_A$, i.e., Avogadro's number, we consider one mole of gas, in which case the volume V has thus to be replaced by the molar volume v, and we have

$$p^{ig}v = RT,\qquad(3.17)$$

where

$$R = N_A k\qquad(3.18)$$

is the gas constant.

The equation of state for an ideal gas is a universal law that can be verified experimentally for real gases in the limit of low density. Thus the significance of the constant k in

$$\beta = \frac{1}{kT}$$

as being the Boltzmann constant is justified, as anticipated in (A 3.17).

EXERCISE 3.1

Calculate the molecular partition function of translation in the semiclassical approximation.

Solution

For the molecular partition function of translation we find from (3.3) that

$$q_{tr} = (N! Q_{tr}^{ig})^{1/N}.$$

Here Q_{tr}^{ig} is the canonical partition function for the translational mode of an ideal gas, which in semiclassical approximation is given by, cf. Exercise 2.2,

$$Q_{tr}^{ig} = \frac{1}{N! h^{N f_{tr}}} \int e^{-H_{tr}^{ig}(\Gamma_{tr})/kT}\,d\Gamma_{tr} = \frac{1}{N! h^{3N}} \int_{-\infty}^{+\infty} e^{-\sum_i^{3N}(p_i^2/2m_i)/kT}\,dp^N dr^N$$

$$= \frac{1}{N!} \Lambda^{-3N} V^N,$$

with

$$\Lambda = \sqrt{\frac{h^2}{2\pi mkT}}.$$

Therefore, the translational partition function becomes

$$q_{tr} = \left(\frac{2\pi mkT}{h^2}\right)^{3/2} V.$$

This is the same result that is found from the quantum mechanical form of the molecular partition function. Clearly, this demonstrates the validity of the semiclassical approximation for this particular case. So, for the translational energy of single molecules, explicit use of the Schrödinger equation results is not necessary. Still, the narrow spacing of the translational energy levels and in particular the correction due to the Heisenberg uncertainty relation have been derived from quantum mechanics. So the impression should not arise that any molecular model could be formulated without a quantum mechanical basis.

At very low temperatures, i.e., $T \to 0$, (3.16) becomes invalid because the two approximations we used break down, i.e., replacing the summation in (3.14) by the integral in (3.15) and using corrected Boltzmann statistics, i.e., (3.3). Although the limit of temperature for replacing the summation by an integral is so small that it is practically irrelevant, cf. Exercise 3.2, the limitation of corrected Boltzmann statistics may well be of practical significance in some selected cases, such as gaseous hydrogen and neon at very low temperatures, say below 30 K. These effects are referred to as quantum effects and will not be discussed in this book.

EXERCISE 3.2

Estimate the temperature at which the approximation of the sum by an integral in the evaluation of the translational partition function becomes questionable for a hydrogen molecule in a container of $V = 1 \, \text{cm}^3$.

Solution

For the approximation to be valid we must have

$$\varepsilon_{n+1} - \varepsilon_n \ll kT,$$

where n stands for any combination of translational quantum numbers. This can be transformed into

$$\frac{h^2}{8mV^{2/3}kT} \ll 1.$$

We thus have

$$\frac{(6.62 \cdot 10^{-34})^2 \, \text{N}^2\text{m}^2\text{s}^2}{8 \cdot \frac{2 \cdot 10^{-3} \, \text{kg}}{6.02 \cdot 10^{23}} \cdot 1.381 \cdot 10^{-23} \, \text{J/K} \cdot 10^{-4} \, \text{m}^2} \ll T.$$

For the temperature we must therefore require

$$T \gg 1.2 \cdot 10^{-14} \, \text{K}.$$

For molecules with larger masses this limiting temperature is even smaller. It is thus practically irrelevant.

When the universal equation of state for the ideal gas is introduced into the well-known general relationships between the caloric state quantities and the equation of state, expressions for the difference of state quantities between a considered state (T, V) or (T, p) and a reference state (T^0, V^0) or (T^0, p^0) result, such as

$$U^{\text{ig}}(T, \{N_j\}) - U^{\text{ig}}(T^0, \{N_j\}) = \int_{T^0}^{T} C_V^{\text{ig}}(T, \{N_j\}) \, dT, \tag{3.19}$$

$$H^{\text{ig}}(T, \{N_j\}) = U^{\text{ig}}(T, \{N_j\}) + NRT, \tag{3.20}$$

$$S^{\text{ig}}(T, V, \{N_j\}) - S^{\text{ig}}(T^0, V^0, \{N_j\}) = \int_{T^0}^{T} \frac{C_V^{\text{ig}}(T, \{N_j\})}{T} \, dT + NR \ln \frac{V}{V^0}, \tag{3.21}$$

$$S^{\text{ig}}(T, p, \{N_j\}) - S^{\text{ig}}(T^0, p^0, \{N_j\}) = \int_{T^0}^{T} \frac{C_p^{\text{ig}}(T, \{N_j\})}{T} \, dT - NR \ln \frac{p}{p^0}, \tag{3.22}$$

$$A^{ig}(T, V, \{N_j\}) - A^{ig}(T^0, V^0, \{N_j\}) = \int_{T^0}^{T} C_V^{ig}(T, \{N_j\}) dT$$

$$- T \int_{T^0}^{T} \frac{C_V^{ig}(T, \{N_j\})}{T} dT - NRT \ln \frac{V}{V^0},$$

$$(3.23)$$

and

$$G^{ig}(T, p, \{N_j\}) - G^{ig}(T^0, p^0, \{N_j\}) = \int_{T^0}^{T} C_p^{ig}(T, \{N_j\}) dT$$

$$- T \int_{T^0}^{T} \frac{C_p^{ig}(T, \{N_j\})}{T} dT + NRT \ln \frac{p}{p^0}.$$

$$(3.24)$$

These expressions contain the heat capacities C_V^{ig} and C_p^{ig}, which depend on molecular species as well as on temperature and, in mixtures, on composition, but not on pressure or volume. They have to be calculated from a detailed evaluation of the total partition function for a molecule. Also, the values of the thermodynamic functions in the reference state, characterized by the suffix 0, are specific to each substance. We learn from (3.19) that the internal energy U of an ideal gas does not depend on volume, in agreement with the behavior of real gases in the limit of low densities. The internal energy, enthalpy, and heat capacity of an ideal gas further do not depend on density and pressure. This is not in agreement with the behavior of real gases in the limit of low densities, because $(\partial u / \partial n)_T$, e.g., is not zero in the limit of zero density, as will be discussed in Section 5.1.

3.5 Mixing Properties

Some general relationships exist for the computation of the thermodynamic functions of ideal gas mixtures from those of the pure components. From the canonical partition function for a mixture of ideal gases, cf. Eq. (3.4),

$$Q^{ig} = \frac{1}{N_\alpha! N_\beta! \cdots N_K!} q_\alpha^{N_\alpha} q_\beta^{N_\beta} \cdots q_K^{N_K}, \tag{3.25}$$

we find for the Helmholtz free energy, by making use of Stirling's formula, cf. App. 2, that

$$A^{ig} - U^0 = -kT \ln Q^{ig} = kT \sum_\alpha (N_\alpha \ln N_\alpha - N_\alpha - N_\alpha \ln q_\alpha)$$

$$= NkT \sum_\alpha x_\alpha (\ln x_\alpha + \ln N) - kT \sum_\alpha (N_\alpha + N_\alpha \ln q_\alpha)$$

$$= NkT \sum_\alpha x_\alpha \ln x_\alpha + \sum_\alpha kT \left(-N_\alpha - N_\alpha \ln \frac{q_\alpha}{N} \right). \tag{3.26}$$

The molecular partition function of component α in the gaseous mixture can generally be written as

$$q_\alpha = f(T)V.$$

If the mixing process occurs at constant pressure and temperature, we have $p = NkT/V = N_\alpha kT/V_{0\alpha}$ and the relationship between the quantities for component α in the mixture and in a pure gas is

$$\frac{q_\alpha}{q_{0\alpha}} = \frac{V}{V_{0\alpha}} = \frac{N}{N_\alpha},$$

which is equivalent to

$$\frac{q_\alpha}{N} = \frac{q_{0\alpha}}{N_\alpha},$$

where $q_{0\alpha}$ is the molecular partition function of pure α at a volume $V_{0\alpha}$, to be calculated from system pressure, temperature, and number of α-molecules. We then get

$$A^{\text{ig}} - U^0 = NkT \sum_\alpha x_\alpha \ln x_\alpha + \sum_\alpha kT(-N_\alpha + N_\alpha \ln N_\alpha - N_\alpha \ln q_{0\alpha})$$

$$= NkT \sum_\alpha x_\alpha \ln x_\alpha + \sum_\alpha (A^{\text{ig}}_{0\alpha} - U^0_{0\alpha}), \tag{3.27}$$

with $A^{\text{ig}}_{0\alpha} = A^{\text{ig}}_{0\alpha}(T, V_{0\alpha}) = A^{\text{ig}}_{0\alpha}(T, p)$. For $N = N_A$, i.e., for one mole of a gaseous mixture, we have

$$a^{\text{ig}} - u^0 = RT \sum_\alpha x_\alpha \ln x_\alpha + \sum_\alpha x_\alpha (a^{\text{ig}}_{0\alpha} - u^0_{0\alpha}). \tag{3.28}$$

Putting together the components separately, i.e., unmixed, to make a total system volume V gives the Helmholtz free energy of the composed system as

$$A^{\text{ig}} - U^0 = \sum_\alpha (A_{0\alpha} - U^0_{0\alpha}), \tag{3.29}$$

where $A_{0\alpha}$ again refers to pure component α at volume $V_{0\alpha}$. Equation (3.29) differs from (3.27) by a term $NkT \sum_\alpha x_\alpha \ln x_\alpha$. Analogous results are obtained for all thermodynamic functions that depend on the factorials of the molecule numbers, such as

$$S^{\text{ig}} = \sum_\alpha S^{\text{ig}}_{0\alpha} - Nk \sum_\alpha x_\alpha \ln x_\alpha \tag{3.30}$$

and

$$G^{\text{ig}} - U^0 = \sum_\alpha (G^{\text{ig}}_{0\alpha} - U^0_{0\alpha}) + NkT \sum_\alpha x_\alpha \ln x_\alpha. \tag{3.31}$$

It is essential to appreciate the physical origin of the mixing terms in Eqs. (3.27) to (3.31). In classical thermodynamics it is usually argued that they follow directly from the definition of the ideal gas. Because the molecules of the ideal gas are independent of each other, each component behaves as if it filled the volume V alone. During the isothermal–isobaric mixing process each component α thus expands from the original volume $V_{0\alpha}$ to the system volume V. Due to this expansion process, the Helmholtz free energy of the component α changes by

$$\Delta A_\alpha^{\text{ig}} = -N_\alpha kT \ln \frac{V}{V_{0\alpha}} = N_\alpha kT \ln x_\alpha.$$

Thus there is a total change of Helmholtz free energy due to the isothermal–isobaric mixing process by

$$\Delta A^{\text{ig}} = \sum_\alpha N_\alpha kT \ln x_\alpha = NkT \sum_\alpha x_\alpha \ln x_\alpha, \tag{3.32}$$

which in combination with (3.29) yields (3.27). No statistical mechanical considerations have entered into this latter derivation. However, in the particular case of mixing two identical pure gases, (3.32) would also be valid, which is obviously wrong (Gibbsian paradox). This indicates the limitations of classical thermodynamics in interpreting fluid phase behavior. The statistical derivation reveals that the Helmholtz free energy of mixing arises due to the distinguishability of the components as documented by the factorials in (3.25) and is not necessarily associated with an expansion. This is conceptually important, because it explains the identical term for an ideal liquid mixture; cf. Section 4.1, where expansion upon mixing does not take place.

For all thermodynamic functions that depend exclusively on a derivative of the logarithm of the canonical partition function, which eliminates the factorials in (3.25), we find simple additive rules for the gas mixture, i.e.,

$$U^{\text{ig}} - U^0 = \sum_\alpha (U_{0\alpha}^{\text{ig}} - U_{0\alpha}^0) \tag{3.33}$$

and

$$H^{\text{ig}} - H^0 = \sum_\alpha (H_{0\alpha}^{\text{ig}} - H_{0\alpha}^0). \tag{3.34}$$

In classical thermodynamics this follows from the fact that U^{ig}, H^{ig} do not depend on volume or pressure. For the pressure, in particular, we find that

$$p^{\text{ig}} = \sum_\alpha \frac{N_\alpha kT}{V} = \sum_\alpha p_\alpha^{\text{ig}}, \tag{3.35}$$

where $p_\alpha^{\text{ig}} = N_\alpha kT/V$ is the partial pressure of component α in the ideal gaseous mixture and (3.35) is referred to as Dalton's law. The general definition of the partial pressure is

$$p_\alpha = x_\alpha p, \tag{3.36}$$

which always gives

$$\sum_\alpha p_\alpha = p. \tag{3.37}$$

Only in the ideal gas mixture does p_α have the clear physical significance following from (3.35).

3.6 Individual Contributions to the Thermodynamic Functions

In the general equations for the thermodynamic functions of an ideal gas, cf. Section 3.4, the individuality of the molecules is contained in the heat capacity and in the thermodynamic functions for the reference state. For these quantities the complete and detailed evaluation of all types of energy of a particular molecule yields further information. Because the final value of a state quantity can be calculated additively from the contributions of the various degrees of freedom it is possible and useful to consider these contributions separately. The molecular parameters needed for their numerical calculation can be taken from established tabulations [1–3], based on the evaluation of molecular spectra with quantum mechanical formulae, or from direct quantum-chemical calculations as discussed in Section 2.5.

3.6.1 Translation

For the molecular partition function of translation we have, according to (3.15),

$$q_{\text{tr}} = \left(\frac{2\pi m k T}{h^2}\right)^{3/2} V. \tag{3.38}$$

From this partition function the translational contributions to the thermodynamic functions are easily derived. As discussed above, the associated zero point energy contribution is negligible. However, it is still included here for formal consistency. For the free energy we get

$$A_{\text{tr}}^{\text{ig}} - U_{\text{tr}}^0 = kT \ln N! - NkT \ln q_{\text{tr}}$$

$$= NkT \left((\ln N - 1) - \ln \left\{ \left(\frac{2\pi m k T}{h^2}\right)^{3/2} V \right\} \right). \tag{3.39}$$

With the values for the universal constants k, h, and N_A (as tabulated in App. 1), we derive the following relationship for the molar Helmholtz free energy of translation in terms of temperature and pressure:

$$\frac{a_{\text{tr}}^{\text{ig}} - u_{\text{tr}}^0}{RT} = 2.6517 - 2.5 \ln(T/K) + \ln(p/\text{bar}) - 1.5 \ln(M/\text{g mol}^{-1}). \tag{3.40}$$

We further have

$$U_{\text{tr}}^{\text{ig}} - U_{\text{tr}}^0 = NkT^2 \left(\frac{\partial \ln q_{tr}}{\partial T}\right)_V = \frac{3}{2} NkT \tag{3.41}$$

$$C_{V,\text{tr}}^{\text{ig}} = \left(\frac{\partial U_{\text{tr}}^{\text{ig}}}{\partial T}\right)_N = \frac{3}{2} Nk \tag{3.42}$$

$$S_{\text{tr}}^{\text{ig}} = Nk\left[-\frac{\ln N!}{N} + \ln q_{tr} + T\left(\frac{\partial \ln q_{tr}}{\partial T}\right)_V\right]$$

$$= Nk\left[-(\ln N - 1) + \ln\left\{\left(\frac{2\pi mkT}{h^2}\right)^{3/2} V\right\} + \frac{3}{2}\right]$$

$$= Nk\left[-\ln N + \ln\left\{\left(\frac{2\pi mkT}{h^2}\right)^{3/2} V\right\} + \frac{5}{2}\right]. \tag{3.43}$$

A simple working equation for the translational molar entropy in terms of temperature and pressure follows from (3.43) as

$$\frac{s_{\text{tr}}^{\text{ig}}}{R} = -1.1517 + 2.5\ln(T/\text{K}) - \ln(p/\text{bar}) + 1.5\ln(M/\text{g mol}^{-1}). \tag{3.44}$$

Finally, we have

$$H_{\text{tr}}^{\text{ig}} - U_{\text{tr}}^0 = U_{\text{tr}}^{\text{ig}} - U_{\text{tr}}^0 + (pV)^{\text{ig}} = \frac{5}{2}NkT \tag{3.45}$$

$$C_{v,\text{tr}}^{\text{ig}} = \left(\frac{\partial H_{\text{tr}}^{\text{ig}}}{\partial T}\right)_N = \frac{5}{2}Nk \tag{3.46}$$

$$G_{\text{tr}}^{\text{ig}} - U_{\text{tr}}^0 = \left(H_{\text{tr}}^{\text{ig}} - TS_{\text{tr}}^{\text{ig}}\right)$$

$$= \frac{5}{2}NkT - NkT\left[-\ln N + \ln\left\{\left(\frac{2\pi mkT}{h^2}\right)^{3/2} V\right\} + \frac{5}{2}\right]$$

$$= NkT\left[\ln N - \ln\left\{\left(\frac{2\pi mkT}{h^2}\right)^{3/2} V\right\}\right], \tag{3.47}$$

and the corresponding working equation for the translational molar Gibbs free energy in terms of temperature and pressure reads

$$\frac{g_{\text{tr}}^{\text{ig}} - g_{\text{tr}}^0}{RT} = 3.6517 - 2.5\ln(T/\text{K}) + \ln(p/\text{bar}) - 1.5\ln(M/\text{g mol}^{-1}). \tag{3.48}$$

The translational contributions to the thermodynamic functions depend on the universal constants k, h and on T, V, N or T, p, N. Molecular individuality is restricted to m, the mass of the molecule. These contributions can thus be computed without any information about the geometry and the internal force field of the molecule. It should be noted again, however, that they become invalid at very low temperature. This can clearly be seen for $S_{\text{tr}}^{\text{ig}}$ from (3.43), which becomes $-\infty$ for $T \to 0$, in contradiction to the third law. Furthermore, the heat capacity in reality falls to zero for $T \to 0$, which is not reproduced by (3.42) and (3.46). The factor $1/N!$ that was introduced in (3.3) to take account of the indistinguishability of identical molecules ensures the proportionality of entropy and all other thermodynamic functions associated with it to N; i.e., it makes these quantities extensive, as they must be.

EXERCISE 3.3

Derive Eq. (3.40).

Solution

We find from (3.39) that

$$
\frac{A_{tr}^{ig}}{NkT} = \frac{a_{tr}^{ig}}{RT} = (\ln N - 1) - \ln \left\{ \left(\frac{2\pi mkT}{h^2} \right)^{3/2} \frac{NkT}{p} \right\}
$$

$$
= \ln N - 1 - \ln N - \ln k - \ln T + \ln p - 1.5 \ln(2\pi) - 1.5 \ln m
$$
$$
- 1.5 \ln k - 1.5 \ln T + 3 \ln h
$$
$$
= -1 - 1.5 \ln(2\pi) - 2.5 \ln T + \ln p - 1.5 \ln M + 1.5 \ln N_A
$$
$$
- 2.5 \ln k + 3 \ln h
$$
$$
= -1 - 1.5 \ln(2\pi) - 2.5 \ln(1.380662 \cdot 10^{-23} \text{ Nm/K})
$$
$$
+ 3 \ln(6.626176 \cdot 10^{-34} \text{ Nm s}) + 1.5 \ln(6.022045 \cdot 10^{23} \text{ mol}^{-1})
$$
$$
- 2.5 \ln T + \ln p - 1.5 \ln M
$$
$$
= -1 - 1.5 \ln(2\pi) - 2.5 \ln(1.380662 \cdot 10^{-23})
$$
$$
+ 3 \ln(6.626176 \cdot 10^{-34}) + 1.5 \ln(6.022045 \cdot 10^{23})
$$
$$
- 2.5 \ln(\text{Nm/K}) + 3 \ln(\text{Nm s}) + 1.5 \ln(\text{mol}^{-1}) - 2.5 \ln T
$$
$$
+ \ln p - 1.5 \ln M - 1.5 \ln(10^{-3} \text{ Ns}^2/\text{m}) + 1.5 \ln g
$$
$$
= -1 - 1.5 \ln(2\pi) - 2.5 \ln(1.380662 \cdot 10^{-23})
$$
$$
+ 3 \ln(6.626176 \cdot 10^{-34}) + 1.5 \ln(6.022045 \cdot 10^{23}) - 1.5 \ln(10^{-3})
$$
$$
+ \ln(10^5) - 2.5 \ln T/\text{K} + \ln p/\text{bar} - 1.5 \ln(M/\text{gmol}^{-1})
$$
$$
= 2.6517 - 2.5 \ln T/\text{K} + \ln p/\text{bar} - 1.5 \ln(M/\text{gmol}^{-1}).
$$

3.6.2 Electronic Energy

The partition function for electronic energy reads

$$
q_{el} = \sum_j g_{el,j} e^{-\varepsilon_{el,j}/kT}
$$
$$
= g_{el,0} + g_{el,1} e^{-\varepsilon_{el,1}/kT} + \cdots . \tag{3.49}
$$

Here, $\varepsilon_{el,j}$ is the energy level j of the electronic energy, and $g_{el,j}$ is its degeneracy. According to our convention for the zero point energy, the energy of the electronic ground configuration $\varepsilon_{el,0}$ is included in the reference value and, therefore, must be set equal to zero here. Thus the first term reduces to $g_{el,0}$, the degeneracy of the electronic ground state. The first excited electronic state has the energy value $\varepsilon_{el,1}$. Its energy is in most cases so large that the second term in (3.49) may be neglected when the temperature is not very high. Reliable calculations must, of course, estimate the effect of electronic energy states. To simplify notation we introduce with $\theta_{el,j} = \varepsilon_{el,j}/k$ the characteristic temperature of the jth excited electronic state.

The levels of electronic energy and their degeneracies have to be determined from quantum-chemical computer codes. Alternatively, these data can be obtained from tables based on spectroscopic experiments, which are interpreted quantum-chemically. A useful tabulation is contained in [1]. For the thermo-dynamic functions we have

$$A_{el}^{ig} - U_{el}^0 = -NkT \ln q_{el} = -NkT \ln \left(\sum_j g_{el,j} e^{-\theta_{el,j}/T} \right) = G_{el}^{ig} - U_{el}^0 \quad (3.50)$$

$$U_{el}^{ig} - U_{el}^0 = NkT^2 \left(\frac{\partial \ln q_{el}}{\partial T} \right) = NkT^2 \frac{1}{q_{el}} \left(\frac{\partial q_{el}}{\partial T} \right)$$
$$= Nk \frac{\sum g_{el,j} \theta_{el,j} e^{-\theta_{el,j}/T}}{\sum g_{el,j} e^{-\theta_{el,j}/T}} = H_{el}^{ig} - U_{el}^0 \quad (3.51)$$

$$S_{el}^{ig} = -\frac{A_{el}^{ig}}{T} + \frac{U_{el}^{ig}}{T}$$
$$= Nk \left[\ln \sum (g_{el,j} e^{-\theta_{el,j}/T}) + \frac{\sum g_{el,j} (\theta_{el,j}/T) e^{-\theta_{el,j}/T}}{\sum g_{el,j} e^{-\theta_{el,j}/T}} \right] \quad (3.52)$$

$$C_{V,el}^{ig} = C_{p,el}^{ig} = \left(\frac{\partial U_{el}^{ig}}{\partial T} \right)_N = \frac{Nk \sum g_{el,j} (\theta_{el,j}/T)^2 e^{-\theta_{el,j}/T}}{\sum g_{el,j} e^{-\theta_{el,j}/T}}$$
$$- \left(\frac{\sum g_{el,j} (\theta_{el,j}/T) e^{-\theta_{el,j}/T}}{g_{el,j} e^{-\theta_{el,j}/T}} \right)^2 Nk. \quad (3.53)$$

EXERCISE 3.4

Calculate the thermodynamic functions of argon in the ideal gas state at $p = 1.01235$ bar and temperatures between 100 and 4000 K.

Solution

Argon consists of monatomic molecules and thus only contributions of translation and electronic energy have to be taken into account.

The following expressions hold for the contributions of translation:

$$\frac{u_{tr}^{ig}}{RT} = \frac{3}{2}$$

$$\frac{h_{tr}^{ig}}{RT} = \frac{5}{2}$$

$$\frac{c_{v,tr}^{ig}}{R} = \frac{3}{2}$$

$$\frac{c_{p,tr}^{ig}}{R} = \frac{5}{2}$$

$$\frac{a_{tr}^{ig}}{RT} = 2.6517 - 2.5 \ln(T/K) + \ln(p/\text{bar}) - 1.5 \ln(M/\text{gmol}^{-1})$$

$$\frac{s_{tr}^{ig}}{R} = -1.1517 + 2.5 \ln(T/K) - \ln(p/\text{bar}) + 1.5 \ln(M/\text{gmol}^{-1}).$$

With $M = 39.948$ g/mol, numerical values are easily obtained. They are summarized in Table E 3.4.1.

Table E 3.4.1. Results

T/K	$(a_{tr}^{ig} - u^0)/RT$	s_{tr}^{ig}/R
100	−14.38	15.88
200	−16.11	17.61
500	−18.40	19.90
1000	−20.14	21.64
2000	−21.87	23.37
3000	−22.88	24.38
4000	−23.60	25.10

To calculate the electronic contribution we need the energy levels and the associated degeneracies. From [1] we find

$$g_{el,0} = 1$$

and

$$\frac{\varepsilon_{el,1}}{hc} = 93{,}143.8 \text{ cm}^{-1}.$$

The ground state is nondegenerate and the characteristic temperature of the first excited electronic state is computed using the transformation of energy units as communicated in App. 1, with the result that

$$\theta_{el,1} = \frac{\varepsilon_{el,1}}{k} = 134{,}013 \text{ K}.$$

A quantum-chemical calculation on the CIS/aug-cc-pvdz level gives $\theta_{el,1} = 145{,}675$ K, and thus roughly confirms this value. Thus, even for the highest temperature of 4000 K, the exponent of the Boltzmann factor of the first excited electronic state is

$$\frac{\theta_{el,1}}{4000} = 33.5$$

and thus so high that the contribution of electronic energy does not have to be taken into account.

[1]   C. E. Moore. *Atomic Energy Levels, Vol.1.* NSRDS-NBS 35, Washington, DC, 1971.

3.6.3 External Rotation

The external rotation molecular partition function reads

$$q_r = \sum_i e^{-\varepsilon_{ri}/kT}. \tag{3.54}$$

The energy values ε_{ri} must be obtained from the Schrödinger equation. To simplify the calculation of the external rotational energy, the molecule has been assumed in Sections 2.3 and 2.5 to be a rigid geometrical structure, with all atomic distances and angles being those of the equilibrium configuration. At normal temperatures this model is in good agreement with reality. Corrections for deviations at high temperatures can be added. We first consider linear molecules.

For the simple model of a linear rigid rotator the solution of the Schrödinger equation gives the energy levels and their associated degeneracies as, cf. (2.203),

$$\varepsilon_{rl} = \frac{l(l+1)h^2}{8\pi^2 I}, l = 0, 1, 2, \ldots, \tag{3.55}$$

with

$$g_{rl} = 2l + 1. \tag{3.56}$$

Here I is the moment of inertia with reference to the center of mass, defined as $I = \sum_i m_i r_i^2$, with m_i as the mass of atom i and r_i as the distance of this atom from the center of mass. The energy of external rotation in the lowest quantum state is zero, and thus no zero energy values have to be considered in the contributions of external rotation to the thermodynamic functions. Using Eqs. (3.55) and (3.56) in Eq. (3.54) for the partition function gives

$$q_r = \sum_l g_{rl} e^{-\frac{l(l+1)h^2}{8\pi^2 IkT}} = \sum_{l=0}^{\infty} (2l+1)e^{-l(l+1)\theta_r/T}. \tag{3.57}$$

Here we have defined

$$\theta_r = \frac{h^2}{8\pi^2 Ik} \tag{3.58}$$

as the characteristic temperature of external rotation. It contains I, i.e., the equilibrium configuration, as a property of a particular molecule, which can be found from spectroscopic tables or from quantum mechanical calculations; cf. Section 2.5. For most molecules we have $\theta_r \ll T$, and successive contributions to the sum differ only slightly from each other. So, for CO, we find $\theta_r = 2.78$ K, which is typical. This indicates that the quantization of the rotational energy is usually not important and we may replace the sum in (3.57) by an integral. Accordingly, as for translation, we can use the semiclassical approximation for the contribution of external rotation to the partition function. We thus make use of the classical mechanical results for the rotation of a linear rigid body; cf. (2.101). In so doing we note that we have to take into account the indistinguishability of identical nuclei by introducing a symmetry number σ_r, with $\sigma_r = 1$ for a heteronuclear and $\sigma_r = 2$ for a homonuclear diatomic molecule. So $\sigma_r = 2$ accounts for the fact that rotation of a homonuclear diatomic molecule by 180° interchanges two identical nuclei. Because the new orientation is indistinguishable from the original one, we have to divide the integral by 2 to avoid overcounting with respect to the quantum mechanical evaluation. The canonical partition function in the semiclassical approximation for a system of linear rigid rotators is

$$Q_r^{ig} = \left[\frac{1}{\sigma_r} \frac{1}{h^{fr}} \int\limits_{p_\theta=-\infty}^{+\infty} \int\limits_{p_\phi=-\infty}^{+\infty} \int\limits_{\theta=0}^{\pi} \int\limits_{\phi=0}^{2\pi} e^{-\frac{1}{2IkT}\left(p_\theta^2 + \frac{p_\phi^2}{\sin^2\theta}\right)} \, dp_\theta \, dp_\phi \, d\theta \, d\phi \right]^N$$

$$= \left[\frac{1}{\sigma_r} \frac{1}{h^2} \sqrt{2IkT\pi} \sqrt{2IkT\pi} \, 2 \cdot 2\pi \right]^N ,$$

where we have transformed the integral into a standard form, cf. App. 6, by introducing the variables

$$x = p_\phi / \sin\theta$$

and

$$y = p_\theta.$$

For the molecular partition function of external rotation of a linear molecule this gives

$$q_r = \frac{1}{\sigma_r} \frac{8\pi^2 IkT}{h^2} = \frac{1}{\sigma_r} \frac{T}{\theta_r}. \tag{3.59}$$

Here θ_r is the characteristic temperature of external rotation; cf. (3.58). We further have introduced $f_r = 2$ as the number of external rotational degrees of freedom for a linear molecule. For the thermodynamic functions we thus get

$$A_r^{ig} = -NkT \ln \frac{T}{\sigma_r \theta_r} = G_r^{ig} \tag{3.60}$$

$$S_r^{ig} = Nk \left(\ln \frac{T}{\sigma_r \theta_r} + 1 \right) \tag{3.61}$$

$$U_r^{ig} = NkT = H_r^{ig} \tag{3.62}$$

$$C_{V,r}^{ig} = Nk = C_{p,r}^{ig}. \tag{3.63}$$

The contributions of external rotation to the internal energy and to the heat capacity of an ideal gas of diatomic molecules do not depend on the molecular identity. Further, the heat capacity is independent of the temperature. This is only true for $T \gg \theta_r$. If the temperature is not high enough for the semiclassical approximation to be used, a series expansion or a direct summation has to be carried out. These quantum effects may become noticeable at low temperatures in diatomic molecules with one hydrogen atom because these have an unusually high characteristic temperature of external rotation, $10 < \theta_r < 90$. For unsymmetrical diatomic molecules direct summation leads to temperature dependence of the rotational contribution to the heat capacity, as shown in Figure 3.1. The maximum results because the exponential terms in the sum decrease while the degeneracies increase. In gases with lower characteristic

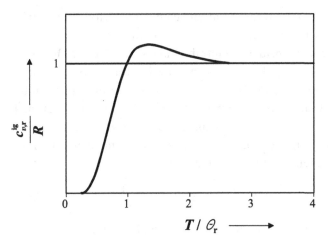

Fig. 3.1. Rotational heat capacity for unsymmetrical linear molecules.

temperatures of external rotation this maximum also exists, without, however, having practical significance, because of the low temperatures where it arises.

The general nonlinear molecule is treated in a similar way. The canonical partition function of a system of N unsymmetrical nonlinear rotators reads, in the semiclassical approximation,

$$Q_r^{ig} = \left[\frac{1}{\sigma_r} \frac{1}{h^{f_r}} \int \cdots \int e^{-E_r^{kin}/kT} dp_\theta dp_\phi dp_\chi d\theta d\phi d\chi \right]^N,$$

where E_r^{kin} is given by (2.102). We have $f_r = 3$ for a nonlinear molecule, and to integrate we substitute

$$x = p_\theta \sin \chi - \frac{\cos \chi}{\sin \theta} (p_\phi - p_\chi \cos \theta)$$

and

$$y = p_\theta \cos \chi + \frac{\sin \chi}{\sin \theta} (p_\phi - p_\chi \cos \theta)$$

with, cf. App. 6,

$$dp_\theta \, dp_\phi = \sin \theta \, dx \, dy.$$

We thus find that

$$Q_r^{ig} = \left[\frac{1}{\sigma_r} \frac{1}{h^3} \int\limits_{x=-\infty}^{+\infty} \int\limits_{y=-\infty}^{+\infty} \int\limits_{p_\chi=-\infty}^{+\infty} \int\limits_{\theta=0}^{\pi} \int\limits_{\phi=0}^{2\pi} \int\limits_{\chi=0}^{2\pi} e^{-\frac{1}{2kT}\left(\frac{x^2}{I_A} + \frac{y^2}{I_B} + \frac{p_\chi^2}{I_C}\right)} \right.$$

$$\left. \sin \theta \, dx \, dy \, dp_\chi \, d\theta \, d\phi \, d\chi \right]^N,$$

which is a product of standard integrals, cf. App. 6, and gives for the molecular partition function of external rotation

$$q_r = \frac{\pi^{1/2}}{\sigma_r} \left(\frac{8\pi^2 I_A kT}{h^2} \right)^{1/2} \left(\frac{8\pi^2 I_B kT}{h^2} \right)^{1/2} \left(\frac{8\pi^2 I_C kT}{h^2} \right)^{1/2}. \tag{3.64}$$

The symmetry number σ_r of external rotation again corrects for the fact that a molecule may have a number of rotational orientations that are indistinguishable. To illustrate, we consider the tetrahedral CH_4 molecule. The three hydrogen atoms in one plane can be oriented in three different ways by rotation of less than 360° without creating a new quantum state. Because this can be done with each of the four hydrogen atoms at the top, there are 12 rotational arrangements associated with one single quantum state; i.e., $\sigma_r = 12$. For many molecules this symmetry number has been tabulated. In compact notation we introduce characteristic temperatures of rotation by

$$\theta_{rA} = \frac{h^2}{8\pi^2 I_A k}; \quad \theta_{rB} = \frac{h^2}{8\pi^2 I_B k}; \quad \theta_{rC} = \frac{h^2}{8\pi^2 I_C k} \tag{3.65}$$

and find for the molecular partition function of external rotation

$$q_r = \frac{\pi^{1/2}}{\sigma_r} \left(\frac{T^3}{\theta_{rA}\theta_{rB}\theta_{rC}} \right)^{1/2}. \tag{3.66}$$

The thermodynamic functions are given by

$$A_r^{ig} = -NkT \ln \left[\frac{\pi^{1/2}}{\sigma_r} \left(\frac{T^3}{\theta_{rA}\theta_{rB}\theta_{rC}} \right)^{1/2} \right] = G_r^{ig} \tag{3.67}$$

$$S_r^{ig} = Nk \left\{ \ln \left[\frac{\pi^{1/2}}{\sigma_r} \left(\frac{T^3}{\theta_{rA}\theta_{rB}\theta_{rC}} \right)^{1/2} \right] + \frac{3}{2} \right\} \tag{3.68}$$

$$U_r^{ig} = \frac{3}{2} NkT = H_r^{ig} \tag{3.69}$$

$$C_{V,r}^{ig} = \frac{3}{2} Nk = C_{p,r}^{ig}. \tag{3.70}$$

These formulae are limited to $T \gg \theta_r$, where θ_r may be any one of the three characteristic temperatures of external rotation. Under this restriction the rotational contributions to the internal energy and heat capacity are again independent of molecular species and temperature. In contrast, the free energy and the entropy do depend on molecular properties, such as the principal moments of inertia. They are calculated from the geometry of the molecule, which in turn is determined from spectroscopic data or quantum-chemical computer codes.

3.6.4 Vibration

During a vibration the atoms vibrate around their equilibrium positions, i.e., the configuration that is associated with the minimum of the intramolecular energy, U_{im}. This intramolecular energy results from the electrostatic interactions of the charges in a molecule. We consider a diatomic molecule in the electronic ground state. The molecular configuration depends only on the distance r between the

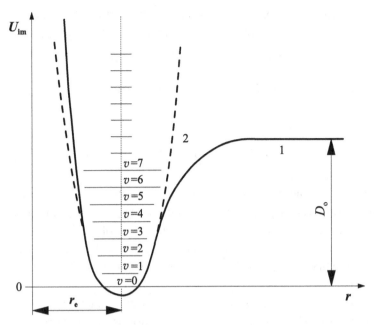

Fig. 3.2. Potential energy of a diatomic oscillator.

two nuclei and the intramolecular energy U_{im} is given by curve number 1 in Figure 3.2. This intramolecular energy determines the work associated with the vibrational motion and thus in particular its frequency. The zero of energy is the energy associated with the lowest possible quantum number, i.e., $v = 0$. We note that at a sufficiently high temperature the molecule will dissociate to the state of separated atoms, i.e., $r \to \infty$. The difference between the energy zero of the molecule and the energy of the separate atoms at $T = 0$ K is referred to as the dissociation energy D_0. To simplify the mechanical analysis, the model of a harmonic oscillator was introduced in Section 2.3. In this model, the actual intramolecular energy is approximated by a parabola, as shown in curve number 2. The harmonic approximation is seen to be adequate at low vibrational quantum numbers v, but becomes insufficient as v increases. For normal temperatures only the first few values of v are relevant, so that the harmonic approximation is sufficient for most applications. In (2.201), cf. Section 2.5, it was shown that the quantum mechanical energy states of the diatomic harmonic oscillator are given by

$$\varepsilon_{vv} = \left(\frac{1}{2} + v\right) h\nu_0, \quad v = 0, 1, 2\ldots. \tag{3.71}$$

Here ν_0 is the fundamental harmonic frequency and v the vibrational quantum number. Each energy level is nondegenerate; i.e., $g_v = 1$. The molecular partition function of vibration for a diatomic molecule thus reads

$$q_v = \sum_v e^{-\varepsilon_{vv}/kT}. \tag{3.72}$$

When we take the zero-point energy ($v = 0$) out of the vibrational energy, in accord with our convention for the thermodynamic zero of energy, cf. Section 3.2, i.e.,

$$(\varepsilon_v - \varepsilon_v^0)_v = \left(\frac{1}{2} + v\right) h\nu_0 - h\nu_0/2 = vh\nu_0, \quad v = 0, 1, 2 \ldots,$$

we arrive at

$$q_v = \sum_{v=0}^{\infty} e^{-vh\nu_0/kT} = \sum_{v=0}^{\infty} e^{-v\theta_v/T} \tag{3.73}$$

with $\theta_v = h\nu_0/k$ as the characteristic temperature of vibration. Physically, it does not make sense to evaluate (3.73) up to $v \to \infty$, because at very high vibrational quantum numbers a real oscillator will dissociate; cf. curve 1 of Figure 3.2. This complication does not have any practical significance, however, because θ_v/T is so large that only the first terms of the sum will make a significant contribution to the molecular partition function.

The vibrational energy levels are so far apart that the summation cannot be replaced by an integral. The semiclassical approximation is thus not applicable to vibration. However, the series $1 + x + x^2 \cdots$ converges to $(1 - x)^{-1}$ for $|x| = e^{-\theta_v/T} < 1$, and we thus have

$$q_v = \frac{1}{1 - e^{-\theta_v/T}}. \tag{3.74}$$

The fundamental frequency ν_0 can be found from spectroscopic data or quantum-chemical computer codes. For the thermodynamic functions we get

$$A_v^{ig} - U_v^0 = -NkT \ln \frac{1}{1 - e^{-\theta_v/T}} \tag{3.75}$$

$$U_v^{ig} - U_v^0 = \frac{\theta_v/T}{e^{\theta_v/T} - 1} NkT = H_v^{ig} - U_v^0 \tag{3.76}$$

$$S_v^{ig} = Nk \left(\ln \frac{1}{1 - e^{-\theta_v/T}} + \frac{\theta_v/T}{e^{\theta_v/T} - 1} \right) \tag{3.77}$$

$$C_{V,v}^{ig} = C_{p,v}^{ig} = Nk \frac{(\theta_v/T)^2 \, e^{\theta_v/T}}{(e^{\theta_v/T} - 1)^2}. \tag{3.78}$$

We note that the contribution of vibration to the heat capacity for a diatomic molecule will lead to Nk for high temperatures. The caloric quantities A_v^{ig} and U_v^{ig} are given with reference to a zero of energy U_v^0, which is the energy of the molecule in the lowest vibrational quantum state.

EXERCISE 3.5

Investigate the occupation numbers of the first five rotational and vibrational energy values for CO at 50, 100, 300, and 1000 K.

Table E 3.5.1. Occupation numbers of rotation

T/K	n_0/n	n_1/n	n_2/n	n_3/n	n_4/n	n_5/n
50	0.0556	0.1492	0.1991	0.1997	0.1646	0.1154
100	0.0278	0.0789	0.1176	0.1394	0.1435	0.1328
300	0.0093	0.0273	0.0438	0.0580	0.0693	0.0772
1000	0.0028	0.0083	0.0137	0.0188	0.0237	0.0281

Solution

The occupation number of the lth rotational energy level, i.e., the number of molecules in the lth level in relation to the total number of molecules, is given for a linear molecule by

$$P_{l,r} = \left(\frac{n_l}{n}\right)_r = \frac{(2l+1)e^{-l(l+1)\theta_r/T}}{T/\theta_r}.$$

The characteristic temperature of external rotation for CO is $\theta_r = 2.78$ K [2]. Introducing this gives the numerical results summarized in Table E 3.5.1.

The occupation number of the vth vibrational energy level is given by

$$P_{v,v} = \left(\frac{n_v}{n}\right)_v = e^{-v\theta_v/T}\left[1 - e^{-\theta_v/T}\right].$$

With $\theta_v = 3120$ K [2] for the characteristic temperature of vibration for CO we obtain the numerical results summarized in Table E 3.5.2.

There is a broad distribution of occupied energy levels for rotation, whereas for vibration, essentially all molecules are in their lowest vibrational state. This latter fact is due to the rather high value of the characteristic temperature of vibration for carbon monoxide. Because the vibrational energy for $v = 0$ has been incorporated into the zero point of energy, there will be no vibrational contribution to the thermodynamic functions from the vibrational ground state, i.e., at low temperatures; cf. Exercises 3.7 and 3.8. Although the results of Table E 3.5.2 do not hold generally, for most molecules at normal temperatures, only the first few vibrational states are occupied.

[1] G. Herzberg. *Molecular Spectra and Molecular Structure. I. Spectra of Diatomic Molecules.* Van Nostrand Reinhold Company, New York, 1950.

EXERCISE 3.6

Calculate the molecular partition function of vibration for a diatomic molecule in the semiclassical approximation.

Solution

The total classical energy of a diatomic harmonic oscillator is given by (2.93) as

$$H_v = E_v^{kin} + U_v = \frac{1}{2m_r}p_r^2 + \frac{1}{2}\kappa(r - r_e)^2,$$

Table E 3.5.2. Occupation numbers of vibration

T/K	n_0/n	n_1/n	n_2/n	n_3/n	n_4/n	n_5/n
50	~1	0	0	0	0	0
100	~1	0	0	0	0	0
300	~1	0	0	0	0	0
1000	0.956	0.042	0.002	~0	0	0

with

$$m_{\mathrm{r}} = \frac{m_1 m_2}{m_1 + m_2}$$

and κ as the constant of the restoring force. The number of vibrational degrees of freedom for a diatomic molecule is $f_v = 1$; symmetry restrictions do not apply. Thus the molecular partition function of harmonic oscillation in the semiclassical approximations is given by

$$q_v = (Q_v^{\mathrm{ig}})^{1/N}$$

with

$$Q_v^{\mathrm{ig}} = \left[\frac{1}{h^{f_v}} \int e^{-H_v(p_r,r)/kT} \mathrm{d}r \mathrm{d}p_r \right]^N$$

$$= \left[\frac{1}{h} \int\limits_{-\infty}^{\infty} \int\limits_{-\infty}^{\infty} e^{-\frac{1}{kT}\left(p_r^2/2m_r + \kappa(r-r_e)^2/2 \right)} \mathrm{d}r \mathrm{d}p_r \right]^N$$

$$= \left[\frac{2\pi kT}{h} \left(\frac{m_r}{\kappa} \right)^{1/2} \right]^N.$$

With $v_0 = \sqrt{\kappa/m_{\mathrm{r}}}/2\pi$, cf. (2.95), this gives the molecular partition function in the semiclassical approximation as

$$q_v = \frac{T}{\theta_v}.$$

The associated contribution to the heat capacity is R. As could be expected, this does not agree with the correct quantum mechanical result (3.74). However, at high temperatures, (3.74) yields to the semiclassical result as postulated by the correspondence principle, because

$$e^{-\theta_v/T} = 1 - \frac{\theta_v}{T} + \frac{1}{2}\left(\frac{\theta_v}{T} \right)^2 - \cdots,$$

and thus

$$\lim_{\theta_v/T \to 0} q_v = \frac{1}{1 - 1 + \theta_v/T - \frac{1}{2}(\theta_v/T)^2 + \cdots} = \frac{T}{\theta_v}.$$

This means that the contribution of vibration to the heat capacity of a diatomic molecule will reach a limit of $1R$ at high temperatures.

With the molecular partition function of vibration obtained from (3.74) and that for rotation from (3.59), very accurate predictions for the ideal gas properties of diatomic molecules can be made.

EXERCISE 3.7

Calculate the individual contributions to the thermodynamic functions of nitrogen in the ideal gas state at $p = 1.01325$ bar at temperatures between 100 and 1000 K.

Solution

Nitrogen is a diatomic homonuclear molecule. We have to consider contributions due to translation, electronic energy, rotation, and vibration.

1. Translation: The evaluation of the appropriate formulae for the translational contributions is analogous to Exercise 3.4. For nitrogen we use $M = 28.0134$ g/mol.

Table E 3.7.1. Contributions of the individual degrees of freedom

T/K	$(a^{ig} - u^0)/RT$			s^{ig}/R			c_v^{ig}/R		
	tr	r	v	tr	r	v	tr	r	v
100	−13.85	−2.84	0.00	15.35	3.84	0.00	1.5	1.0	0.00
200	−15.58	−3.53	0.00	17.08	4.53	0.00	1.5	1.0	0.00
500	−17.87	−4.45	0.00	19.37	5.45	0.01	1.5	1.0	0.06
1000	−19.60	−5.14	−0.04	21.10	6.14	0.16	1.5	1.0	0.42

2. Electronic energy: From [1] we find $\varepsilon_{el,1}/hc = 69{,}290$, $g_{el,1} = 3$, and $g_{el,0} = 1$. The characteristic temperature of the first excited electronic state is thus

$$\theta_{el,1} = 99{,}693 \text{ K},$$

   which is essentially confirmed by a quantum-chemical calculation on a perturbed CIS-level. This temperature is so high that a contribution from excited electronic states may be neglected. Because the electronic ground state is nondegenerate, the electronic energy does not make any contribution apart from that to the zero-point energy.

3. Rotation: From [1] we find a characteristic temperature of rotation of $\theta_r = 2.89$ K; in [2] the corresponding value is $\theta_r = 2.92$ K. A quantum-chemical calculation on the MP2/cc-pvtz level gives $\theta_r = 2.81$ K. These differences lead to only minor uncertainties in the thermodynamic functions. The symmetry number is $\sigma_r = 2$.

4. Vibration: From [1] we find a characteristic temperature of vibration of $\theta_v = 3394$ K; [2] the corresponding value is $\theta_v = 3352$ K. A quantum-chemical calculation on the MP2/cc-pvtz level gives $\theta_v = 3161$ K. Again, these differences are insignificant for the thermodynamic functions.

Table E 3.7.1 summarizes the contributions of the individual degrees of freedom with the rotational and vibrational characteristic temperatures taken from [2]. The calculated heat capacities agree with experimental data [3] between 250 and 350 K to within their accuracy of 0.3%. We note that there is practically no contribution of vibration to the thermodynamic functions for this simple molecule at normal temperatures.

[1]   G. Herzberg. *Molecular Spectra and Molecular Structure. I. Spectra of Diatomic Molecules*. Van Nostrand Reinhold, New York, 1950.

[2]   Landolt-Börnstein. *Zahlenwerte und Funktionen aus Physik, Chemie, Geographie und Technik*, Volume 2,4. Springer, Berlin, 1961.

[3]   W. Lemming. *VDI-Fortschritt-Ber.*, 19(32): 1813, 1989.

For polyatomic molecules we get, again introducing our definition of the zero-point energy,

$$q_{vj} = \sum_v e^{-\varepsilon_{vvj}/kT} = \sum_v e^{-vh\nu_{0j}/kT} = \frac{1}{1 - e^{-\theta_{vj}/T}} \qquad (3.79)$$

for the molecular partition function of the jth vibrational normal mode and

$$q_v = \prod_j^{\substack{3n-5 \\ 3n-6}} q_{vj} = \prod_j^{\substack{3n-6 \\ 3n-5}} \frac{1}{1 - e^{-\theta_{vj}/T}} \qquad (3.80)$$

for the total vibrational contribution with $\theta_{vj} = h\nu_{0j}/k$ as the characteristic temperature of the jth normal mode. The thermodynamic functions are analogous to those in the case of diatomic molecules. Because the logarithm of q is relevant, the total vibrational contribution is simply equal to the sum of the vibrational contributions of the normal modes. The frequencies of the normal modes have to be found from tabulated spectroscopic data or quantum chemical computer codes. Useful compilations for simple molecules are contained in [1,3].

EXERCISE 3.8

Calculate the individual contributions to the thermodynamic functions of ammonia in the ideal gas state at $p = 1.01325$ bar at temperatures between 100 and 1000 K.

Solution

Ammonia consists of four atoms. We consider contributions due to translation, electronic energy, rotation, and vibration.

1. Translation: The evaluation of the formulae for the translational contributions is analogous to the earlier examples. We use $M = 17.031$ g/mol.
2. Electronic energy: In [1] we find $\varepsilon_{el,1}/hc = 46{,}136$, $g_{el,1} = g_{el,0} = 1$. The characteristic temperature of the first excited electronic state is thus

$$\theta_{el,1} = 66{,}379 \text{ K},$$

again confirmed by a quantum-chemical calculation on a perturbed CIS-level. This temperature is so high that the contribution of electronic energy is fully taken into account by its contribution to the zero of energy.
3. Rotation: Ammonia belongs to the class of symmetrical tops. Two of its principal moments of inertia are identical. In [1] we find $\theta_{r,A} = \theta_{r,B} = 14.87$ K and $\theta_{r,C} = 9.45$ K; in [2] the corresponding values are $\theta_{r,A} = \theta_{r,B} = 14.5$ K and $\theta_{r,C} = 9.34$ K, whereas the results of quantum-chemical calculations are reproduced in Table 2.1 of Section 2.5. Application of these various sets of parameters leads to only minor discrepancies in the thermodynamic functions. The symmetry number is $\sigma_r = 3$.
4. Vibration: As a nonlinear molecule with four atoms, ammonia has six vibrational degrees of freedom. In [1] we find $\theta_{v,1} = 2340$ K, $\theta_{v,2} = 1342$ K, $\theta_{v,3} = 4801$ K, and $\theta_{v,4} = 4955$ K. The corresponding values in [2] are $\theta_{v,1} = 2344$ K, $\theta_{v,2} = 1365$ K, $\theta_{v,3} = 4794$ K, and $\theta_{v,4} = 4961$ K. The first and fourth normal modes are twofold degenerate. A quantum-chemical set of vibration frequencies is shown in Table 2.1 of Section 2.5. Again these differences in the molecular data are of minor significance.

Table E 3.8.1 presents a summary of the contributions of the individual degrees of freedom with the molecular data taken from [2]. Very accurate statistical calculations for NH_3 with a much more elaborate molecular model were performed by Haar [3], who quotes agreement of better than 0.1% with experimental data. The values calculated here by the rigid rotator–harmonic oscillator model for the heat capacity deviate from those in [3] by 0.3% at 240 K up to 1.1% at 420 K. We note that the vibrational contribution now becomes relevant even at normal temperatures. At 1000 K it is the dominant single contribution.

Table E 3.8.1. Contributions of the individual degrees of freedom

	$(a^{ig} - u^0)/RT$			s^{ig}/R			c_v^{ig}/R		
T/K	tr	r	v	tr	r	v	tr	r	v
100	−13.10	−2.59	0.00	14.60	4.09	0.00	1.5	1.5	0.00
200	−14.83	−3.63	0.00	16.33	5.13	0.01	1.5	1.5	0.05
500	−17.12	−5.00	−0.09	18.62	6.50	0.37	1.5	1.5	0.99
1000	−18.86	−6.04	−0.52	20.36	7.54	1.60	1.5	1.5	2.70

[1] G. Herzberg. *Molecular Spectra and Molecular Structure. III. Electronic Spectra and Electronic Structure of Polyatomic Molecules.* Van Nostrand Reinhold, New York, 1966.
[2] Landolt-Börnstein. *Zahlenwerte und Funktionen aus Physik, Chemie, Geographie und Technik*, volume 2,4. Springer, Berlin, 1961.
[3] L. Haar. NSRDS-NBS 19, 1968.

As shown in Exercise 3.8, even for NH_3, the contribution of vibration is again moderate at room temperature, far below the high-temperature value of $6R$. The latter is typical for all molecules, but for larger molecules the relative contribution of vibration will tend to be dominant even at moderate temperatures due to the large number of vibrational modes; cf. Exercise 3.9. The major challenge in predicting ideal gas properties of large molecules is therefore a sound knowledge of vibration frequencies. Tabulated spectroscopic values are frequently available. If not, frequencies can be obtained from quantum-chemical computer codes, unless the molecules become exceptionally large, such as some of those shown in Figure 1.14.

As a further illustration of the prediction of ideal gas properties from a combination of statistical mechanics and quantum chemistry, we consider the standard heat of formation of gaseous NH_3. Table 3.1 shows results for the standard heat of formation of NH_3 with all energy values being given in hartree; cf. App. 1. The quantum-chemical computer code gives the energy $E(0)$ of the ground state at 0 K with reference to the nuclei and electrons being infinitely separated. So the enthalpy of formation for NH_3 at 0 K is obtained from the difference of the ground state energies, according to

$$\Delta h_{NH_3}^f(0) = E_{NH_3}(0) - \left(\frac{1}{2}E_{N_2}(0) + \frac{3}{2}E_{H_2}(0)\right),$$

and the transformation to the standard temperature at 25°C, i.e., $\Delta h_i = h_i(298.15) - h_i(0)$, is done with the methods discussed above using the moments of inertia and the vibration frequencies as calculated from the computer code; cf. Table 2.1 of Section 2.5. It turns out that an accurate calculation of the electronic energy is rather demanding with respect to the quantum-chemical level, because the final result for the enthalpy of formation is a small number obtained from the difference of large numbers. A coupled cluster (CC) calculation with a MP-calculation for the temperature dependence gives close

Table 3.1. Standard enthalpy of formation of NH_3 in the ideal gas state in hartree from the Gaussian 03 computer code ($\Delta h_{NH_3,exp}^{f,0}(g) = -46,11$ kJ/mol $= -0.01756$ hartree, cf. App. 1)

	B3LYP 6-31 + g(d,p)	MP2 aug-cc - pVDZ	MP2 aug-cc - pVTZ	CCSD(T) aug-cc - pVTZ
$E_{NH_3}(0)$	−56.56699	−56.40489	−56.46054	−56.48054
$E_{N_2}(0)$	−109.52978	−109.28065	−109.36480	−109.38059
$E_{H_2}(0)$	−1.17854	−1.15622	−1.16502	−1.17262
Δh_{NH_3}	0.03819	0.03820	0.038392	0.038392[*]
Δh_{N_2}	0.00890	0.00822	0.00829	0.00829[*]
Δh_{H_2}	0.01348	0.01347	0.01360	0.01360[*]
$\Delta h_{NH_3}^{f,0}(g)$	−0.02077	−0.01635	−0.01676	−0.01747

[*] = MP2 -aug-cc - pVTZ.

agreement to the experimental value. It can be seen that a coupled cluster calculation is necessary for the electronic energy, whereas for the temperature correction a lower level appears to be sufficient, as also found in Table 2.1.

3.6.5 Internal Rotation

The energy mode of internal rotation was discussed within the framework of classical mechanics in Section. 2.3. We show here how we can evaluate the molecular partition function. The intramolecular potential energy shown in Figure 2.8 is essentially determined by $U_{ir,max}$, referred to as the barrier of internal rotation. We now classify three different cases with respect to the magnitude of $U_{ir,max}/kT$.

1. $U_{ir,max}/kT$ Is Large

If $U_{ir,max}/kT$ is on the order of 10 or more, i.e., for large potential energies in combination with low temperatures, the molecule does not have enough energy to surpass the potential barrier of internal rotation. The actual motion is then rather a torsional oscillation around the position of the minimum of U_{ir} with an angle ϕ that is very small. This allows us to approximate $\cos 3\phi$ in (2.103) by $1 - (3\phi)^2/2$. We then find $U_{ir} = U_{ir,max}(3\phi/2)^2$ and the Hamilton functions for harmonic vibration (2.93) and torsional oscillation (2.107) become formally identical, when the reduced mass m_r in (2.93) is replaced by the reduced moment of inertia I_{ir} and the potential energy U_{im} of vibration by that of internal rotation, i.e., U_{ir}. The Schrödinger equation for internal rotation thus reads, by comparison with (2.199),

$$\frac{d^2\psi}{d\phi^2} + \frac{8\pi^2 I_{ir}}{h^2}\left(\varepsilon_\phi - U_{ir,max}\frac{9}{4}\phi^2\right)\psi = 0.$$

With the abbreviations $\lambda = 8\pi^2 I_{ir}/h^2\varepsilon_\phi$ and $\alpha = \frac{\pi}{h}3\sqrt{2I_{ir}U_{ir,max}}$ this equation becomes formally identical to (2.199) for the harmonic oscillator; i.e.,

$d^2\psi/dx^2 + (\lambda - \alpha^2 x^2)\,\psi = 0$. Replacing the reduced mass and the frequency in $\alpha = 4\pi^2 m_r \nu_0 / h$ for the harmonic oscillator by the corresponding quantities for internal rotation leads to $\alpha = 4\pi^2 I_{ir}\nu_\phi / h$. We thus find for the torsional frequency $\nu_\phi = (3/2\pi)\sqrt{U_{ir,max}/2I_{ir}}$, which can be calculated from $U_{ir,max}$ and I_{ir}. Torsional oscillation is frequently associated with double bonds between carbon atoms, as in ethylene. If the frequency is known one does not have to worry about the nature of this oscillation and can just treat it like the other normal modes of vibration.

2. $U_{ir,max}/kT$ Is Small

If $U_{ir,max}/kT$ is so small that we can neglect it, i.e., for low potential energies in combination with high temperatures, only the kinetic contribution to the energy of internal rotation remains to be considered. The associated Hamilton function is analogous to that for external rotation, although constrained to just one coordinate, ϕ. As there, it is found again that quantization of the energy levels becomes unimportant. Then, we just need the classical mechanical expression for the kinetic energy of free internal rotation (2.106) to evaluate the molecular partition function in the semiclassical approximation and find that

$$
\begin{aligned}
q_{ir,free} &= \frac{1}{h\sigma_{ir}} \int\limits_{p_\phi=-\infty}^{\infty} \int\limits_{\phi=0}^{2\pi} e^{-\frac{1}{2I_{ir}kT}p_\phi^2}\,dp_\phi d\phi \\
&= \frac{1}{\sigma_{ir}}\left(\frac{8\pi^3 I_{ir}kT}{h^2}\right)^{1/2} = \frac{1}{\sigma_{ir}}\left(\frac{T}{\theta_{ir}}\right)^{1/2},
\end{aligned}
\tag{3.81}
$$

where σ_{ir} is the symmetry number of internal rotation with a significance analogous to that of external rotation and $\theta_{ir} = h^2/(8\pi^3 kI_{ir})$ is the characteristic temperature of internal rotation. The contributions of free internal rotation to the thermodynamic functions are

$$
\begin{aligned}
A_{ir,free}^{ig} &= -NkT \ln q_{ir,free} \\
&= -NkT\left[-\ln\sigma_{ir} + \frac{1}{2}\ln\left(\frac{T}{\theta_{ir}}\right)\right]
\end{aligned}
\tag{3.82}
$$

$$
S_{ir,free}^{ig} = -\left(\frac{\partial A_{ir,free}^{ig}}{\partial T}\right)_V = Nk\left[-\ln\sigma_{ir} + \frac{1}{2}\ln(T/\theta_{ir}) + \frac{1}{2}\right]
\tag{3.83}
$$

$$
U_{ir,free}^{ig} = A_{ir,free}^{ig} + TS_{ir,free}^{ig} = \frac{1}{2}NkT = H_{ir,free}^{ig}
\tag{3.84}
$$

$$
C_{p,\,ir,free}^{ig} = \left(\frac{\partial H_{ir,free}^{ig}}{\partial T}\right)_N = C_{V,ir,free}^{ig} = \left(\frac{\partial U_{ir,free}^{ig}}{\partial T}\right)_N = \frac{1}{2}Nk.
\tag{3.85}
$$

3. $U_{ir,max}/kT$ Is Neither Large Nor Small

If $U_{ir,max}/kT$ is neither large nor small, the rotational motion is neither a torsional vibration nor a free internal rotation. We refer to it as a hindered internal rotation. The semiclassical approximation is then no longer valid and the associated energy values must be obtained from the Schrödinger equation. In this equation we can frequently formulate the angular dependence of the periodic potential energy in terms of a Fourier series $U_{ir} = \frac{1}{2} \sum_n U_{ir,max}^{(n)} [1 - \cos(n\phi)]$, where in a typical case n goes from 1 to 6 and some terms may be zero due to symmetry of the molecule. The values of $U_{ir,max}^{(n)}$ are found from spectroscopic data or from quantum-chemical computer codes. The solution of the Schrödinger equation has been obtained numerically [4] and the energy eigenvalues can be summed to give the molecular partition function. In many simple cases the potential energy can be replaced by one single term such as $U_{ir} = \frac{1}{2} U_{ir,max}^{(n)} [1 - \cos(n\phi)]$, e.g., for ethane with $n = 3$, cf. Section 2.3, which is also approximately valid for many other molecules. For this particular case numerical contributions to the thermodynamic functions are tabulated in App. 7 [5]. It should be noted that the n in the expression for U_{ir} is not necessarily equal to the symmetry number of internal rotation although it is obviously so for ethane. We note from the tables that the general case of hindered internal rotation yields to the limiting cases of torsional oscillation and free internal rotation for $U_{ir,max}/kT \to \infty$ and $U_{ir,max}/kT \to 0$, respectively. So, for $U_{ir,max}/kT \to 0$, i.e., free internal rotation, we find a contribution of $R/2$ to the heat capacity independent of $1/q_{ir}$. For $U_{ir,max}/kT > 0$ we find a spectrum of values depending on q_{ir}. We note that $1/q_{ir}$ becomes large at small values of (T/θ_{ir}), i.e., large values of I_{ir} and small values of T. At constant $U_{ir,max}/kT$, an increase of $1/q_{ir}$ reduces the effect of internal rotation because the motions tend to die off. At constant $1/q_{ir}$, increasing $U_{ir,max}/kT$ first increases the contribution to the heat capacity, even above $1R$, the classical value of torsional vibration, and then reduces it to essentially zero for the same reason.

EXERCISE 3.9

Calculate the contribution of internal rotation as well as the other contributions to the thermodynamic functions of methanol (CH_4O) in the ideal gas state at $p = 1.01325$ bar at temperatures between 100 and 1000 K and compare to data [1].

[1] E. Strömsoe, H. G. Rönne, and A. L. Lydersen. *J. Chem. Eng. Data*, 15:286, 1970.

Solution

Methanol is a nonlinear molecule with six atoms. Its geometry has been shown in Figure E 2.3.1, which is reproduced in Figure E 3.9.1. It has 12 internal degrees of freedom. In [1] we find the following principal moments of inertia:

$$I_A \cdot I_B = 233.9 \cdot 10^{-80} \text{ g}^2\text{cm}^4$$

and

$$I_C = 35.31 \cdot 10^{-40} \text{ g cm}^2.$$

Further, we find the following characteristic temperatures of vibration:

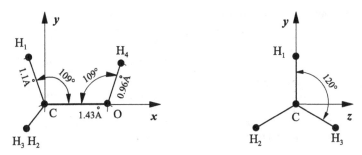

Fig. E 3.9.1. Molecular geometry of the methanol molecule.

$\theta_{v1} = 5295 \text{ K}; \quad \theta_{v2} = 4279 \text{ K}; \quad \theta_{v3} = 4093 \text{ K}; \quad \theta_{v4} = 2127 \text{ K};$
$\theta_{v5} = 2052 \text{ K}; \quad \theta_{v6} = 1936 \text{ K}; \quad \theta_{v7} = 1485 \text{ K}; \quad \theta_{v8} = 1546 \text{ K};$
$\theta_{v9} = 4279 \text{ K}; \quad \theta_{v10} = 2092 \text{ K}; \quad \theta_{v11} = 1769 \text{ K}.$

A 12th vibrational temperature is not given because there is obviously an internal rotation around the CO bond with $\sigma_{ir} = 3$. The moment of inertia for internal rotation of the methanol molecule has been calculated under simplifying assumptions in Exercise 2.3, with the result that

$$I_{ir} = \frac{I_1 I_2}{I_1 + I_2} = 1.100 \cdot 10^{-40} \text{ g cm}^2.$$

With this value the molecular partition function of internal rotation can easily be evaluated. We find that

$$q_{ir} = \frac{1}{\sigma_{ir}} \sqrt{\frac{T}{\theta_{ir}}} = \frac{1}{3} \sqrt{\frac{T}{11.7}}.$$

In [2] a value of $u_{ir,\max} = 4486 \text{ J/mol}$ from direct spectroscopic information can be found. We thus find, at $T = 1000 \text{ K}$,

$$\frac{1}{q_{ir}} = 0.323$$

$$\frac{u_{ir,\max}}{RT} = 0.54$$

and from the tables of App. 7

$$\frac{c_{v,ir}^{ig}}{R} \cong 0.529$$

$$\frac{s_{ir}^{ig}}{R} \cong \frac{s_{ir,free}^{ig}}{R} - 0.016 = \left[\frac{1}{2} - \ln \sigma_{ir} + \frac{1}{2} \ln \frac{T}{\theta_{ir}} \right] - 0.016 = 1.61.$$

Table E 3.9.1 summarizes the contributions of internal rotation. An uncertainty of about 5% should be associated with these values due to the uncertain value of the potential barrier and the approximate nature of the tables in App. 7 for the molecule methanol. The contribution of internal rotation to the thermodynamic functions of methanol is not negligible. For the heat capacity at $T = 200 \text{ K}$ we have

$$\frac{c_v^{ig}}{R} = 1.5 + 1.5 + 0.086 + 0.702 = 3.788,$$

which shows the contribution of internal rotation to be about 20%. In view of the uncertainty of this contribution, the overall error in the heat capacity will be about 1%.

Table E 3.9.1. Contributions of internal rotation

T/K	$1/q_{ir}$	$u_{ir,max}/RT$	$(a/RT)^{ig}_{ir}$		$(s/R)^{ig}_{ir}$		$(c_v/R)^{ig}_{ir}$	
			free	hindered*	free	hindered*	free	complete*
200	0.72	2.70	−0.32	0.125	0.82	−0.201	0.5	0.702
500	0.46	1.08	−0.78	0.114	1.28	−0.056	0.5	0.594
1 000	0.32	0.54	−1.13	0.071	1.63	−0.016	0.5	0.529

*Additional contribution due to hindered rotation.

Table E 3.9.2. Comparison of calculated heat capacity with experimental data ($p = 1$ atm)

$t/°C$	$c_{p,exp}/$(J/mol K)	$c_p^{ig}/$ (J/mol K)	$\frac{c_{p,exp}-c_p^{ig}}{c_{p,exp}}100\%$
158.3	55.77	54.03	3.1
169.0	56.02	54.90	2.0
184.2	56.99	56.14	1.5
204.6	57.24	57.78	−0.9
211.9	56.57	58.36	−3.1
225.8	60.12	59.46	1.1
248.2	61.55	61.21	0.6
282.8	63.93	63.84	0.1
308.2	66.36	65.72	1.0
312.2	66.81	66.00	1.2

At high temperatures the contribution of the vibrational modes becomes dominant. At 1000 K we find

$$\frac{c_v^{ig}}{R} = 1.5 + 1.5 + 6.235 + 0.529 = 9.764,$$

and the contribution of internal rotation has fallen to 5%.

Table E3.9.2 shows a comparison of the calculated heat capacity with experimental data that claim an accuracy of 0.3%. The comparison between experimental and ideal gas heat capacities reveals somewhat unsystematic errors between zero and up to 3%. It appears that the accuracy claimed for the data is too optimistic. This is supported by a comparison with the correlation of [3] from which the deviations of the present calculations are less than 1%. Using the more accurate formula (2.108) for the reduced moment of inertia leads to $I_{ir} = 1.03 \cdot 10^{-40}$ g cm^2 with only minor effects on the calculated total ideal gas heat capacities.

[1] Landolt-Börnstein. *Zahlenwerte und Funktionen aus Physik, Chemie, Geographie und Technik*, volume 2,4. Springer, Berlin, 1961.

[2] Landolt-Börnstein. *Zahlenwerte und Funktionen aus Naturwissenschaft und Technik. Neue Serie. Gruppe II, Bd. 4*. Springer, Berlin, 1967.

[3] B. E. Poling, J. M. Prausnitz, and J. P. O'Connell. *The Properties of Gases and Liquids*, 5th ed. McGraw-Hill, New York, 2001.

To illustrate the prediction of ideal gas properties for somewhat more complicated fluids we compare experimental and predicted ideal heat capacities of

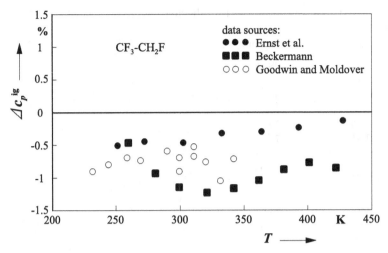

Fig. 3.3. Experimental and predicted heat capacity of CF_3-CH_2F (R134a). Reference line: heat capacity calculated from B3LYP/cc-pVDZ//MP2/cc-pVTZ. For references to data cf. [6].

R134a (CF_3-CH_2F) [6] in Figure 3.3. The reference (zero percent deviation) is based on the B3LYP/cc-pVDZ quantum chemical level for the calculation of the structure and the vibrational frequencies along with the MP2/cc-pVTZ level for the potential energy of hindered internal rotation; cf. Section 2.5. It can be seen that the predicted heat capacities are about 0.5% to 1% higher than the experimental ones, which in turn differ by about the same order of magnitude. This is a typical result for this type of fluids, as shown by many calculations performed in [6]. It demonstrates that reliable predictions for ideal gas thermodynamic functions can be made on the basis of quantum and statistical mechanics without support of experimental data.

An even more challenging problem arises when there are several single bonds around which internal rotations are possible. Then the various internal rotations, in principle, are not independent of each other, because the moments of inertia of the rotating groups, as well as the force field, depend on the rotational angles of the other rotating groups. In such cases the approach discussed above for one rotational group can only be applied to the whole molecule when all internal rotations are considered to be independent as an approximation. This is frequently done. A more adequate approach, however, is to consider the gas as a mixture of those conformations, obtained from internal rotations, which are favored due to their low energy among the many that are combinatorially possible. Each conformer is then treated on the rigid rotator–harmonic oscillator level. The moments of inertia, as well as the vibration frequencies, are calculated quantum-chemically for each conformer, the torsional vibration being treated in the standard harmonic way. As an illustration, we consider the molecule n-hexane. A quantum-chemical calculation on the same level as in Figure 3.3 gives the structures and the relative energies of the first 10 different conformers. They are ordered in terms of energy and displayed in Figure 1.12.

Table 3.2. Predicted ideal gas properties for n-hexane

$T/(\text{K})$	$s^{\text{ig}}/(\text{J/mol/K})$	$c_{\text{p}}^{\text{ig}}/(\text{J/mol/K})$	$c_{\text{p,exp}}^{0}/(\text{J/mol/K})$
200	341.98	108.56	110
250	367.56	122.08	125
298.15	390.47	139.11	143
300	391.34	139.82	144
350	414.36	159.65	162
400	437.01	179.88	181
450	459.33	199.48	199
500	481.32	217.93	217
550	502.90	235.07	233
600	524.04	250.90	249
650	544.71	265.52	263
700	564.89	279.02	275

Table 3.2 shows the prediction of the ideal gas heat capacity and entropy, obtained from n-hexane being treated as a mixture of the 10 conformers with a composition calculated from Boltzmann factor averaging [7]. It can be seen that the agreement between the molecular model predictions and experiment for the heat capacity [8] is again very satisfactory, with a general difference below 1%, although a difference of about 2% is found around room temperature. The accuracy of the experimental data is estimated to be on the order of 1%, although this is not clearly established in the literature. Clearly, the computational effort increases considerably with increasing size of the molecule, when more conformations have to be taken into account.

3.6.6 Corrections

The calculations in the preceding paragraphs were based on the rigid rotator–harmonic oscillator model augmented by the contribution of internal rotation. The interatomic distances were considered to be constant for the evaluation of the rotational partition function and the true intramolecular potential energy for vibration was approximated by a parabola. Clearly, in this approximation the rotational and vibrational energy states are independent of each other. The rigid rotator–harmonic oscillator model gives surprisingly good agreement with experimental data at ordinary temperatures. For improved accuracy, in particular at high temperatures, corrections can be applied that take the following three effects into account:

1. Centrifugal Stretching

When there is very fast rotation the molecule will stretch and therefore increase its moment of inertia. For the rigid diatomic rotator we have

$$\varepsilon_{\text{r}} = \frac{h^2}{8\pi^2 I} j(j+1), \quad j = 0, 1, 2, \ldots,$$

or, in a somewhat different notation,

$$\frac{\varepsilon_r}{hc} = B_e j(j+1), \tag{3.86}$$

with $B_e = h/8\pi^2 Ic$ and c as the speed of light. Here B_e refers to that molecular configuration where the electronic energy is at its minimum, i.e., the equilibrium configuration; cf. Section 2.5.4. In order to take account of the centrifugal stretching of the molecule one formally sets

$$\frac{\varepsilon_r}{hc} = B_e j(j+1) - D_e j^2(j+1)^2 + \cdots, \quad j = 0, 1, 2, \ldots. \tag{3.87}$$

D_e is a parameter that can be obtained from spectroscopic data or quantum-chemical calculations. The correction term will decrease the value of ε_r/hc with respect to the rigid rotator assumption as an increased moment of inertia would do and thus is able to correct for centrifugal stretching.

2. Anharmonic Vibration

In strong vibrations it will become apparent that the true potential energy deviates from a simple parabolic form; cf. Figure 3.2. For the harmonic oscillator we have

$$\varepsilon_v = h\nu_0 \left(v + \frac{1}{2}\right), \quad v = 0, 1, 2, \ldots \tag{3.88}$$

or, in a somewhat different notation,

$$\frac{\varepsilon_v}{hc} = \tilde{\nu}_0 \left(v + \frac{1}{2}\right), \quad v = 0, 1, 2, \ldots \tag{3.89}$$

with

$$\tilde{\nu}_0 = \frac{\nu_0}{c}.$$

The anharmonicity can be taken into account by formally setting

$$\frac{\varepsilon_v}{hc} = \tilde{\nu}_0 \left(v + \frac{1}{2}\right) - x_e \tilde{\nu}_0 \left(v + \frac{1}{2}\right)^2 + \cdots, \quad v = 0, 1, 2, \ldots. \tag{3.90}$$

Here x_e is the anharmonicity constant, which is generally small ($\sim 10^{-2}$) and can be obtained from spectroscopic tables or quantum chemical calculations. The correction term with the anharmonicity constant has the effect of reducing the increase of ε_v/hc with increasing vibrational quantum number and thus gives a better match with the true intramolecular potential curve.

3. Rotational–Vibrational Coupling

In high vibrational states the average distance between two atoms increases due to the anharmonicity of the intramolecular potential function. Thus the intramolecular distance becomes a function of the vibrational state, which leads to a coupling between rotation and vibration. We can take this coupling into account by replacing the rotational constant B_e in (3.87) by a quantity B_v depending on the vibrational state, according to

$$B_v = B_e - \alpha \left(v + \frac{1}{2} \right) + \cdots , \tag{3.91}$$

where α is a coupling constant, which again has to be taken from spectroscopic data or quantum-chemical calculations. An analogous replacement can be applied to the constant D_e in Eq. (3.87), although this is frequently unnecessary.

The energy levels of rotation and vibration for a diatomic molecule including the effects of centrifugal stretching, anharmonic vibration, and rotational–vibrational coupling can thus be written as

$$\left(\frac{\varepsilon_{r,v}}{hc} \right)_{j,v} = \tilde{\nu}_0 \left(v + \frac{1}{2} \right) - x_e \tilde{\nu}_0 \left(v + \frac{1}{2} \right)^2 + B_e j(j+1)$$
$$- \alpha \left(v + \frac{1}{2} \right) j(j+1) - D_e j^2 (j+1)^2, \quad j, v = 0, 1, 2, \ldots . \tag{3.92}$$

By subtracting the zero-point energy for vibration ($v = 0$) and using the substitutions

$$\tilde{\nu} = \tilde{\nu}_0 (1 - 2x_e),$$
$$x = x_e / (1 - 2x_e),$$

and

$$B = B_e - 1/2\alpha,$$

this transforms into

$$\left(\frac{\varepsilon_{r,v} - \varepsilon_{r,v}^0}{hc} \right)_{j,v} = \tilde{\nu} v - \tilde{\nu} x v (v - 1) + B j(j+1) - \alpha v j(j+1)$$
$$- D_e j^2 (j+1)^2, \quad j, v = 0, 1, 2, \ldots . \tag{3.93}$$

Here, the frequency $\tilde{\nu} = \tilde{\nu}_0 (1 - 2x_e)$ is the one that is actually observed in spectroscopic experiments. This can be seen from the energy quantum ($\varepsilon = h\nu$) absorbed to move a realistic anharmonic oscillator from the ground vibrational state to the first excited vibrational state in an absorption experiment, which follows from (3.90) as

$$h\tilde{\nu}(0 \to 1) = h \left[\tilde{\nu}_0 \left(1 + \frac{1}{2} \right) - \tilde{\nu}_0 x_e \left(1 + \frac{1}{2} \right)^2 - \frac{1}{2} \tilde{\nu}_0 + \frac{1}{4} \tilde{\nu}_0 x_e \right]$$
$$= h\tilde{\nu}_0 (1 - 2x_e).$$

The molecular partition function for rotation and vibration including corrections thus reads, for a diatomic molecule,

$$
\begin{aligned}
q_{r,v} &= \frac{1}{\sigma_r} \sum_{v=0}^{\infty} \sum_{j=0}^{\infty} (2j+1) e^{-\left(\frac{\varepsilon_{r,v} - \varepsilon_{r,v}^0}{kT}\right)_{j,v}} \\
&= \frac{1}{\sigma_r} \frac{1}{(hcB/kT)} \frac{1}{1 - e^{-\theta_v/T}} \\
&\quad \cdot \left[1 + 2\frac{D_e}{B} \frac{1}{(hcB/kT)} + \frac{\alpha}{B} \frac{1}{(e^{\theta_v/T} - 1)} + 2x \frac{\theta_v/T}{(e^{\theta_v/T} - 1)^2} \right].
\end{aligned}
\tag{3.94}
$$

Here $\theta_v = hc\tilde{v}/k = hv/k$ is the characteristic temperature of vibration in terms of the observed spectroscopic frequency. The final form of (3.94) has been found by expanding the exponential up to first order and replacing the sum over j by an integral, as before. Clearly, (3.94) reduces to the earlier result of the rigid rotator–harmonic oscillator model for $x_e = \alpha = D_e = 0$. Extending the summations up to infinity strictly does not make sense because $\varepsilon_{r,v} - \varepsilon_{r,v}^0$ will become negative at large quantum numbers. This inconsistency is cured by the truncated expansion of the exponential.

The contribution of rotation and vibration including corrections to the heat capacity, as derived from (3.94), is, for a diatomic molecule,

$$
\begin{aligned}
\left(\frac{c_p^{ig}}{R} \right)_{r,v} &= \left(\frac{c_v^{ig}}{R} \right)_{r,v} = 1 + \frac{4D_e}{B^2} \frac{kT}{hc} + \frac{(\theta_v/T)^2 e^{\theta_v/T}}{(e^{\theta_v/T} - 1)^2} \\
&\quad + 4x \frac{(\theta_v/T)^2 e^{\theta_v/T} \left[2(\theta_v/T) e^{\theta_v/T} - 2e^{\theta_v/T} + \theta_v/T + 2 \right]}{(e^{\theta_v/T} - 1)^4} \\
&\quad + \frac{\alpha}{B} \frac{(\theta_v/T)^2 e^{\theta_v/T} (e^{\theta_v/T} + 1)}{(e^{\theta_v/T} - 1)^3}.
\end{aligned}
\tag{3.95}
$$

These corrections are easily applied to diatomic molecules, for which sufficient molecular data are generally available. For polyatomic molecules, too, corrections have to be introduced into the rigid rotator–harmonic oscillator model at high temperatures for the same physical reasons. Very few polyatomic molecules, however, have been investigated thoroughly enough, either spectroscopically or quantum-chemically, to provide the necessary molecular data about the intramolecular force field. Rather complicated expressions for the energy levels have been derived in particular cases. For most polyatomic molecules one is limited to performing the statistical thermodynamical calculations in the rigid rotator–harmonic oscillator model. The accuracy of the computed thermodynamic functions in the ideal gas state is then usually satisfactory and limited not so much by a failure of that approximation, but rather by a lack of accurate data for vibrational and internal rotational parameters.

EXERCISE 3.10

Evaluate the contribution of the corrections to the rigid rotator–harmonic oscillator model for the ideal gas heat capacity of Cl_2 at temperature between 200 and 1000 K.

Solution

In [1] we find the following molecular data for Cl_2:

$$B_e = 0.2408 \text{ cm}^{-1}$$
$$\alpha = 0.0017 \text{ cm}^{-1}$$
$$\tilde{\nu}_0 = 561.1 \text{ cm}^{-1}$$
$$\tilde{\nu}_0 x_e = 4.0 \text{ cm}^{-1}.$$

A convenient estimate of the centrifugal distortion constant D_e follows from [2] as

$$D_e = 4\frac{B_e^3}{\tilde{\nu}_0^2} = 0.1774 \text{ cm}^{-1}.$$

Using the above definitions we find for the molecular constants to be used in the corrected formula for the partition function

$$\tilde{\nu} = \tilde{\nu}_0(1 - 2x_e) = 553.1 \text{ cm}^{-1}$$
$$\theta_v = hc\tilde{\nu}/k = 795.8 \text{ K}$$
$$x = x_e/(1 - 2x_e) = 0.7232 \cdot 10^{-2}$$
$$B = B_e - 1/2\alpha = 0.2399 \text{ cm}^{-1}.$$

The results are summarized in Table E 3.10.1. We note that the differences between the rigid rotator–harmonic oscillator model and the complete model including corrections are below 1% for Cl_2 at ordinary temperatures. The relevance of corrections depends strongly on the molecules considered and increases with increasing temperature.

Table E 3.10.1. The contribution of corrections to the heat capacity of Cl_2

T/K	$c_{p,\text{RRHO}}^{\text{ig}}/\text{J/mol K}$	$c_p^{\text{ig}}/\text{J/mol K}$	$\left(c_{p,\text{RRHO}}^{\text{ig}} - c_p^{\text{ig}}\right) \cdot 100\%/c_{p,\text{RRHO}}^{\text{ig}}$
200	31.58	31.67	−0.4
500	25.82	36.09	−0.7
1000	36.98	37.50	−1.4

[1] F. P. Incropera. *Introduction to Molecular Structure and Thermodynamics.* Wiley, New York, 1974.

[2] G. Herzberg. *Molecular Spectra and Molecular Structure. I. Spectra of Diatomic Molecules.* Van Nostrand Reinhold, New York, 1950.

3.7 Equilibrium Constant

A particularly fruitful application of the ideal gas thermodynamic functions is the calculation of equilibrium constants for gas phase reactions. The second law of thermodynamics leads to the condition for the equilibrium state in a single chemical reaction, cf. (2.13),

$$\sum \nu_j \mu_j = 0, \qquad (3.96)$$

where μ_j is the chemical potential of component j in the mixture and ν_j the stoichiometric coefficient of this component in a given reaction equation. For a mixture of ideal gases the condition for reaction equilibrium can be transformed into

$$\ln K = -\frac{\sum \nu_j \mu_{0j}^{\text{ig}}(T, p^0)}{RT} = \ln \prod_j \left(\frac{x_j p}{p^0}\right)^{\nu_j}. \tag{3.97}$$

If we can calculate the equilibrium constant K, we can also calculate the composition of the gas in equilibrium. This also applies to real gases, if $(x_j p)$ is replaced by the fugacity f_j. The equilibrium constant is defined in terms of the pure ideal gas chemical potentials μ_{0j}^{ig} as long as the ideal gas state is an adequate reference state.

The Gibbs free energy of pure component α in the ideal gas state is given, cf. (3.27), by

$$G_{0\alpha}^{\text{ig}} - U_{0\alpha}^0 = A_{0\alpha}^{\text{ig}} - U_{0\alpha}^0 + N_\alpha kT = -N_\alpha kT \left[\frac{1}{N_\alpha} \ln \frac{q_{0\alpha}^{N_\alpha}}{N_\alpha!} - 1\right]$$

$$= -N_\alpha kT \ln q_{0\alpha} + kT(N_\alpha \ln N_\alpha - N_\alpha) + N_\alpha kT$$

$$= -N_\alpha kT \ln \frac{q_{0\alpha}}{N_\alpha}. \tag{3.98}$$

From this the molar chemical potential of pure component α in the ideal gas state at the standard pressure p^0 can be expressed as

$$\mu_{0\alpha}^{\text{ig}}(T, p^0) - u_{0\alpha}^0 = -RT \ln \frac{q_{0\alpha}(T, p^0)}{N_A}. \tag{3.99}$$

Here the energy reference value $u_{0\alpha}^0 = u_{0\alpha}^{\text{ig}}(T = 0) = u_{0\alpha}^{\text{ig}}(0)$ is particularly important because chemical reactions transform molecules and thus the reference values of energy do not cancel in the equations. We thus arrive at

$$K = \left(e^{-\Delta u^0/RT}\right) \prod_\alpha \left(\frac{q_{0\alpha}}{N_A}\right)^{\nu_\alpha} \tag{3.100}$$

with

$$\Delta u^0 = \sum_\alpha \nu_\alpha u_{0\alpha}^{\text{ig}}(0). \tag{3.101}$$

The partition function $q_{0\alpha}$ can be calculated from the equations derived before. In order to calculate the equilibrium constant the quantity Δu^0 is needed, which, according to the zero point energy convention, may be interpreted as the enthalpy of reaction in the ideal gas state at $T = 0$ K. To evaluate Δu^0, all molecules must be energetically referred to a state that cancels in a chemical reaction. An appropriate state is the lowest quantum state of the separated atoms, which is obtained by adding the energy of dissociation D_0 to the zero energy convention introduced in Section 3.2; cf. Figure 3.2. Thus for a molecule A_aB_b one has the energy balance $au_{0A}^{ig}(T=0) + bu_{0B}^{ig}(T=0) = u_{0A_aB_b}^{ig}(T=0) + D_{0A_aB_b}$ and it can easily be seen that

$$-\Delta u^0 = \sum_{\alpha} \nu_{\alpha} D_{0\alpha}, \qquad (3.102)$$

because atoms are conserved in chemical reactions. Dissociation energies can frequently be found in tables of molecular properties. They are obtained from spectroscopic data or quantum-chemical calculations. Alternatively, we can use enthalpies of formation, which we either find tabulated in the standard state or can calculate quantum mechanically, as shown in Table 3.1.

EXERCISE 3.11

Calculate the equilibrium constant of the oxygen dissociation reaction

$$O_2 \rightleftharpoons 2O$$

at 4000 K.

Solution

According to (3.100), the equation to be evaluated is

$$K = e^{-\Delta u^0/kT} \prod_{\alpha} \left(\frac{q_{0\alpha}}{N_A} \right)^{\nu_{\alpha}} = e^{-\frac{D_{0,O_2}}{kT}} \frac{(q_{0,O}/N_A)^2}{(q_{0,O_2}/N_A)}.$$

The partition function of the oxygen atom consists of contributions due to translation and electronic energy. With $M = 15.999$ g/mol, $\theta_{el,1} = 22,831.2$ K, and $g_{el,0} = 9$ [1] for atomic oxygen we find that

$$\ln\left(\frac{q_{0,O}}{N_A}\right)_{tr} = \ln\left[\left(\frac{2\pi m_O kT}{h^2}\right)^{3/2} \frac{v}{N_A}\right] = \ln\left[\left(\frac{2\pi M_O kT}{N_A h^2}\right)^{3/2} \frac{v}{N_A}\right]$$

$$= \ln\left[\left(\frac{2\pi M_O kT}{N_A h^2}\right)^{3/2} \frac{kT}{p^0}\right] = 21.2292,$$

where v is the molar volume of the system, and

$$\ln(q_{0,O})_{el} = 2.1972.$$

Similarly with $\theta_{el,1} = 11,392.5$ K, $g_{el,0} = 3$, $g_{el,1} = 2$, $\theta_r = 2.08$ K, and $\theta_v = 2280$ K for molecular oxygen [2] we find that

$$\ln\left(\frac{q_{0,O_2}}{N_A}\right)_{tr} = 22.2690$$

$$\ln(q_{0,O_2})_{el} = 1.1365$$

$$\ln(q_{0,O_2})_r = 6.8685$$

$$\ln(q_{0,O_2})_v = 0.8336.$$

The dissociation energy of molecular oxygen is $D_{0,O_2} = 494,013.4$ J/mol [2] and we thus find for the equilibrium constant

$$K = 2.44.$$

[1] C. E. Moore. *Atomic Energy Levels, vol.1.* NSRDS-NBS 35, Washington, DC, 1971.
[2] G. Herzberg. *Molecular Spectra and Molecular Structure. I. Spectra of Diatomic Molecules.* Van Nostrand Reinhold, New York, 1950.

Quite analogous are calculations for ionization reactions, where ΔU_0 is given by the ionization potential. Such calculations yield the compositions of high-temperature gases and plasmas, as displayed in Figure 1.3.

EXERCISE 3.12

Calculate the equilibrium constant of the ionization reaction

$$\text{Ar} \rightleftharpoons \text{Ar}^+ + e^-$$

at $10,000$ K.

Solution

According to (3.100) the equilibrium constant can be calculated from

$$K = e^{-\Delta u^0/RT} \prod_\alpha \left(\frac{q_{0\alpha}}{N_A}\right)^{\nu_\alpha} = e^{-I/RT} \frac{(q_{0,\text{Ar}^+}/N_A)(q_{0,e}/N_A)}{(q_{0,\text{Ar}}/N_A)},$$

where I is the energy to remove one electron from the argon atom, i.e., the ionization potential. According to [1], the relevant molecular properties are

- argon: $g_{\text{el},0} = 1$, $\frac{\varepsilon_{\text{el},1}}{hc} = 93,143.8$ cm^{-1}

- argon$^+$: $g_{\text{el},0} = 4$, $\frac{\varepsilon_{\text{el},1}}{hc} = 1432$ cm^{-1}
  $g_{\text{el},1} = 2$, $\frac{\varepsilon_{\text{el},2}}{hc} = 108,722.5$ cm^{-1}

The ionization potential is $I = 15.755$ eV, so that

$$I/k = 182,834 \text{ K}.$$

The masses of the argon atom and the argon ion are practically equal and thus their translational contributions cancel. The partition functions due to electronic energy are

$$(q_{0,\text{Ar}})_{\text{el}} = g_{\text{el},0} = 1$$

$$(q_{0,\text{Ar}^+})_{\text{el}} = g_{\text{el},0} + g_{\text{el},1}e^{\varepsilon_{\text{el},1}/kT} = 4 + 2e^{-2060/10,000} = 5.6276.$$

The partition function of the electron has a translational and an electronic contribution due to degeneracy and reads

$$\left(\frac{q_{0,e^-}}{N_A}\right) = 2\left(\frac{2\pi m_e kT}{h^2}\right)^{3/2}\frac{v}{N_A} = 2\left(\frac{2\pi m_e kT}{h^2}\right)^{3/2}\frac{kT}{p^0} = 6582.$$

Here the factor of 2 has been introduced to take into account the degeneracy of the electron's electronic energy due to the two possible spin orientations. For the equilibrium constant we thus find

$$K = 0.00043.$$

[1]   C. E. Moore. *Atomic Energy Levels, vol.1*. NSRDS-NBS 35, Washington, DC, 1971.

Again analogous, but with a somewhat more complicated procedure to arrive at ΔU_0, are the calculations for reaction equilibria with polyatomic gases; cf. Exercise 3.13.

EXERCISE 3.13

Calculate the equilibrium constant of the reaction

$$CO + H_2O \rightleftharpoons CO_2 + H_2$$

at 500, 1000, and 2000 K.

Solution

The equation for the equilibrium constant reads, cf. (3.100),

$$K = \left(e^{-\Delta u^0/RT}\right) \prod_\alpha \left(\frac{q_{0\alpha}}{N_A}\right)^{v_\alpha}.$$

Taking logarithms, we find

$$\ln K = -\Delta u^0/RT + \ln \frac{q_{0,CO_2} q_{0,H_2}}{q_{0,CO} q_{0,H_2O}}$$

$$= -\Delta u^0/RT + \ln \left(\frac{q_{0,CO_2} q_{0,H_2}}{q_{0,CO} q_{0,H_2O}}\right)_{tr} + \ln \left(\frac{q_{0,CO_2} q_{0,H_2}}{q_{0,CO} q_{0,H_2O}}\right)_r + \ln \left(\frac{q_{0,CO_2} q_{0,H_2}}{q_{0,CO} q_{0,H_2O}}\right)_v,$$

where we have made use of the fact the electronic contribution is fully taken into account by the zero of energy for the components considered.

The contribution of translation becomes

$$\ln \left(\frac{q_{0,CO_2} q_{0,H_2}}{q_{0,CO} q_{0,H_2O}}\right)_{tr} = \frac{3}{2} \ln \frac{M_{CO_2} M_{H_2}}{M_{CO} M_{H_2O}} = -2.60745,$$

where the temperature T and the reference pressure p^0 cancel, because all components have stoichiometric coefficients of identical magnitude.

To evaluate the contribution of external rotation we need the characteristic temperatures θ_r. We find in [1]

$$\theta_{r,H_2} = 85.29 \text{ K}; \quad \theta_{r,CO} = 2.815 \text{ K}; \quad \theta_{r,CO_2} = 0.57 \text{ K};$$
$$\theta_{rA,H_2O} = 39.4 \text{ K}; \quad \theta_{rB,H_2O} = 21.0 \text{ K}; \quad \theta_{rC,H_2O} = 13.7 \text{ K}.$$

This gives for the contribution of external rotation

$$\ln \left(\frac{q_{0,CO_2} q_{0,H_2}}{q_{0,CO} q_{0,H_2O}}\right)_r = \ln \left(\frac{\sigma_{r,CO} \sigma_{r,H_2O}}{\sigma_{r,CO_2} \sigma_{r,H_2}}\right) - \frac{1}{2} \ln(\pi T/K) + \ln \left(\frac{\theta_{r,CO} \left(\theta_{r,A} \theta_{r,B} \theta_{r,C}\right)_{H_2O}^{1/2}}{\theta_{r,CO_2} \theta_{r,H_2}}\right)$$

$$= 0.55335 - \frac{1}{2} \ln(T/K).$$

For the contribution of vibration we need the characteristic temperatures of vibration, for which we find in [1]

$$\theta_{v,H_2} = 5995 \text{ K}; \quad \theta_{v,CO} = 3080.7 \text{ K};$$
$$\theta_{v1,CO_2} = 961 \text{ K}; \quad \theta_{v2,CO_2} = 961 \text{ K}; \quad \theta_{v3,CO_2} = 1924 \text{ K}; \quad \theta_{v4,CO_2} = 3379 \text{ K};$$
$$\theta_{v1,H_2O} = 5258.8 \text{ K}; \quad \theta_{v2,H_2O} = 2293.0 \text{ K}; \quad \theta_{v3,H_2O} = 5400.8 \text{ K}.$$

We thus find for the contribution of vibration

T/K	$\ln \left(q_{0,CO_2} q_{0,H_2}/q_{0,CO} q_{0,H_2O}\right)_v$
500	0.32670
1000	0.99602
2000	1.89598

We finally can calculate the energy of reaction at 0 K from tabulated dissociation energies D_0 of the molecules. In [1] and [2] we find

$$D_{0,CO} = 11.1 \text{ eV} = 1{,}071{,}039 \text{ J/mol}$$

$$D_{0,H_2} = 4.476 \text{ eV} = 431{,}889 \text{ J/mol}$$

$$D_{0,CO\text{-}O} = 5.453 \text{ eV} = 526{,}160 \text{ J/mol}$$

$$D_{0,H\text{-}OH} = 5.113 \text{ eV} = 493{,}353 \text{ J/mol}$$

$$D_{0,O\text{-}H} = 4.390 \text{ eV} = 423{,}591 \text{ J/mol}.$$

This gives

$$-\Delta u^0 = \sum \nu_\alpha D_{0\alpha} = 41{,}105 \text{ J/mol}.$$

Alternatively, we can use enthalpies of formation as obtained from quantum mechanical computer codes; cf. Table 3.1.

Calculating the equilibrium constant for the temperatures considered with Δu^0 obtained from the dissociation energies, we find the results in the following table. The experimental values are from [3]:

T/K	K	K_{exp}
500	156.48	117.22
1000	1.540	1.315
2000	0.226	0.207

There are notable differences between the calculated and the experimental values. The view is taken that the calculated equilibrium constants, although very sensitive to any inaccuracies of the molecular data, are more reliable than those obtained from experiments, which are back-calculated from measured compositions. The equilibrium compositions calculated by the two sets of equilibrium constants are not very different, except for components of high dilution.

[1] Landolt-Börnstein. *Zahlenwerte und Funktionen aus Physik, Chemie, Geographie und Technik*, volume 2,4. Springer, Berlin, 1961.

[2] G. Herzberg. *Molecular Spectra and Molecular Structure. III. Electronic Spectra and Electronic Structure of Polyatomic Molecules*. Van Nostrand Reinhold, New York, 1966.

[3] I. Barin and O. Knacke. *Thermodynamical Properties of Inorganic Substances*. Springer, Berlin, 1977.

3.8 Summary

The ideal gas is a model system that, although not identical with a real gas under any conditions, still accurately reproduces many aspects of real gas thermodynamic behavior in the limit of low density. Thus the ideal gas is an extremely useful model and its thermodynamic functions find wide applications in fluid phase calculations.

The thermodynamic properties of ideal gases can be calculated with good accuracy from molecular models. Such calculations are the basis of many tabulated ideal gas standard data, which are frequently used in practical calculations. The geometrical structure of the molecule gives the moments of inertia and the intramolecular force field permits the determination of the complete set of vibration frequencies and of the barriers of internal rotation. Generally, the molecular properties can be obtained either from tables of spectroscopic data or from quantum-chemical computer codes. The agreement of calculated values for the ideal gas state with experimental data of simple gases is typically on the order of 1% for the heat capacity. For complicated molecules, molecular data are rare, and the reliability of the results obtained from the molecular models discussed in this chapter has not been tested yet on a broad scale. Generally, it depends on the reliability of the vibrational frequencies and on the adequate treatment of internal rotation. Calculations are most conveniently based on molecular data obtained from quantum-chemical computer codes. Very large molecules, for which neither spectroscopic nor quantum mechanical data may be available, do not often play an important part in applications of ideal gases because their vapor pressures are generally small and their thermodynamic behavior thus is primarily interesting in the liquid state.

3.9 References to Chapter 3

[1] C. E. Moore. *Atomic Energy Levels*, volume 1. NSRDS-NBS 35, Washington, DC, 1971.

[2] G. Herzberg. *Molecular Spectra and Molecular Structure. I. Spectra of Diatomic Molecules*. Van Nostrand Reinhold, New York, 1950.

[3] G. Herzberg. *Molecular Spectra and Molecular Structure. III. Electronic Spectra and Electronic Structure of Polyatomic Molecules*. Van Nostrand Reinhold, New York, 1966.

[4] J. D. Lewis, T. B. Malloy, T. H. Chao, and J. Laane. *J. Mol. Struct.*, 12:427, 1972.

[5] B. J. McClelland. *Statistical Thermodynamics*. Chapman and Hall, London, 1973.

[6] M. Speis, U. Delfs, and K. Lucas. *Int. J. Thermophys.*, 22:1813, 2001.

[7] A. Schäfer. BASF AG. Personal communication.

[8] B. E. Poling, J. M. Prausnitz, and J. P. O'Connell. *The Properties of Gases and Liquids*, 5th ed. McGraw-Hill, New York, 2001.

4 Excess Function Models

The fluid phase behavior of dense fluids, e.g., liquids, is not described adequately by the ideal gas molecular model. Thus, our interest now turns to configurational properties, i.e., those that are determined by the intermolecular potential energy. It is this potential energy that controls the most important aspects of fluid phase behavior, such as phase equilibria, and reaction equilibria in solutions.

A particularly simple approach to fluid phase behavior of dense fluids is offered by excess function models. The thermodynamic relations for computing phase and reaction equilibria as well as heat effects from excess functions are well established; cf. Section 2.1. They are basically rather straightforward and even allow evaluation by hand in many simple applications. The excess function approach is particularly suited to liquid mixtures made up of large molecules. The fluid phase behavior of such fluids tends to be interesting only over a narrow liquid density range at normal pressure, such that it may be formulated in terms of constant density and constant pressure mixing effects. Such effects are adequately formulated in terms of excess functions. Restriction to excess functions for liquid mixtures implies two important simplifications in the molecular models. First, we will not need explicit potential models over a large range of intermolecular distances and orientations. Because the molecules in liquids can be considered to be densely packed, the interactions between them can be modeled by contact energies of nearest neighbors. In this sense the potential model adequate for excess functions is the strict opposite to that of an ideal gas, where the distance between interacting molecules was assumed to be infinite. Second, we will not have to deal with density-dependent internal motions of the molecules, as they appear in the general formulation of the partition function. Their effect may be assumed as an approximation to be the same in the pure fluids and in the mixture and will thus cancel in the excess functions. The penalty for this simplified treatment is a restriction in generality. There are many important applications that cannot be covered adequately by excess function models.

4.1 General Properties

Some general aspects of excess function models can be obtained from statistical mechanics without yet introducing any simplifying assumptions about the geometry of the molecules, the interactions between them, and molecular structure.

4.1.1 Repulsive and Attractive Contribution

The first general property of excess function models is their separability into a repulsive and an attractive contribution. To see this we remember that the free energy of a molecular system is related to the canonical partition function via, as shown in (2.60),

$$A(T, V, \{N_j\}) = -kT \ln Q(T, V, \{N_j\}).$$

We consider a mixing process in which N_1 molecules of liquid component 1 and N_2 molecules of liquid component 2 are mixed together at constant temperature and volume to give a liquid mixture of $N = N_1 + N_2$ molecules. For simplicity, the mixture and the pure components are assumed to have the same temperature T and the same number density $N/V = N_1/V_1 = N_2/V_2$. The free energy of mixing associated with this process can then be related to the canonical partition functions of the mixture and the two pure components via

$$
\begin{aligned}
\Delta A^{\mathrm{M}} &(T, V, N, x_1) \\
&= A(T, V, N, x_1) - N_1 a_{01} (T, V_1/N_1) - N_2 a_{02} (T, V_2/N_2) \\
&= A(T, V, N, x_1) - x_1 N a_{01} (T, V_1/N_1) - x_2 N a_{02} (T, V_2/N_2) \\
&= A(T, V, N, x_1) - x_1 A_{01} (T, V, N, x_1 = 1) \\
&\qquad - x_2 A_{02} (T, V, N, x_1 = 0) \\
&= -kT \ln \frac{Q}{Q_{01}^{x_1} Q_{02}^{x_2}}.
\end{aligned}
\tag{4.1}
$$

Here Q is the canonical partition function of N molecules of the binary mixture and Q_{01} the canonical partition function of N molecules of component 1, with an analogous interpretation for Q_{02}, all at temperature T and volume V. Noting that the canonical partition function in the semiclassical approximation reads, for a mixture of monatomic components, cf. (2.72),

$$
\begin{aligned}
Q &= \frac{1}{\prod_\alpha N_\alpha! h^{N_\alpha f_\alpha}} \int_{r^N} \int_{p^N} e^{-H(r^N p^N)/kT} dr^N dp^N \\
&= \frac{1}{\prod_\alpha N_\alpha! h^{N_\alpha f_\alpha}} \int_{p^N} e^{-H(p^N)/kT} dp^N Q^{\mathrm{C}},
\end{aligned}
$$

we realize that the contributions due to Planck's constant as well as the kinetic part of the phase integral cancel in (4.1), because they do not depend on a particular configuration of the N molecules in either a mixture or a pure component. However, the configurational, i.e., the potential energy part of the phase integral and the permutational factor taking account of the indistinguishability of the molecules remain. This latter contribution, i.e., the term with the factorials, is transformed, for two components 1 and 2, using Stirling's formula, cf. App. 2, as

$$\ln \frac{\frac{1}{N_1! N_2!}}{\left(\frac{1}{N!}\right)^{x_1} \left(\frac{1}{N!}\right)^{x_2}} = -\ln N_1! - \ln N_2! + \ln N!$$

$$= -(N_1 \ln N_1 - N_1) - (N_2 \ln N_2 - N_2)$$
$$+ N \ln N - N$$
$$= -N_1 \ln x_1 - N_2 \ln x_2,$$

and we finally arrive at the result

$$\Delta A^M (T, V, N, x_1) = NkT [x_1 \ln x_1 + x_2 \ln x_2]$$
$$- kT \ln \frac{Q^C}{\left(Q_{01}^C\right)^{x_1} \left(Q_{02}^C\right)^{x_2}}. \tag{4.2}$$

For more complicated molecules (4.2) still applies because the nonconfigurational contributions of the rotational and internal degrees of freedom do not contribute, being represented by their ideal gas values and thus strictly canceling in (4.1). So we consider (4.2) to be a general statistical mechanical expression for the free energy of mixing in a binary mixture related to the mixing process defined above.

To introduce the excess free energy we consider the special case of a mixture of molecules with identical force fields between them. Then the intermolecular potential energy is the same in the mixture and in the pure components. The configurational partition functions thus cancel in (4.2) and we find

$$\Delta A^{M,is} = NkT [x_1 \ln x_1 + x_2 \ln x_2], \tag{4.3}$$

which is the free energy of mixing in an ideal solution; cf. Section 2.1. The ideal solution model is generally not realistic. However, it will be close to reality for isotope mixtures and also approximately true for mixtures formed from similar components, such as benzene and toluene. Deviations from the properties of an ideal solution are described in terms of excess functions. Thus, the statistical mechanical formula for the Helmholtz excess free energy of a binary mixture is

$$A^E(T, V, N_1, N_2) = \Delta A^M - \Delta A^{M,is}$$
$$= -kT \ln \frac{Q^C}{\left(Q_{01}^C\right)^{x_1} \left(Q_{02}^C\right)^{x_2}}. \tag{4.4}$$

We now separate the partition function into a repulsive and an attractive part. To do this we make use of the Gibbs–Helmholtz relation in terms of A and U, cf. Section 2.1, and find

$$\left(\frac{A^C}{kT}\right)_{T=\infty} - \left(\frac{A^C}{kT}\right)_T = \ln Q^C(T, V, N) - \ln Q^C(T = \infty, V, N)$$

$$= \int_{T=\infty}^{T} \frac{U^C(T, V, N)}{kT^2} dT. \tag{4.5}$$

At $T = \infty$ the thermodynamic properties of a fluid are free from attractive intermolecular forces and exclusively determined by hard body effects. The attractive potential energy is contained in the integral term in (4.5). We thus write

$$Q^C = Q^C_{rep} Q^C_{att} \tag{4.6}$$

and arrive at molecular models for the excess free energy of the general form

$$A^E = A^E_{rep} + A^E_{att}. \tag{4.7}$$

The first term results from a contribution due to molecular size and shape effects; i.e., it introduces the molecular geometry. These effects are frequently summarized as steric effects and are modeled by hard bodies. Their notation is then changed from A^E_{rep} to A^E_h. They contribute to the excess free energy even when there is no difference between the attractive interactions in the mixture and in the pure components. Because this term does not depend on temperature, as long as the hard core volume is constant, it will not contribute to the excess enthalpy. It is therefore frequently termed the athermal part, where an athermal solution is one with $h^E = 0$. Because the athermal part will contribute only to the excess entropy, it is sometimes also referred to as the entropic part. This classification is misleading, however, since differences in the intermolecular attractive forces, i.e., nonsteric effects, also produce an entropic contribution. The second term is the contribution from attractive intermolecular forces. It is frequently denoted as the residual contribution (res). Here, we prefer to refer to it as the attractive contribution (att) because this appears to be more descriptive and, also, the term residual has been used before for formulating the deviation from ideal gas behavior; cf. Section 2.1. We note, though, that the contact energies between molecules may well assume positive and negative values, so that the term "attractive" is meant in an average sense for the whole fluid.

4.1.2 Nonrandomness

The second general property of excess function models is related to the appearance of local compositions as a major mixing effect. The intermolecular interactions result in the formation of a nonhomogeneous microscopic composition in a liquid, as a result of the mixing process. As a consequence, the

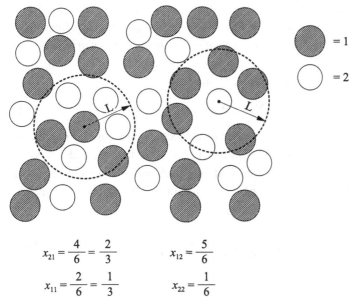

$$x_{21} = \frac{4}{6} = \frac{2}{3} \qquad x_{12} = \frac{5}{6}$$

$$x_{11} = \frac{2}{6} = \frac{1}{3} \qquad x_{22} = \frac{1}{6}$$

Fig. 4.1. The concept of local compositions ($x_1 = 0.6$, $x_2 = 0.4$).

local composition deviates significantly from the bulk composition. Figure 4.1 illustrates this phenomenon for a system of simple spherical molecules in one particular configuration, i.e., arrangement of molecules in space. We focus on two different regions of the fluid, one with the (dark) molecule of type 1 at the center and the other with the (light) molecule of type 2 at the center. Although in the total fluid the bulk composition is $x_1 = 0.6$ and $x_2 = 0.4$, we find local compositions that are quite different from the bulk values. Their definition is

$$x_{ji} = \frac{N_{ji}}{\sum_k N_{ki}}, \qquad (4.8)$$

where N_{ji} is the number of molecules j in a sphere of radius L around a particular central molecule i. There are general closure relationships between the local compositions that follow from their definitions in (4.8). So we find for a binary mixture

$$x_{12} + x_{22} = 1 \qquad (4.9)$$

and

$$x_{21} + x_{11} = 1. \qquad (4.10)$$

Because the intermolecular potential energy is directly related to interactions between the molecules, it is dependent on the local compositions. Any excess function model will thus have to be formulated in terms of them, and thus, we need their relation to the bulk composition. Clearly, a direct computation of local compositions from their definition (4.8) is not possible, because the extent of the region around a central molecule, i.e., L, is not physically defined. However, we can derive a formal expression for them by looking at time-averaged values in an arbitrarily specified region around an arbitrary central molecule. We then can express the time-averaged number of molecules j around an arbitrary molecule of type i formally as

$$\overline{N}_{ji} = \overline{N}_j \overline{g}_{ji}, \tag{4.11}$$

where $\overline{N}_j$ is the time-averaged number of molecules of component j in the considered region, as it would be found without interactions, and $\overline{g}_{ji}$ is the time-averaged relative probability of finding a molecule j around a molecule i in the considered region with respect to a situation without interactions. In a binary mixture we then have, for example,

$$\overline{x}_{11} = \frac{\overline{N}_1 \overline{g}_{11}}{\overline{N}_1 \overline{g}_{11} + \overline{N}_2 \overline{g}_{21}} = \frac{x_1}{x_1 + x_2 G_{21}}, \tag{4.12}$$

where

$$G_{21} = \frac{\overline{g}_{21}}{\overline{g}_{11}} \tag{4.13}$$

is a nonrandomness factor indicating deviations from random mixing due to intermolecular interactions. For $G_{21} = 1$ we have $x_{11} = x_1$; i.e., the local and the bulk compositions are equal. This happens when all interactions in the mixture are equal, as would be the case in an ideal solution. When $G_{21} > 1$ we have $x_{11} < x_1$, indicating a stronger local attraction between unlike molecules than between pairs of type 1 molecules. The opposite is true for $G_{21} < 1$. Analogously, we find

$$\overline{x}_{21} = \frac{\overline{N}_2 \overline{g}_{21}}{\overline{N}_2 \overline{g}_{21} + \overline{N}_1 \overline{g}_{11}} = \frac{x_2 G_{21}}{x_2 G_{21} + x_1} \tag{4.14}$$

and thus

$$\frac{\overline{x}_{21}}{\overline{x}_{11}} = \frac{x_2}{x_1} G_{21}. \tag{4.15}$$

Equivalently, we have

$$\frac{\overline{x}_{12}}{\overline{x}_{22}} = \frac{x_1}{x_2} G_{12}, \tag{4.16}$$

where

$$G_{12} = \frac{\overline{g}_{12}}{\overline{g}_{22}} \tag{4.17}$$

is a second nonrandomness factor with $G_{12} \neq G_{21}$. The general relation between the local compositions and the nonrandomness factors follows from (4.15) and (4.16) as

$$\frac{\overline{x}_{21} \overline{x}_{12}}{\overline{x}_{11} \overline{x}_{22}} = G_{21} G_{12}. \tag{4.18}$$

Clearly the nonrandomness factors will in general depend on temperature, on density, and, in particular, on composition. So there will be no simple expressions derivable for them. Still, most of the semiempirical models for the excess free energy in use today apply simple ad hoc formulae for G_{21} and G_{12}.

Some guidance about the form of the nonrandomness factors can be obtained from statistical mechanics. As a first reasonable approximation we may assume that

$$\overline{g}_{ij} = e^{-\overline{\phi}_{ij}/kT}, \tag{4.19}$$

with $\overline{\phi}_{ij}$ as a time-averaged value for the potential energy of an i, j-pair in the considered region, because the relative probability of finding an i, j-pair should be related to the associated Boltzmann factor; cf. Section 2.2. This gives for $\overline{\phi}_{ij} = -\varepsilon_{ij}$, where ε_{ij} is the attractive energy parameter of an i, j-interaction,

$$\frac{\overline{x}_{21}}{\overline{x}_{11}} = \frac{x_2}{x_1} G_{21} = \frac{x_2}{x_1} e^{(\varepsilon_{21}-\varepsilon_{11})/kT} \tag{4.20}$$

and also

$$\frac{\overline{x}_{12}}{\overline{x}_{22}} = \frac{x_1}{x_2} G_{12} = \frac{x_1}{x_2} e^{(\varepsilon_{12}-\varepsilon_{22})/kT}. \tag{4.21}$$

Computer simulations have revealed that the nonrandomness factors of (4.20) and (4.21), although exact in the low-density limit, give too much order at high densities [1]. This is plausible in view of the fact that the simple Boltzmann factor approach of (4.19), i.e., postulating that the probability of a particular binary pair depends only on the associated potential energy, can only be reasonable for rather dilute systems; cf. Section 2.4. In dense systems other interactions should also influence $\overline{g}_{ij}$. We shall show this in Section 5.4.4. In particular, we find from (4.20) and (4.21) that

$$\frac{\overline{x}_{21}\overline{x}_{12}}{\overline{x}_{11}\overline{x}_{22}} = G_{21}G_{12} = e^{-\omega_{12}/kT}, \tag{4.22}$$

where

$$\omega_{12} = \varepsilon_{12} + \varepsilon_{21} - \varepsilon_{11} - \varepsilon_{22} \tag{4.23}$$

is the so-called exchange energy. We shall see later that (4.22) with (4.23) is indeed valid for liquid mixtures under well-defined conditions; not, however, (4.20) and (4.21).

Summarizing, we realize that any molecular model for the excess free energy should be made up of two separate contributions, a repulsive part associated with hard body size and shape effects of the molecules, and an attractive part associated with attractive intermolecular interactions. In most systems the attractive part is the more important one. Because the intermolecular interactions are formulated in terms of pair contacts, they will depend on the local structure in the liquid. Thus, an essential part of any molecular model for the excess functions in liquid mixtures will be a model for the relation between the unknown local compositions and the known bulk compositions in terms of the intermolecular potential energy.

4.2 Intermolecular Potential Energy

Excess function models are designed to describe the properties of liquid mixtures with reference to the pure component properties. Generally, a liquid has a very complicated molecular structure to be discussed in more detail in Section 5.4. Here, we wish to introduce a simplified treatment of mixing effects.

4.2.1 Simplified Liquid Models

Because the mixing effects occur at high density we assume, as a reasonable approximation, that the molecular system is represented by a space-filling arrangement of molecules, interacting via surface contacts. There is thus no free space between the molecules, contrary to what would be the situation in a gas. Interaction energies in this approximation are thus binary surface–surface contact energies, i.e., numbers. Such surface interaction energies are independent of intermolecular distance and orientation, although the resulting complete potential energy between two molecules obviously depends on their mutual orientation. Because the distance dependence has been suppressed, the contact energies must generally be assumed to be temperature- and density-dependent. Although the density dependence is not taken into account in the simplified liquid models considered here, a temperature dependence may basically be introduced. The total potential energy is obtained from summing up the energies associated with all binary surface contacts in the system. To calculate the excess functions we need in particular those contributions to the potential energy that differ in the mixture and the pure components.

To fix ideas, we may assume that the molecular system of a liquid is represented by a regular lattice. The molecules are then placed on the various lattice sites in a space-filling manner; cf. Figure 4.2 in two-dimensional representation. We want to treat relatively large and complicated molecules, which do not have an obvious center and which may be flexible and exist in various conformations in the liquid state. To incorporate such molecules into the lattice we may consider a molecule to be a structure consisting of a number of volume segments, which have spherical shapes and equal sizes; cf. Figure 4.3. Here, the number of volume segments is $r = 8$. Each volume segment fills one lattice site. The intermolecular interactions between two such molecules are then visualized as occurring via surface contacts between the various segments of different molecules. There is no free volume in the lattice, and so all density effects are neglected. As a consequence, the volume of mixing, and thus also the excess volume v^E, is zero and we do not have to make a difference between

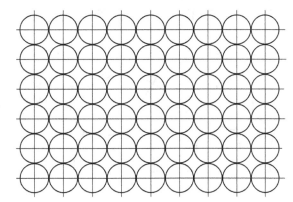

Fig. 4.2. Liquid model in the lattice picture.

the excess energy and the excess enthalpy. Clearly, this molecular system can only be a reasonable model for a densely packed liquid and even then it is oversimplified, because in real liquids a molecule may leave its position and move to other regions. Also, compressibility effects may not be negligible in some applications, such as liquid–liquid equilibria between components with large size differences. Still, it is well known and will be shown below that the lattice model is a useful basis for developing excess function models.

As an alternative to the lattice picture, the interactions between molecules in a liquid may also be visualized as occurring between polyhedral surface segments representing the molecular shapes of the molecules and contacting each other in a space-filling manner; cf. Figure 4.4 for a two-dimensional representation [2]. A regular lattice is not assumed. Each surface segment of one molecule is in contact with one particular surface segment of another molecule. With reference to molecule i, in the situation shown in Figure 4.4, there are interactions between the segments of that molecule and those of the molecules j, k, l, m, and n. The particular geometric arrangement is not constant, but will change continuously during observation time. Although the two visualizations of the liquid, i.e., molecules situated on the site of a regular lattice and molecules being considered as polyhedral geometries in contact, call for different forms of potential energy models, they will be shown to lead to the same model equation for the excess free energy.

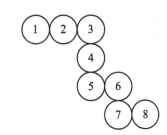

Fig. 4.3. Model of a structured molecule ($r = 8$).

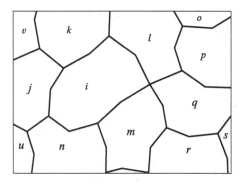

Fig. 4.4. Liquid model for polyhedral surface interactions.

4.2.2 The Free Segment Approximation

In Figure 4.5 we consider a two-dimensional lattice system composed of two types of molecules, one with eight segments ($r_1 = 8$) and one with two segments ($r_2 = 2$). There are two molecules of the large type and six molecules of the small type. Each segment is capable of four contacts with a neighboring segment. It is important to differentiate between external and internal contacts. External contacts between segments of different molecules contribute to the intermolecular potential energy and are represented by dots in Figure 4.5. Internal contacts between the segments of the same molecule represent the intramolecular connectivities and do not contribute to the intermolecular potential energy. They are denoted by bars in Figure 4.5. External and internal contacts may also be referred to as nonbonded and bonded, respectively. We note that in the special case of ring formation during a change of conformation, nonbonded contacts may also occur between segments of the same molecule and then contribute to the intermolecular potential energy. There are 36 external interaction sites and 20 internal surface connections visible in the figure. The intermolecular potential energy is found from adding up all pair interaction

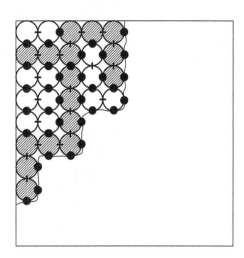

Fig. 4.5. Molecular interaction model of a liquid system composed of two components ($r_1 = 8, r_2 = 2$); (•) interaction sites (36); (·) internal surface connections (20).

energies of the segments, i.e., all external or nonbonded contact energies. In general, this energy will clearly depend on the particular geometrical arrangement considered, e.g., the configuration displayed in Figure 4.5. As different geometrical arrangements arise, different external surface contacts will be the consequence, and thus a different intermolecular potential energy of the whole system will be found.

A considerable simplification in applying this pair segment contact model for the intermolecular potential energy is achieved by the assumption that the molecular segments interact independently with each other. This is referred to as the free segment approximation. A particular segment will then interact with another particular segment irrespective of its actual position in a real molecule. So, in this approximation, the interacting molecular segments have lost their memories about their original associations with particular molecules. Because relatively few types of molecular segments are sufficient to build up the enormous number of molecules of technical interest, the free segment approximation is a crucial step toward the development of practical potential models for liquids. Also, for such potential models of independently interacting segments, simple statistical models for the excess functions can be derived, as shown below.

To visualize the free segment approximation in the lattice picture, we consider in Figure 4.6 the same number of white and dark segments as in Figure 4.5 with the same number of interaction sites, but the external contacts, i.e., the contacts contributing to the intermolecular potential energy, are now distributed in a manner different from that in Figure 4.5. They are again shown by points. With bars we here again indicate internal contact sites, although there are no real internal contacts any more because the molecules have been broken into their segments. The bars here rather indicate contacts that do not contribute to the

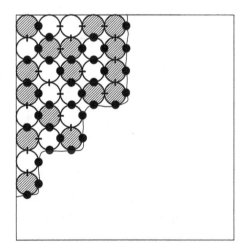

Fig. 4.6. Molecular interaction model of a liquid assumed to be statistically identical to that of Fig. 4.5 in the free segment approximation: (•) interaction sites (36); (-) internal surface connections (20).

intermolecular potential energy of the system. It is clear that the intermolecular potential energies of the particular configurations in Figures 4.5 and 4.6 are different, because, e.g., the number of white–white contacts in Figure 4.5 is five, whereas it is only two in Figure 4.6. The free segment approximation, however, implies that, in the statistical average, relevant to evaluating the partition function, a system of molecules 1 and 2, for which one particular configuration is shown in Figure 4.5, has the same intermolecular potential energy as a system of independent segments, for which one particular configuration is shown in Figure 4.6. In the polyhedral surface interaction picture of Figure 4.4 the free segment approximation applies to surface contacts, as shown in a schematic way for molecules visualized as squares in the two-dimensional representation of Figure 4.7. Here four different surface properties are considered, allowing for 10 different interaction types between neighboring segments. In the free segment approximation these surface contacts can combine independently, i.e., without the constraints of their attachment to a particular molecular segment, and their addition is still assumed to give the correct intermolecular potential energy. The 10 types of different free segment contacts are also shown in Figure 4.7.

Clearly, the free segment approximation neglects any influence of the geometry of the molecules and of the coherent charge distribution over the molecular surface on the potential energy of the system. Therefore, it cannot be generally accurate. However, many of the standard systems in chemical technology appear to follow the free segment approximation quite closely. This can be

Attached
Segments   Free Segments

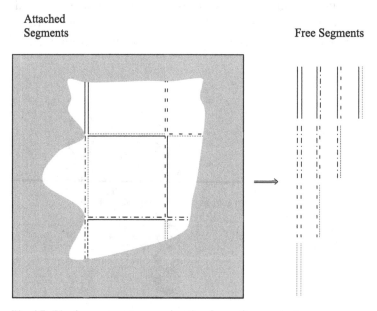

Fig. 4.7. The free segment approximation for surface contacts.

concluded from the successful applications that have been made on this basis and is also supported by computer simulations to be demonstrated below. We note that the geometrical effects of size and shape, although neglected in the attractive part, are taken into account by the repulsive contribution to the excess functions. Also, the external surfaces of the molecules will be shown to appear in the final equations for the attractive contribution.

4.2.3 Group Interaction Models

One rather popular interaction model based on the free segment approximation in the lattice picture is the group interaction model. Group interaction models have proven to be very simple, but, at the same time, very useful representations of the intermolecular potential energy in liquids. They consider the intermolecular potential energy to be added up from independently interacting groups of atoms, so-called functional groups, located in the molecules. These groups are the volume segments of the molecules to which the free segment approximation is applied in the lattice picture. Currently, the most popular group interaction model for predicting fluid phase equilibria is UNIFAC [3]. In UNIFAC, e.g., the polar molecule butanone with the chemical formula C_4H_8O is broken up into three functional groups, CH_2, CH_3, and CH_3CO. The nonpolar molecule hexane with the chemical formula C_6H_{14} is broken up into CH_3-groups and CH_2-groups. By breaking up a molecule into functional groups we arrive at certain group frequencies for the pure components. For butanone each of the three groups appears only once, whereas in hexane, there are two CH_3-groups and four CH_2-groups. In a mixture the group mole fractions are calculated from those of the pure components and their associated group frequencies by

$$x_\alpha = \frac{\sum_i x_i n_{\alpha,i}}{\sum_i x_i n_i}, \tag{4.24}$$

where i is the index for the molecules and α the index for the functional groups. Further, n_i is the number of groups in species i and $n_{\alpha,i}$ is the number of α-groups in a molecule of component i. The intermolecular potential energy is now calculated from a summation over all pair contact energies between the different groups in the mixture. The interaction energy $\phi_{\alpha\beta}$, i.e., the contact energy between a group α and a group β, cannot yet be calculated theoretically from quantum mechanics and so must be determined by fitting to experimental data. Clearly, it should be temperature-dependent. It has been shown [3] that, by using a large database and a large number of functional groups, values for $\phi_{\alpha\beta}$ can be obtained that have a considerable predictive capacity for the macroscopic behavior of fluid mixtures. This is due to the fact that the number of groups is much less than the number of components that can be synthesized from them.

EXERCISE 4.1

Calculate the group frequency profile of an equimolar mixture of butanone with hexane for the UNIFAC model.

Solution

For an equimolar mixture of butanone with hexane we have for the CH_3 group from (4.24)

$$x_{CH_3} = \frac{0.5 \times 1 + 0.5 \times 2}{0.5 \times 3 + 0.5 \times 6} = 0.3333,$$

with analogous results for the other groups, leading to a group frequency profile as shown in Figure E 4.1.1.

In spite of the remarkable success of fluid phase calculations based on group contributions, it is obvious that there are a number of principal drawbacks associated with this approach. First, there is no theoretical basis for choosing the functional groups and associating a particular interaction energy with a group contact. Thus, different group divisions have been reported, and the individual group interactions have to be found from fitting the models to data. As a consequence, group contribution methods are restricted in application to molecules that are composed of those groups that have already been parameterized. Any molecule containing a group not yet considered in the database

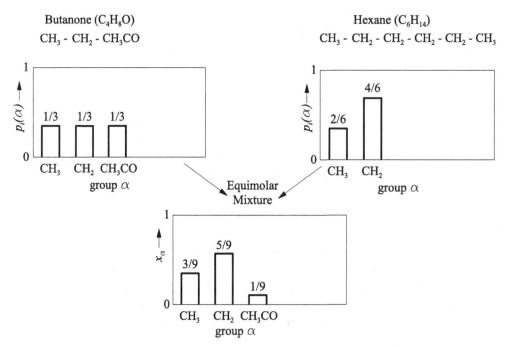

Fig. E 4.1.1. Frequency profile of functional groups in the system butanone–hexane ($x = 0.5$).

cannot be treated. This is a serious disadvantage in a situation where technological development proceeds to ever new compounds. Also, even if all group interaction energies have been parameterized, group contribution methods suffer from the principal inability to distinguish between isomers consisting of identical groups, unless taken into account by particular group definitions. Also, they have found to be inaccurate when compounds with several strong nonalkyl functional groups are considered. This becomes clear when it is noted that modeling the intermolecular potential energy by group contributions implies that each contact between two groups α and β is associated with a unique interaction energy. This energy is assumed to incorporate all different types of interactions, such as electrostatics, induction, dispersion, and hydrogen bonding. In reality, any two groups can form a variety of different contacts, depending on their location in the molecule, i.e., the intramolecular environment. For example, an OH-group capable of hydrogen bonding will have quite different effects on the intermolecular potential energy depending upon whether it is freely available to contact another molecule or almost unavailable due to an intramolecular hydrogen bond. Because different macroscopic phenomena differ in sensitivity with respect to such an ad hoc combination of effects, one usually finds, e.g., that group contributions obtained from vapor–liquid equilibrium are unable to predict the liquid–liquid equilibrium, even at the same temperature. Finally, no satisfactory temperature dependence of the interaction parameter can be derived from the model. As a consequence, quantities such as the excess enthalpy cannot be predicted from parameters fitted to vapor–liquid equilibrium data unless the temperature dependence is fitted separately for each group. So, summarizing, a potential model based on group contributions reflecting chemical functionality is essentially a useful interpolation tool for a database with predictive capacity for compounds composed of the same groups already considered as long as intramolecular environment effects are not severe. It does not, however, represent a realistic picture of the intermolecular interaction resulting from the surface properties of the molecules.

4.2.4 Surface Charge Interaction Models

An alternative and more theoretically founded approach to the intermolecular potential energy in liquids can be designed in the form of a surface charge interaction model. From the continuum solvation model COSMO discussed in Section 2.4 [4], and in combination with quantum-chemical calculations, we obtain a distribution of screening charges along the inner surface of a cavity representing the geometry of a dissolved molecule in an electrical conductor. This so-called COSMO file is available from quantum-chemical computer codes. In such calculations, the charge distribution and the geometry of the single molecule in a vacuum are used to start an iterative computation that finally gives the geometry and the screening charge, i.e., an effective surface charge

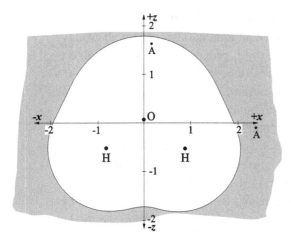

Fig. 4.8. COSMO cavity surface for H_2O.

distribution of the molecule dissolved in a dielectric continuum. As an illustration, Figure 4.8 shows the shape of the cavity in the COSMO model for H_2O in a two-dimensional representation. As compared to the vacuum geometry, there are some minor differences for this molecule. In the quantum-chemical level usually applied to COSMO calculations the angle between the two OH-bonds in the vacuum is found to be 104.07°, whereas it is slightly reduced to 103.5° in the conductor. Also, the OH-bond length is 0.97123 Å in the vacuum and 0.97242 Å in the conductor. In other molecules the differences may be much larger. Figure 4.9 shows a qualitative representation of the screening charge density distribution on the inner surface of a cavity for an H_2O molecule embedded in an electrical conductor. We see that the positively charged hydrogen atoms lead to an accumulation of negative screening charges at the bottom of the cavity, whereas there is an accumulation of positive charges screening the

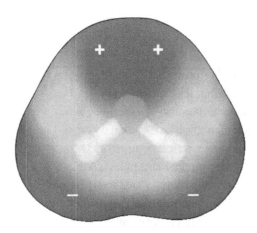

Fig. 4.9. Screening charge distribution for H_2O (schematic).

electronegative oxygen atom at the top. In between there is a broad region of essentially zero charge.

The screening charges can be used as a basis to establish the intermolecular potential model due to electrostatic effects in a real solvent. This generalization of COSMO is referred to as COSMO-RS. In the COSMO-RS (real solvent) potential model [2] the interactions between molecules are visualized to occur between the surface segments of polyhedral surfaces contacting each other in a space-filling manner; cf. Figure 4.4. Each surface segment is associated with a characteristic electrostatic charge density. The segmentation of a molecule's surface into surface patches can be made fine enough to reproduce the real charge distribution properly. An order of magnitude of several hundred segments is typical for COSMO. So, contrary to the group contribution models, the pair contact interactions are not assumed to occur between coarse functional groups. Rather, the characteristic charge density distribution over the molecule's surface is taken into account. The notion of groups is irrelevant for this model. Clearly, a much more realistic representation of the interaction properties of a molecule is now possible than with a group contribution model.

The COSMO-RS model replaces the conductor by real solvent molecules. It thus abandons the limitations of a continuum model. When it does so, a new reference state is created, defined as that of molecule i being ideally screened electrostatically by solvent molecules, as if it were dissolved in a conductor. The intermolecular contacts in this reference state occur via pairwise interactions of surface segments of solute and solvent molecules with COSMO surface charge densities. Clearly, this reference state has a system-specific potential energy, because the effect of the electrostatic response of the solvent and solute molecules to the solution process is taken into account, at least in the approximation of the continuum solvation model. Thus, it is generally much closer to reality for molecules in solution than the ideal gas state, i.e., of molecules in vacuum without any potential energy, and also than the pure liquid state, which neglects all solution effects. It is close in philosophy, although not identical, to the ideal dilute solution reference state, with a degree of approximation of experimental reality as shown in Table 2.2. In a real solution it will not be possible to screen the solute molecule perfectly by solvent molecules. Thus, there will be energetic deviations from the reference state. Only the potential energy associated with such deviations from the reference state enters into the calculation of the excess functions.

Figure 4.10 explains the COSMO reference state in more detail [5]. We consider a solute molecule i and four solvent molecules 1, 2, 3, and 4 belonging to different components, all embedded in an electrical conductor. For the sake of simplicity of the argument all molecules are considered to be of cubic shape but projected onto a plane. Thus all molecules are modeled as squares interacting with each other via their four sides. All five molecules are perfectly screened by surface charges on the associated inner surfaces of the cavities; cf. Section 2.4.

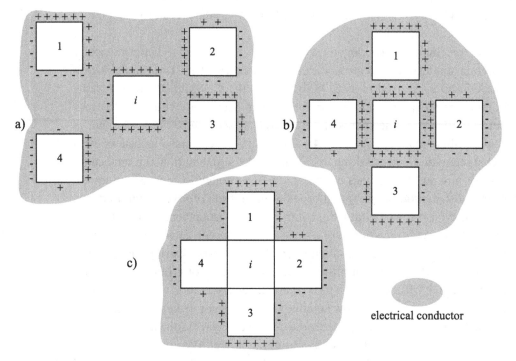

Fig. 4.10. COSMO reference state of solute molecule i in a solvent.

The screening charges are obtained from a COSMO calculation and are shown as arrangements of positive and negative charges with equal magnitude in Figure 4.10. Thus, the solute molecule i is screened by six positive charges on two opposite sides and also six negative charges on the remaining two opposite sides. The solvent molecules are screened by six positive and six negative charges on two opposite sides and various positive and negative charges on the remaining two opposite sides. We first consider the arbitrary configuration in Figure 4.10a, i.e., the typical COSMO configuration of molecules embedded in an electrical conductor. The potential energy of this molecular system is available from five COSMO calculations. Because the molecules are ideally screened in the conductor they do not interact with each other. They can thus be moved about without any change in electrostatic energy. In particular, we can, while keeping the solute molecule in the fixed position of Figure 4.10a, move the four solvent molecules in such a way that the configuration in Figure 4.10b appears, without a change of electrostatic energy with respect to the situation in Figure 4.10a. This new configuration is characterized by the surface charges of the solute molecule being ideally matched by opposite surface charges of the solvent molecules. Because the net charge around each of the four sides of the solute molecule is then zero it does not make any difference whether there is a conducting medium around it or not, and we may thus remove the electrical conductor without any change in electrostatic energy; cf. Figure 4.10c. As a result the solute molecule

i is now perfectly screened by solvent molecules, not any more by a conductor, but still has the same electrostatic energy as in Figure 4.10a. We note that the situation in Figure 4.10 requires that the solvent molecules surfaces dispose of equal but opposite surface charges for all surfaces of the solute molecule. Only under this condition are the solvent molecules able to replace the electrical conductor without any change of energy. The COSMO reference state for solute i is thus defined by a situation where all interacting surface segments of the solute are matched electrostatically in such an ideal manner by solvent molecules. Its potential energy is fully described by the conductor-like screening charges on the surfaces of the molecule, i.e., by a COSMO calculation.

In a real liquid mixture most surface pairs will not meet this perfect fit. This results in a less perfect screening of the solute molecule i by the solvent molecules, and removing the electric conductor between them is associated with a change in electrostatic energy, referred to as a misfit energy. It is this misfit energy that is responsible for deviations from the COSMO reference state due to local electrostatic effects and thus will be needed to calculate excess functions and activity coefficients. The potential energy associated with the reference state is the same for a solute i dissolved in a solvent or for a molecule in the environment of pure liquid i, because in both cases it is modeled by a hypothetical environment of ideally matched surface segments. It is, therefore, independent of the particular environment in either the mixture or the pure components and thus will cancel in the excess functions. The misfit energy can be computed from the conductor-like screening charges. Figure 4.11 shows again the liquid model adapted to take into account surface charge interactions in the framework of

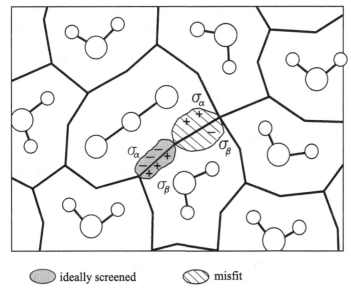

Fig. 4.11. The COSMO-RS interaction model.

space-filling surface segments. Now, the polyhedral structures contain schematically the molecular geometries that they are meant to represent. The contacting surface segments are made of grounded electrically conducting surfaces, each being represented by a constant surface charge density σ_α or σ_β, respectively. In a real liquid mixture these electrically conducting surfaces have to be removed. Those surfaces with a perfect fit of the positive and negative surface charges can be removed without any change of the system's electrostatic energy. Such a surface is represented by the shaded region in Figure 4.11. Other surface pairs, such as that belonging to the dashed area in Figure 4.11, do not have such a perfect fit. Thus, a misfit energy must be accounted for when the electrical conducting surface is removed. An expression for this misfit energy can be derived from visualizing the process of eliminating the electrically conducting surface as one of adding the opposite of the net charge on the unbalanced surface. The surface is then balanced and can be taken away without any change of electrostatic energy. The energy penalty for this is the energy necessary for charging the surface to be eliminated additionally with its net charge. We found in Section 2.4 that the energy associated with charging a surface is proportional to the square of the charge density. This leads to an expression for the interaction energy contribution due to a misfit between the associated surface segments of the form

$$(\phi_{\alpha\beta})_{\text{misfit}} = A_{\alpha\beta}\frac{\alpha}{2}(\sigma_\alpha + \sigma_\beta)^2, \tag{4.25}$$

where $A_{\alpha\beta}$ is the surface area of the segment, transforming the charge densities $\sigma_\alpha, \sigma_\beta$ into charges, and α is an electrostatic misfit energy coefficient for converting a charge into an energy. Both are adjustable parameters of the model. We note that the misfit pair potential energy vanishes for $\sigma_\alpha = -\sigma_\beta$, as it must. It is assumed that the total intermolecular potential energy due to the electrostatic energy difference between the system in the reference state and the real system is given by a sum over the misfit energies over all interacting surface pairs.

Although local electrostatic effects in the fluid are quite well modeled by the misfit energy, there will be further contributions of electrostatic origin to the intermolecular potential energy of liquids. One of particular significance in many systems is that related to hydrogen bonding. Hydrogen bonding is observed when a hydrogen atom covalently bonded to an electronegative atom such as oxygen approaches another electronegative atom in the same or in a second molecule, which again may be oxygen. A rather strong attractive interaction then occurs in the O-H $\cdots$ O system with a bond angle of about 180°, having significant consequences for the structural and thermodynamic properties of the system. In a simplified visualization, the electronegativity of the oxygen atom will make the O-H bond polar and the proton of the H atom will be exposed unshielded. Due to the smallness of the H atom, the O atom of the second molecule can approach the O-H bond closely before repulsive forces become important. Thus the electrons of this O atom can interact strongly with the H nucleus. So, in other words, in a hydrogen bond a proton acts as a link between two electronegative atoms. Hydrogen bonding is found not only with the oxygen atom but also frequently with nitrogen or fluorine. It can either occur within a molecule as an intramolecular effect or contribute to the intermolecular potential energy of a system. A large part of the hydrogen bond interaction is electrostatic in origin, although some covalent contribution may also be involved. All in all, hydrogen bonding reflects the interaction energy resulting from the interpenetration of atomic electron densities when two sufficiently polar pieces of surface of opposite polarity are in contact. To account for hydrogen bonding effects within a surface interaction model beyond those contained in the misfit energy and in the COSMO reference state, it has been suggested to take the difference between a threshold value of local polarity and the screening charge density σ as a measure of the hydrogen bonding effects. The function chosen is [2]

$$(\phi_{\alpha\beta})_{\mathrm{hb}} = A_{\alpha\beta} C_{\mathrm{hb}} \max\left[0, \sigma_{\mathrm{acc}} - \sigma_{\mathrm{hb}}\right] \min\left[0, \sigma_{\mathrm{don}} + \sigma_{\mathrm{hb}}\right], \qquad (4.26)$$

with $\sigma_{\mathrm{don}} = \min(\sigma_\alpha \sigma_\beta)$ and $\sigma_{\mathrm{acc}} = \max(\sigma_\alpha \sigma_\beta)$ being the donor and the acceptor part of the hydrogen bond, respectively. The screening charge for a positive molecular region is negative, which makes the negative σ the donor part and the positive σ the acceptor part. The parameter σ_{hb} is a threshold for hydrogen bonding and C_{hb} is a strength coefficient, which should be temperature-dependent. Both are universal parameters of the model that have to be fitted to a database. From the structure of (4.26), the hydrogen bonding pair potential will only become effective when $|\sigma_{\mathrm{don}}| > \sigma_{\mathrm{hb}}$ and $\sigma_{\mathrm{acc}} > \sigma_{\mathrm{hb}}$ and then it is proportional to the product of the two excess charge densities.

Finally, there are nonelectrostatic parts of the intermolecular potential energy, notably dispersion effects, which are known to be important, as a rule even dominant, in nonpolar fluids. They depend on the polarizability of a molecule, cf. Section 2.5, and thus are related to the molecular volume or the surface. In the simplest way they can thus be modeled by simple proportionality to the surface area of an atom and an element-specific strength parameter as

$$\phi_{\mathrm{disp}} = \sum_i x_i \left(\sum_k A_{k,i} \gamma_k\right), \qquad (4.27)$$

where k denotes the atoms of a considered molecule, $A_{k,i}$ is the surface of atom k in molecule i, and γ_k is an associated energy coefficient universal for the element k to be fitted to data. This formulation of the nonelectrostatic contribution does not depend on segment–segment interactions. It is rather an energy contribution to the reference state. Thus, it does not contribute to the excess functions, although it will be crucial for the ideal dilute solution properties, such as Henry coefficients. It is clear that the above model for the dispersion interactions is no more than an ad hoc concept and certainly is a major source of uncertainty when the COSMO-RS model is applied to the prediction of fluid phase properties.

So far, the electrostatic part of the COSMO-RS model is based on the interaction of molecular surface patches α and β, with a full account of their association with the molecular geometry, as obtained from the COSMO files. A crucial simplification is now possible by invoking the free segment approximation. Then the geometrical information contained in the COSMO files is irrelevant and statistical information about the frequencies of particular charge densities in a molecule, the so-called σ-profile [2], is sufficient. The σ-profile of a molecule is obtained by a suitable statistical averaging of the charge density distribution over the effective contact areas. It will basically be a histogram in which discrete σ-values are related to associated surface areas. On the order of 60 discrete segments representing the σ-profile has in most cases been found to be sufficient. The average surface charge density σ_m on segment m is obtained from the COSMO charge densities σ_n by [6]

$$\sigma_m = \frac{\sum_n \sigma_n \frac{r_n^2 r_{\mathrm{av}}^2}{r_n^2 + r_{\mathrm{av}}^2} \exp\left(-\frac{d_{mn}}{r_n^2 + r_{\mathrm{av}}^2}\right)}{\sum_n \frac{r_n^2 r_{\mathrm{av}}^2}{r_n^2 + r_{\mathrm{av}}^2} \exp\left(-\frac{d_{mn}}{r_n^2 + r_{\mathrm{av}}^2}\right)},$$

where r_n is the radius of the much smaller actual COSMO surface segment, assumed to be circular, r_{av} is the averaging radius, and d_{mn} is the distance between the two segments. The sum is over all COSMO surface segments. Generally, r_{av} is a universal parameter to be found from fitting to data. It has a significant influence on the final results. A very small value leads to segments that are highly correlated whereas a large value suppresses much of the charge structure over the surface. A value of $r_{\mathrm{av}} = 0.5$ Å has been found to be reasonable. As an illustration, Figure 4.12 shows the smoothed, i.e., continuous σ-profiles of some molecules, hexane, water, benzene, toluene, and xenon, obtained in this way from COSMO. The charge density σ is given in e/Å^2 and $A_{\sigma,i}$ is the surface area associated with a particular charge density σ in a molecule of component i. Summing up all $A_{\sigma,i}$ values gives the total external surface area of a molecule of component i. It can be seen that the σ-profile of the water molecule is very broad, with surface charge densities between about -0.02 and $+0.02$ e/Å^2, reflecting the strongly polar character of this molecule. The σ-profile is further almost symmetrical with one broad negative peak at about -0.015 e/Å^2 and one broad positive peak at about $+0.015$ e/Å^2. The peak on the negative side results from the screening of the two positively charged hydrogen atoms, whereas the positive peak screens the two lone electron pairs of the negative oxygen atom; cf. Figure 4.9. Due to the symmetric nature of the σ-profile, almost any surface segment can find its ideal electrostatic partner. So there is little necessity

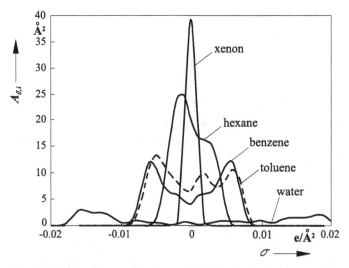

Fig. 4.12. Surface charge density profiles for some molecules.

for the segments of pure water molecules to form nonideal pairs, and the surface segments of water have a high electrostatic affinity to each other. Thus, much energy is required to vaporize water, resulting in a high boiling point. Further, water has surface segments with a broad spectrum of charges. Thus, it can offer the opposite ideal surface charge density for all segments of a solute molecule and thus is a good solvent for many components. The σ-profile of the hexane molecule is quite different. It is rather narrow, the available surface charge densities lying between -0.005 and $+0.005$ e/Å^2. The peaks that can be detected result from the hydrogens at the negative and from the carbons at the positive side. The small range of surface charge densities explains the nonpolar character of hexane, as opposed to water. Because the σ-profiles of the two molecules do not fit to each other, these components are not very mutually soluble. The σ-profile of benzene reflects the benzene ring and the hydrogen atoms at the positive and the negative side, respectively. The additional peak in the σ-profile of toluene represents the additional CH_3-group. We note the similarity of the σ-profiles of benzene and toluene, which indeed are known to form an ideal solution to a good approximation. The σ-profile for the noble gas xenon is extraordinary, peaked around $\sigma = 0$, which results from its strictly spherical character. The small positive and negative charges are artifacts of the smoothing process. Thus, we find that the σ-profiles indeed reflect the essential electrostatic interaction properties of the molecules and should be a sensible basis for modeling the associated intermolecular interactions.

In surface charge interaction models the total intermolecular potential energy adds up from the contributions of all surface patches. In the free segment approximation these surface patches are no longer associated with particular molecules. It is therefore convenient to express the σ-profile of a component in terms of surface fractions of the various discrete σ-values, such as, for a σ-value of α,

$$\theta_{\alpha,i} = \frac{A_{\alpha,i}}{\sum_{\beta} A_{\beta,i}} = \frac{A_{\alpha,i}}{A_i}, \qquad (4.28)$$

where $\sum_{\beta} A_{\beta,i} = A_i$ is the total surface area of a molecule of component i. In a mixture we need the σ-profile of the mixture. The surface fraction of a segment of type α in a mixture is calculated from the surfaces of the components associated with it, analogously to (4.24), via

$$\theta_{\alpha} = \frac{\sum_i x_i A_{\alpha,i}}{\sum_i x_i A_i}, \qquad (4.29)$$

where x_i is the mole fraction of component i in the mixture.

When comparing the COSMO-RS potential model with the UNIFAC potential model, one finds that the σ-profile is analogous to the frequency distribution of the particular groups. Because, in the free segment approximation, simple statistical mechanical formulae can be derived for the excess free energy which can be solved rapidly on a computer, it has been possible to find the parameters of both models by fitting to a large database. The intermolecular potential energy of any mixture can thus be calculated when the molecules can adequately be treated in the free segment approximation. We note that the demanding quantum mechanical COSMO calculations in the COSMO-RS model have to be performed only once for each new molecule considered.

4.3 Simple Model Molecules

To introduce the basic concepts of deriving excess function models in the free segment approximation, we first treat simple model molecules. In particular, we consider spherical molecules of equal sizes and uniform surface properties. There will thus be no hard core contribution to the excess free energy, because the excess entropy for a system of hard spheres of equal diameters is obviously zero. Also, the attractive contact energies are numbers associated with the various pairs of molecules without the necessity of structuring the molecules into volume segments or surface patches. The liquid model adopted for the statistical mechanical derivation is that of a regular lattice, although identical results can be derived in the polyhedral surface contact picture.

4.3.1 The Partition Function

We consider a mixture of two components. The molecules of the two components 1 and 2 are spheres of equal size. They are arranged on the sites of a regular lattice in a space-filling manner, as shown in the two-dimensional case in Figure 4.13 for a pure liquid of component 1. The type of lattice considered

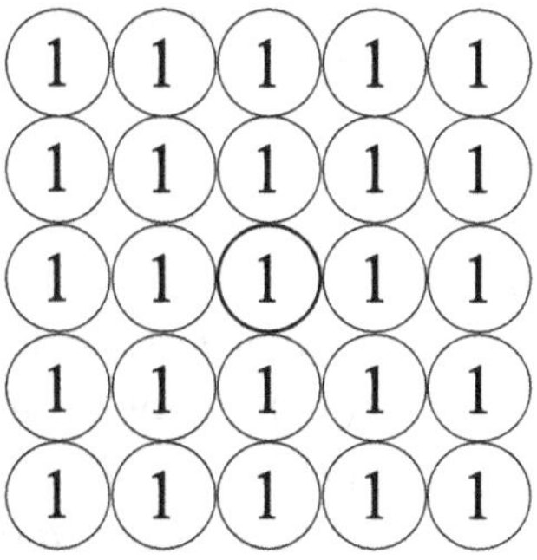

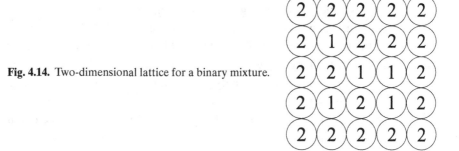

Fig. 4.13. Two-dimensional lattice for pure component 1.

defines a so-called coordination number z, which represents the number of nearest neighbors with which a selected molecule will interact. For the two-dimensional lattice considered in Figure 4.13 the coordination number is $z = 4$. Depending on the type of packing in three dimensions it may have values between 6 and 12. Figure 4.14 shows a two-dimensional lattice for a binary system of components 1 and 2. The energetic information about this system is fully contained in the numbers of contacts between unlike and like molecules, which are restricted by the conservation equations

$$zN_1 = N_{11} + N_{21} \tag{4.30}$$

and

$$zN_2 = N_{22} + N_{12}. \tag{4.31}$$

Here zN_1 is the number of nearest neighbor molecules around all molecules 1, N_{11} is the number of nearest neighbor molecules 1 around all central molecules 1, and N_{21} is the number of nearest neighbor molecules 2 around all central molecules 1. An analogous notation applies for N_{22} and N_{12}, and by symmetry we have $N_{12} = N_{21}$. We further note that the numbers of nearest neighbor contacts of the 11- or 22-type are $N_{11}/2$ and $N_{22}/2$, respectively, whereas the number of nearest neighbor contacts of the 12-type is N_{12}. Clearly, when N_{12} is given, all nearest neighbor binary contacts are fully defined in a system of given composition. A particular value of N_{12} thus defines a particular molecular energy

Fig. 4.14. Two-dimensional lattice for a binary mixture.

of the lattice. We can use N_{11}, N_{12}, and N_{22} to define mean local compositions by

$$x_{ij} = \frac{N_{ij}}{\sum_k N_{kj}} = \frac{N_{ij}}{zN_j}, \quad i,j = 1,2. \tag{4.32}$$

Although formally defined in the same manner as the local compositions of (4.8), the two definitions are clearly different. Whereas the mean local compositions in a lattice take only nearest neighbors into account and are average values for the whole lattice, those of (4.8) are truly local. Besides the closure conditions (4.9) and (4.10), mean local compositions in a lattice, by their definition and the conservation equations, also fulfill the symmetry relation

$$x_1 x_{21} = x_2 x_{12}, \tag{4.33}$$

which is not obeyed by the local compositions of (4.8). Note that the conservation equations apply to an infinite lattice and thus cannot generally be verified by a small sample such as the one shown in Figure 4.14.

In view of the molecular energy of a lattice being defined by a particular value of N_{12} and because the various possible configurations arise by specific arrangements of the molecules over the lattice sites, the configurational partition function (2.75) for a lattice transforms to a sum over all N_{12}; i.e.,

$$Q^C = \sum_{N_{12}} g(N_1, N_2, N_{12}) e^{-U(N_1, N_2, N_{12})/kT}. \tag{4.34}$$

Here, $U(N_1, N_2, N_{12})$ is the intermolecular potential energy associated with a configuration defined by N_{12}. Further, $g(N_1, N_2, N_{12})$ represents the degeneracy of a particular energy level. The degeneracy takes into account that the configurational partition function (2.75) is an integral over all different configurations, whereas in (4.34) the sum extends over all numbers of unequal pair contacts between different molecules and thus over all energy levels. The same number of N_{12} may be associated with a large number of different configurations. Thus, the degeneracy is the number of configurations with an identical intermolecular potential energy and is given by the combinatorial factor, equal to the number of distinguishable ways of arranging N_1 molecules of type 1 and N_2 molecules of type 2 such that there are N_{12} unlike contacts between them. For a pure component there is only a single type of interaction and only one distinguishable way of arranging the indistinguishable molecules, giving $g = 1$. In a mixture, however, there are very many distinguishable configurations leading to the same number of N_{12}-contacts. Due to the large numbers involved there is one particular energy level, i.e., one particular number of N_{12}-contacts, associated with an overwhelmingly large statistical weight. The sum in (4.34) thus can be replaced by this one term as an excellent approximation. This is called the maximum term method; cf. App. 3. We thus write for the configurational partition function

$$Q^C = g(N_1, N_2, N_{12}) e^{-U(N_1, N_2, N_{12})/kT}, \tag{4.35}$$

where it is understood that N_{12} here and in the following indicates the particular energy level that is found with a dominant probability because it is associated with a dramatically larger number of configurations than any other energy level. We shall turn to the determination of N_{12} later. We note in particular that the degeneracy and thus Q^C here contains the ideal mixing contribution due to the permutational factor in the canonical partition function, because it is obtained from considering combinatorial permutations. This must be taken into account in calculating the mixing functions and excess functions.

4.3.2 The Excess Free Energy

The intermolecular potential energy of a lattice system is

$$U = \frac{1}{2}N_{11}\phi_{11} + \frac{1}{2}N_{22}\phi_{22} + N_{12}\phi_{12}, \tag{4.36}$$

where ϕ_{ij} is the intermolecular potential energy of an ij-contact. Using the conservation equations (4.30) and (4.31), the intermolecular potential energy of the lattice in a state characterized by N_{12} can be written as

$$U = \frac{z}{2}N_1\phi_{11} + \frac{z}{2}N_2\phi_{22} + \frac{N_{12}}{2}\omega_{12}, \tag{4.37}$$

where

$$\omega_{12} = \phi_{12} + \phi_{21} - (\phi_{11} + \phi_{22}) \tag{4.38}$$

is referred to as the exchange energy for the binary system; cf. (4.23). On the premise that the only statistically relevant energy level of the lattice is the one characterized by N_{12}, (4.37) along with (4.38) gives the configurational part of the macroscopic internal energy of the system. The excess internal energy then follows as

$$U^E = N_{12}\omega_{12}/2. \tag{4.39}$$

It is possible to transform this result for a binary mixture into a more symmetrical form for a mixture of an arbitrary number of components $1, 2, \ldots$ as

$$U^E = \frac{1}{4}Nz\sum_i\sum_j x_i x_{ji}\omega_{ji}, \tag{4.40}$$

where

$$\omega_{ji} = \phi_{ji} + \phi_{ij} - (\phi_{ii} + \phi_{jj}) \tag{4.41}$$

is the exchange energy of the ij-pair and x_{ji} is the mean local composition of j-molecules around central i-molecules in the lattice; cf. (4.32).

To arrive at an expression for the excess free energy we remember that, cf. (2.60),

$$A = U - TS = -kT\ln Q.$$

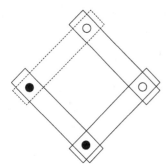

Fig. 4.15. Correlation of pair interactions in a lattice.

This gives the configurational entropy, i.e., the entropy determined by the intermolecular contact interactions, for the only relevant molecular energy level of the lattice characterized by N_{12} from (4.35) as

$$S^C = k \ln g(N_1, N_2, N_{12}).\qquad(4.42)$$

Thus, we need the degeneracy of the state characterized by N_{12}.

There is no exact solution to calculating this degeneracy in the general case. However, a solution is possible on the assumption that the pair interactions of nearest neighbors can be treated as independent of each other [7]. This assumption is not strictly correct in a lattice as visualized in Figure 4.15, from which it is clear that in the situation shown only three pairs can be arranged independently, whereas the fourth pair, the dashed one, results from the lattice conditions, and thus must be a black–white interaction. However, we feel justified using this approximation in the statistical average in view of the good results obtained from it in comparison with computer simulations, as demonstrated below. Thus we do not fix the fourth pair to be a black–white interaction but rather leave it free. This approximation implies that the total intermolecular potential energy, and thus also the distribution of pair interactions, does not depend significantly on the nature of a particular interacting pair, at least in an average sense. With this approximation, we can conclude that each pair of interactions can be occupied in four ways, i.e., 11, 22, 12, and 21, with equal probabilities. From the conservation equations (4.30) and (4.31), the total number of nearest neighbor contacts, i.e., $N_{11}/2 + N_{22}/2 + N_{12}$, is $(z/2)(N_1 + N_2)$. The combinatorial problem is then one of arranging this total number of contacts over the four independent categories we have. This has been addressed before, and the result is, cf. (A 3.6), for an arbitrary distribution N_{12}

$$h(N_1, N_2, N_{12}) = \frac{\left[\frac{z}{2}(N_1 + N_2)\right]!}{(N_{11}/2)!(N_{22}/2)!(N_{12}/2)!(N_{21}/2)!},\qquad(4.43)$$

where the notation of the N_{ij} is as discussed following (4.30) and (4.31). Because the independent pair approximation will count many configurations that are actually impossible, (4.43) does not satisfy the ideal mixing condition

$$\sum_{N_{12}} g(N_1, N_2, N_{12}) = \frac{(N_1 + N_2)!}{N_1! N_2!} \neq \sum_{N_{12}} h(N_1, N_2, N_{12}),\qquad(4.44)$$

which is the number of ways a total of $(N = N_1 + N_2)$ molecules can be arranged so that N_1 are in one group and N_2 in the other one, irrespective of N_{12} and all other local pair contacts. To enforce the ideal mixing condition we normalize (4.43) by a normalization factor that depends only on the bulk composition. Thus we find

$$g(N_1, N_2, N_{12}) = C(N_1, N_2)h(N_1, N_2, N_{12}), \tag{4.45}$$

where C has to be determined from (4.44); i.e.,

$$\sum_{N_{12}} g(N_1, N_2, N_{12}) = C(N_1, N_2) \sum_{N_{12}} h(N_1, N_2, N_{12})$$
$$= \frac{(N_1 + N_2)!}{N_1! N_2!}. \tag{4.46}$$

The details are worked out in App. 8. As a result we find the degeneracy as, cf. (A 8.6),

$$g(N_1, N_2, N_{12}) = \left(\frac{N!}{N_1! N_2!}\right)^{1-z} \cdot \frac{(\frac{z}{2}N)!}{(N_{11}/2)! \, (N_{22}/2)! \, [(N_{12}/2)!]^2} \tag{4.47}$$

and its logarithm, cf. (A 8.7), as

$$\ln g = -N(x_1 \ln x_1 + x_2 \ln x_2)$$
$$+ z\frac{N}{2}\left(x_{11}x_1 \ln \frac{x_1}{x_{11}} + x_{21}x_1 \ln \frac{x_1}{x_{21}} + x_{12}x_2 \ln \frac{x_2}{x_{12}} + x_{22}x_2 \ln \frac{x_2}{x_{22}}\right). \tag{4.48}$$

Here we realize that (4.48) is an exact result for $z = 1$, because in that case the independent pair approximation becomes irrelevant and the general formula for distributing $1/2(N_1 + N_2)$ pairs over N_{11}, N_{22}, and N_{12} is exactly given by (4.43). In general, it is an approximation.

For the contribution to the excess entropy resulting from attractive intermolecular interactions we then get from (4.42), using the symmetry relations (4.33) between x_{12} and x_{21} [8],

$$S_{att}^E = \Delta S^M - \Delta S^{M,is} = S^C - S^{C,is}$$
$$= \frac{z}{2}kN\left(x_{11}x_1 \ln \frac{x_1}{x_{11}} + x_{21}x_1 \ln \frac{x_2}{x_{21}} + x_{12}x_2 \ln \frac{x_1}{x_{12}} + x_{22}x_2 \ln \frac{x_2}{x_{22}}\right). \tag{4.49}$$

For random mixing, when $x_{11} = x_{12} = x_1$ and $x_{21} = x_{22} = x_2$, we recover the ideal mixing entropy, i.e., $S_{att}^E = 0$. Generalization to multicomponent mixtures is straightforward. For the excess entropy we get

$$S_{att}^E = -\frac{z}{2}kN \sum_i \sum_j x_i x_{ji} \ln \frac{x_{ji}}{x_i}. \tag{4.50}$$

Combining the equation (4.49) for the excess entropy with (4.40) for the excess internal energy of a binary mixture gives the Helmholtz excess free energy of a binary mixture due to attractive interactions as

$$A_{att}^{E} = \frac{1}{4} Nz \left[x_1 x_{11} \left(\omega_{11} + 2kT \ln \frac{x_{11}}{x_1} \right) + x_1 x_{21} \left(\omega_{21} + 2kT \ln \frac{x_{21}}{x_1} \right) \right.$$
$$\left. + x_2 x_{12} \left(\omega_{12} + 2kT \ln \frac{x_{12}}{x_2} \right) + x_2 x_{22} \left(\omega_{22} + 2kT \ln \frac{x_{22}}{x_2} \right) \right]. \quad (4.51)$$

Generalization to multicomponent mixtures leads to

$$A_{att}^{E} = \frac{1}{4} \sum_i \sum_j N_{ji} \left(\omega_{ji} + 2kT \ln \frac{x_{ji}}{x_i} \right)$$
$$= \frac{1}{4} zN \sum_i \sum_j x_i x_{ji} \left(\omega_{ji} + 2kT \ln \frac{x_{ji}}{x_i} \right). \quad (4.52)$$

Due to the symmetry relations (4.33), the mole fraction in the logarithmic term may also be changed to x_j in all of the above equations. We note that the excess free energy is formulated in terms of the local compositions as anticipated in Section 4.1.2.

4.3.3 Local Compositions

The mean local compositions x_{ij} or, alternatively, the contact numbers N_{ij} are not yet known. They represent the particular molecular distribution the system will most likely assume. This distribution will determine the maximum term in the sum of (4.34) and is thus found, for a binary system, by maximizing

$$g(N_1, N_2, N_{12}) e^{-U(N_1, N_2, N_{12})/kT},$$

i.e., working out (cf. App. 3 for the justification for going over to logarithms)

$$\frac{\partial}{\partial N_{12}} [\ln g(N_1, N_2, N_{12}) - U(N_1, N_2, N_{12})/kT] = 0.$$

We note that in these equations N_{12} is now a variable and not the particular value associated with the maximum number of configurations. Using the expression (4.37) for the lattice energy, the maximum condition gives

$$\frac{\partial \ln g}{\partial N_{12}} = \omega_{12}/2kT;$$

i.e., with (A 8.1) and

$$\partial \ln g / \partial N_{12} = \partial \ln h / \partial N_{12},$$

we now find that value of N_{12} with the maximum number of configurations from

$$\frac{N_{12}^2}{N_{11} N_{22}} = \frac{x_{12} x_{21}}{x_{11} x_{22}} = \tau^2, \quad (4.53)$$

where

$$\tau^2 = e^{-\omega_{12}/kT}. \tag{4.54}$$

The same result would be obtained by minimizing A_{att}^E as obtained from (4.51) for a binary mixture with respect to x_{12}. As expected, the dominant molecular distribution depends on the intermolecular interactions. Due to the formal analogy of (4.53) with the equilibrium condition of a chemical reaction

$$11 + 22 \rightleftharpoons 2(12),$$

this result, together with the expression for the excess free energy A_{att}^E, is referred to as the quasichemical approximation (QUAC) [7]. For $\omega_{12} = 0$ we have complete random mixing. In this case the local compositions become equal to the bulk compositions and we return to an ideal solution.

For multicomponent mixtures the mean local compositions have to be calculated numerically from the relations

$$\frac{x_{ji}x_{ij}}{x_{ii}x_{jj}} = e^{-\omega_{ji}/kT} = \tau_{ji}^2 \tag{4.55}$$

along with the closure condition

$$\sum_j x_{ji} = 1 \tag{4.56}$$

and the symmetry conditions

$$x_i x_{ji} = x_j x_{ij}. \tag{4.57}$$

Equation (4.52) for the excess free energy of a system of equal-sized molecules, based on a clear statistical mechanical derivation, can alternatively be written in a simpler form in terms of activity coefficients. For this purpose we realize that the local compositions can be expressed as [9,10]

$$x_{ij} = \tau_{ij} x_i \gamma_i \gamma_j, \tag{4.58}$$

where τ_{ij} is defined as in (4.55) and γ_i, γ_j are the activity coefficients of the components i and j, respectively; cf. Section 2.1. Substituting (4.58) into the quasichemical equations (4.55) to (4.57) shows that they are satisfied, with the closure condition $\sum_j x_{ji} = 1$ leading to

$$1/\gamma_i = \sum_j \tau_{ji} x_j \gamma_j. \tag{4.59}$$

If we now introduce (4.58) into (4.52) for the excess free energy in the quasi-chemical approximation we find, in molar units, after again using the symmetry relation (4.57) and the closure condition, cf. (4.56),

$$a^E = \frac{1}{4} z \sum_i \sum_j x_i x_{ji} \left(\omega_{ji} + 2RT \ln(\tau_{ji} \gamma_i \gamma_j) \right)$$

$$= \frac{1}{2} z RT \sum_i \sum_j x_i x_{ji} \ln(\gamma_i \gamma_j)$$

$$= \frac{1}{2} z RT \left[\sum_i \sum_j x_i x_{ji} \ln \gamma_i + \sum_i \sum_j x_j x_{ij} \ln \gamma_j \right]$$

$$= z RT \sum x_i \ln \gamma_i. \tag{4.60}$$

Because z is the number of contacts of one molecule with its neighbors in a lattice, we find from an appeal to classical thermodynamics that (4.59) is equivalent to (4.52) along with (4.55). The QUAC equations to be solved scale with m^2 for m components, whereas (4.59) scales with m. We thus find that (4.59) is simpler than QUAC and thus an attractive formulation of the quasichemical approximation.

The quasichemical approximation is not an exact solution to the problem of mixing in a lattice because the independent pair approximation is clearly incorrect there. It is not clear, however, to what extent this theory is adequate as a model for the excess functions. Computer simulations on a lattice demonstrate that the quasichemical approximation performs quite well for the excess free energy, although there is some deterioration for the excess energy and more so for the excess entropy; cf. Figure 4.16. Here, we show two sets of comparisons

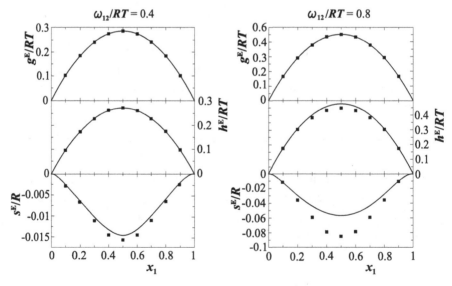

Fig. 4.16. Excess functions: comparison between QUAC (—) and computer simulations (■■■) [11].

between the results obtained from (4.51), along with (4.53) and (4.54), and Monte Carlo simulations on a lattice [11]. We remember that $a^E = g^E$ and $u^E = h^E$ in a lattice. It must be noted that whereas the system at $\omega_{12}/RT = 0.4$ is at a typical, moderately nonideal dense fluid state, that at $\omega_{12}/RT = 0.8$ is close to the liquid demixing condition. It is clear that the simple model of the quasichemical approximation is less accurate close to the liquid stability condition, where complicated cooperative phenomena become noticeable, than far away from it. Still, the performance is surprisingly good for g^E and also h^E, whereas significant errors manifest themselves in s^E. This is not surprising because the approximations have gone primarily into the derivation of a model for s^E. The local compositions, too, have been calculated from the simulations and the comparison is shown in Figure 4.17 in terms of the quantity $-\ln[(x_{12}x_{21})/(x_{11}x_{22})] = \omega_{12}/RT$. Although the quasichemical approximation leads to a constant value for this expression, it turns out to be composition-dependent in the simulations [11]. Again the deviations increase with increasing value of ω_{12}/RT. The criticism has frequently been advanced that the quasichemical approximation in the form of (4.50) can only produce negative excess entropies, independent of the value of ω_{12}/RT, whereas in real liquid mixtures positive excess entropies are frequently found. In the context of the limitation to spherical molecules of equal sizes this is not really a concern, because such simple systems indeed have negative excess entropies, as also shown in the simulations; this holds irrespective of the sign of ω_{12}/RT. More complex molecules need corrections to the simple model of this section that will allow for negative and positive excess entropies, depending on the molecules considered. Such corrections have frequently been introduced by making the exchange energies temperature-dependent or applying other essentially empirical ingredients. A more straightforward extension to real molecules will be given in the next section. All in all, we conclude that the quasichemical approximation is a sensible basis for designing excess function models. We can further expect that the error of the approximation is even less in a real liquid, which does not have the rigid structure of a lattice.

It can be shown that the results of the quasichemical approximation can be derived without the constraint of a lattice if the independent pair approximation is introduced and the molecules in the liquid are assumed to be space-filling, i.e., subject to contacts with surfaces of neighboring molecules in the picture of Figure 4.4; cf. App. 9. This also holds for the more complicated molecules to be treated below under the free segment approximation. The excess free energy models derived in this chapter thus are not restricted to being lattice models. They are all based on the assumption of independently interacting molecules or molecular segments, and the notions of quasichemical and free segment approximation will be used interchangeably.

The excess free energy model in the quasichemical approximation (4.52) is formulated in terms of local compositions and thus belongs to the large class of local composition models. Local composition models for the excess free

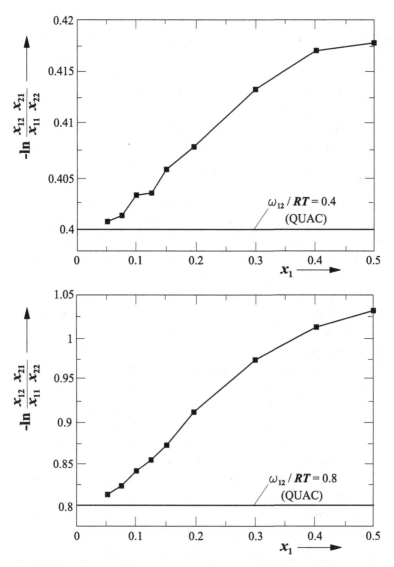

Fig. 4.17. Local compositions: comparison between QUAC (ω_{12}/RT = const.) and computer simulation (■■■) [11].

energy have become very popular in chemical engineering phase equilibrium calculations and many semiempirical versions have been formulated. One of the better known models of this type is the Wilson equation [12]. For a binary mixture it reads

$$\frac{g^{\mathrm{E}}}{RT} = -x_1 \ln(x_1 + G_{21}x_2) - x_2 \ln(G_{12}x_1 + x_2), \qquad (4.61)$$

where G_{21}, G_{12} are the nonrandomness factors defined in Section 4.1.2. The model adopted for the nonrandomness factors by Wilson is

$$G_{21} = \frac{v_2}{v_1} e^{-(\lambda_{12}-\lambda_{11})/kT} \qquad (4.62)$$

and

$$G_{12} = \frac{v_1}{v_2} e^{-(\lambda_{12} - \lambda_{22})/kT}, \tag{4.63}$$

where v_1, v_2 are the molar volumes of pure liquids 1 and 2, respectively, and λ_{12}, λ_{11}, and λ_{22} attractive interaction energy parameters of the 11, 12, and 22 contacts. To show the relation of the Wilson equation to the quasichemical approximation we first note that at constant pressure p we have in general, not restricted to a regular lattice,

$$A^E[T, V(T, p, \{N_j\}), \{N_j\}] \cong G^E(T, p, \{N_j\}), \tag{4.64}$$

because the excess volume in a liquid mixture tends to be very small [13]. Then we transform (4.51) to give

$$
\begin{aligned}
\left(\frac{g^E}{RT}\right)_{\text{att}} &= \frac{1}{4}z\left[x_1 x_{11} 2 \ln \frac{x_{11}}{x_1} + x_1 x_{21}\left(\frac{\omega_{21}}{kT} + 2 \ln \frac{x_{21}}{x_2}\right)\right. \\
&\quad \left. + x_2 x_{12}\left(\frac{\omega_{12}}{kT} + 2 \ln \frac{x_{12}}{x_1}\right) + x_2 x_{22} 2 \ln \frac{x_{22}}{x_2}\right] \\
&= -\frac{1}{2}z\left[x_1 \ln(x_1 + G_{21}x_2) + x_2 \ln(G_{12}x_1 + x_2)\right] \\
&\quad + \frac{1}{4}z\left[x_1 x_{21}\left(\frac{\omega_{21}}{kT} + 2 \ln G_{21}\right) + x_2 x_{12}\left(\frac{\omega_{12}}{kT} + 2 \ln G_{12}\right)\right]. \tag{4.65}
\end{aligned}
$$

Here we use (4.10) and (4.15) as well as (4.9) and (4.16) for the relations between bulk and local compositions of the molecules and the nonrandomness factors. We realize that the first bracket is formally identical to the Wilson equation. The second bracket is zero because we have the symmetry condition $x_1 x_{21} = x_2 x_{12}$ and the quasichemical solution for the local compositions gives, cf. (4.18) and (4.53),

$$\frac{x_{21} x_{12}}{x_{11} x_{22}} = G_{21} G_{12} = e^{-\omega_{21}/kT}$$

with $\omega_{12} = \omega_{21}$. So the formal expression of the Wilson equation is identical to the quasichemical approximation for the excess free energy. The nonrandomness factors G_{12}, G_{21}, however, do not appear in (4.51). Instead, local compositions enter in a more complicated way. We further note that in the Wilson equation size effects, which have not been considered in this section, are taken into account in an ad hoc manner through the ratios of molar volumes in the nonrandomness factors. The major difference of the Wilson equation from the quasichemical approximation is the simple expressions adopted for the non-randomness factors in the former. Due to the separate appearance of G_{21} and G_{12}, each with different energy parameters, $(\lambda_{12} - \lambda_{11})$ and $(\lambda_{12} - \lambda_{22})$, respectively, we note that there are two energy parameters in the Wilson equation for a binary mixture as compared to only one in the quasichemical approximation. Also, the symmetry condition $x_1 x_{21} = x_2 x_{12}$ is not obeyed in the Wilson equation. Although not adequate in the general case, the Wilson nonrandomness factors agree with the exact low-density result for equal molar volumes of the components; cf. Section 4.1.2.

4.4 Complex Model Molecules

The excess free energy model (4.52), although rigorous under the assumption of independent pairs and supported by computer simulations is not yet adequate to describe the mixing behavior of real liquids. It is restricted to a mixture of spherical molecules of equal sizes and uniform surface interaction properties. Thus, there are no differences in sizes and shapes of the molecules and also there is no distribution of nonuniform interaction sites over a molecule's surface. Real liquid mixtures owe their thermodynamic behavior to a more complicated potential energy arising from interaction properties that are generally distributed nonuniformly over a molecule's surface and also to size and shape effects. In this section we generalize (4.52) to include size and shape as well as surface effects.

An essential step forward to practically useful models for the excess functions of real liquid mixtures is to consider a molecule as possessing an arbitrary shape and size to which a particular interaction active surface belongs. We choose to base the following derivations again on the lattice model, although the same results are obtained from the more general polyhedral surface contact model; cf. App. 9. Consistent with the lattice model, the molecules are visualized as structures of spherical volume segments of equal volumes. Each lattice site can then be considered to be occupied with one volume segment. The number of volume elements of a molecule defines its total volume. Such a molecule in a particular shape is shown in Fig. 4.18 and it is characterized by the number of volume segments r_i or the nondimensional volume r_i defined as

$$r_i = V_i / V_{\text{ref}}, \tag{4.66}$$

and the number of external, i.e., nonbonded surface contacts, defined as

$$N_i = A_i / A_{\text{ref}}, \tag{4.67}$$

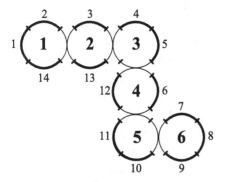

Fig. 4.18. Model of a structured molecule with a uniform external surface.

where V_i, A_i are the volume and the external surface area of molecule i and V_{ref}, A_{ref} are some suitable reference values. In the simple case considered in Figure 4.18, V_{ref} equals the volume of one volume segment and A_{ref} is the surface of one external contact. The number of external surface contacts via which interactions with other molecules can take place in a lattice can also be expressed in terms of a nondimensional external surface, by

$$q_i = N_i/z. \tag{4.68}$$

For the molecule shown in Figure 4.18, we have $r_i = 6$ and $N_i = zq_i = 14$. For a two-dimensional lattice such as the one in Figure 4.13, $z = 4$ and thus $q_i = 3.5$. The surface areas associated with the contacts are considered to be equal in Figure 4.18. For molecules without ring formation we find that the values of r and q are related by $qz = rz - 2(r-1)$, which defines a maximum value for q.

Generally, the interaction properties of a real molecule will not be distributed uniformly over its surface. Figure 4.19 shows a generalization of the simple model of Figure 4.18 to incorporate this nonuniformity in a simple manner. The molecule here consists of six equal-sized volume segments with two types of surface segments, μ and ν. In the case shown in Figure 4.19 the number of μ-surface segments is 6 and that of ν-surface segments 8. The total number of external surface segments, is $N_i = zq_i = 14$ as before in Figure 4.18. In what follows we refer to the volume segments as groups. So in Figure 4.19 we have two groups of type μ and four groups of type ν. The sizes of the external group surfaces may in general vary from group to group. The type of surface, i.e., the interaction property of the surface segments, is assumed to be unique for a particular group. A group α then is characterized by a surface α, a different group β by a surface β, different in size and interaction property from α, and so on.

For a real molecule the geometrical structure is not generally given by the simple picture of Figures 4.18 or 4.19 associated with the lattice model. In the gas phase it may be obtained from quantum-chemical computer codes; cf. Section 2.5. Modifications in a liquid environment can be taken into account by a continuum solvation model; cf. Section 2.4. The volume and the external surface of a molecule may thus be calculated if some reasonable cutoff of

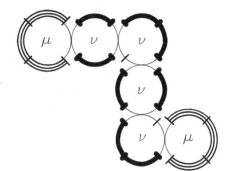

Fig. 4.19. Model of a structured molecule with a nonuniform external surface.

the electron density is assumed. Alternatively, tabulated values for r_i and q_i based on experimental data are available [14]. In these tables, so-called van der Waals volumes $V_{VDW,\alpha}$ and surfaces $A_{VDW,\alpha}$ are given for various functional groups α. A molecule i is synthesized from the groups it contains and we have $r_i = V_{VDW,i}/V_{ref}$, $q_i = A_{VDW,i}/A_{ref}$. Here V_{ref} and A_{ref} are usually and arbitrarily calculated from an effective sphere representing the CH_2 monomer, with the results $V_{ref} = 15.17$ cm^3/mol and $A_{ref} = 2.50344 \cdot 10^9$ cm^2/mol, and $z = 10$. So, generally, r_i is just a nondimensional volume parameter, not necessarily the even number of volume elements or lattice sites. Accordingly, q_i is a nondimensional surface parameter and the functional groups generally have different volumes. Usually r_i will be a noninteger number and q_i will be smaller than the maximum value. We assume in the following that the volume and the external surface of a molecule are known. The results from quantum-chemical calculations and from the tables differ, but not strongly so. So, although we shall base the statistical mechanical derivations for the excess functions on the lattice model, we take the freedom to extract the geometry of a molecule more generally from experimental evidence or quantum-chemical calculations.

4.4.1 Size and Shape Effects

As a first step to formulating models for excess functions of complex molecules, we realize that we now have a hard core repulsive contribution to the excess free energy. So, for a binary mixture of spheres of unequal sizes, there are different distinguishable ways of arranging the molecules on the lattice for the mixture and for the pure components. This will result in a contribution to the entropy of mixing, not present in mixtures of equal-sized spheres. Similar but more complicated effects appear for the complex model molecules of Figure 4.18. The distinguishable arrangements then depend on the internal connectivities of the molecular volume segments, and so this contribution will also contain the external molecular surface parameter q_i, in addition to the volume parameter r_i. We refer to these phenomena as size and shape effects. They are to be evaluated without taking notice of the attractive intermolecular forces, because the effect of those is taken into account in the attractive part. The arrangements to be considered are thus purely random and appear as an additional degeneracy in the partition function (4.35). We refer to the repulsive contribution in this context as the combinatorial excess entropy, because it is due to combinatorial effects related to the positioning of the molecular segments in the lattice. We note that any density dependence generally associated with the hard core repulsive contribution, cf. Section 5.5, is eliminated in the lattice framework. The contribution to the excess entropy will therefore be independent of the molecular state, except for the composition. We find that

$$A_{comb}^E = -T S_{comb}^E, \tag{4.69}$$

with

$$S_{\text{comb}}^{\text{E}} = \Delta S_{\text{comb}}^{\text{M}} - \Delta S^{\text{M,is}}, \tag{4.70}$$

where $\Delta S^{\text{M, is}}$ is the ideal entropy of mixing, i.e., of the ideal solution.

To derive an expression for $\Delta S_{\text{comb}}^{\text{M}}$ we consider the mixing process of N_1 identical molecules of type 1 with N_2 identical molecules of type 2, both with an arbitrary number of volume segments r_1 and r_2, respectively. The model of a structured molecule is thus that of Figure 4.18 and the liquid model that of a regular lattice. Due to the absence of energetic effects, the mixing process is random, and thus the associated combinatorial entropy of mixing is given by the formula derived for the microcanonical ensemble, i.e., valid for fixed (zero) energy; cf. Exercise 2.1,

$$\Delta S_{\text{comb}}^{\text{M}} = k \ln \frac{\Omega(N_1, N_2)}{\Omega(N_1, 0)\Omega(0, N_2)}, \tag{4.71}$$

where $\Omega(N_1, N_2)$ is the number of distinguishable arrangements of N_1, N_2 molecules on the total number of the lattice sites, whereas $\Omega(N_1, 0)$ is that for the N_1 molecules on $N_1 r_1$ sites and $\Omega(0, N_2)$ that for the N_2 molecules on $N_2 r_2$ sites. The total number of sites is $Nr = N_1 r_1 + N_2 r_2$ with $r = x_1 r_1 + x_2 r_2$ and d the diameter of a molecular segment. Because each lattice site is assumed to be occupied by either a segment of molecule 1 or a segment of molecule 2, the mixing process is considered at constant packing fraction, a quantity defined as $v/(\frac{1}{6}\pi r d^3)$, where v is the volume per molecule of the lattice and $(1/6)\pi r d^3$ the hard core volume per molecule.

There is no exact expression available for the combinatorial entropy of mixing for molecules of different sizes and shapes in a lattice. Here, we state two frequently used approximate results. The details of the derivation are discussed in App. 10. The first is the Flory–Huggins equation [15,16], cf. (A 10.2) and (A 10.3),

$$\frac{S_{\text{comb}}^{\text{E}}}{R} = -\sum_i N_i \ln \frac{\phi_i}{x_i}, \tag{4.72}$$

where

$$\phi_i = \frac{x_i r_i}{\sum_k x_k r_k} = \frac{x_i V_i}{\sum_k x_k V_k} \tag{4.73}$$

is the volume fraction and N_i here is the number of moles of component i. The Flory–Huggins model takes size effects into account, but it does not contain any parameter related to molecular shape. In particular, it does not take the internal connectivities of the segments of a molecule into account. An improved expression, due to Guggenheim and Stavermann [7,17], differentiating explicitly between external surfaces q_i and internal contacts of the molecules, is, cf. (A 10.20) and (A 10.22),

$$\frac{S_{\text{comb}}^{\text{E}}}{R} = -\sum_i \left[N_i \ln \frac{\phi_i}{x_i} + \frac{z}{2}\left(N_i q_i \ln \frac{\theta_i}{\phi_i} \right) \right], \tag{4.74}$$

with θ_i as the surface fraction of a molecule of type i, defined as

$$\theta_i = \frac{x_i q_i}{\sum_k x_k q_k} = \frac{x_i A_i}{\sum_k x_k A_k}. \tag{4.75}$$

Evidently, the Guggenheim–Stavermann equation, which can differentiate between ring shapes and no-ring shapes by the relation of r_i to q_i, is an extension of the Flory–Huggins expression taking the difference between external and internal surface segments into account. When this distinction is omitted, $q_i = r_i$ and (4.74) reduces to (4.72). This also holds for the ratio q_i/r_i, as a rough measure of the shape of a molecule, being independent of its identity. To apply these expressions for the combinatorial excess entropy we need values for the volume and the surface of a molecule. When these parameters are identical for all molecules in a mixture the combinatorial excess entropy is zero. Size and shape effects thus become more important as the geometrical differences between the molecules increase. We note that the nondimensional surface q_i appears in the combination zq_i, which is the number of external contacts associated with a molecule of component i.

EXERCISE 4.2

Calculate the combinatorial part of the excess free energy for the system 2-butanone(1)–hexane(2) at $x = 0.5$ and $t = 60°$C. The size and surface parameters of the molecules as found from tables are [1]

$$r_1 = 3.2479, \quad q_1 = 2.8760$$

$$r_2 = 4.4998, \quad q_2 = 3.8560.$$

The coordination number is $z = 10$.

Solution

The Helmholtz excess free energy, which in a lattice is identical to the Gibbs excess free energy, is given, according to the Guggenheim–Stavermann equation (4.74), by

$$g_{\text{comb}}^{\text{E}} = -Ts_{\text{comb}}^{\text{E}} = RT \left[x_1 \ln \frac{\phi_1}{x_1} + x_2 \ln \frac{\phi_2}{x_2} + \frac{z}{2} \left(x_1 q_1 \ln \frac{\theta_1}{\phi_1} + x_2 q_2 \ln \frac{\theta_2}{\phi_2} \right) \right].$$

We have

$$\phi_1 = \frac{x_1 r_1}{x_1 r_1 + x_2 r_2} = \frac{0.5 \times 3.2479}{0.5 \times 3.2478 + 0.5 \times 4.4998} = 0.4192$$

$$\theta_1 = \frac{x_1 q_1}{x_1 q_1 + x_2 q_2} = \frac{0.5 \times 2.8760}{0.5 \times 2.8760 + 0.5 \times 3.8560} = 0.4272$$

and also $\phi_2 = 0.5808$ and $\theta_2 = 0.5728$. Evaluating the Guggenheim–Stavermann equation leads to

$$\begin{aligned}
g_{\text{comb}}^{\text{E}} &= 2769 \, [0.5 \ln 0.8384 + 0.5 \ln 1.1616 + 5 \\
&\quad \times (0.5 \times 2.8760 \ln 1.0191 + 0.5 \times 3.8560 \ln 0.9862)] \\
&= 2769 \, [-0.0881 + 0.0749 + 5(0.0272 - 0.0268)] \\
&= 2769 \, [-0.0132 + 0.002] = -31.0 \text{ kJ/kmol}.
\end{aligned}$$

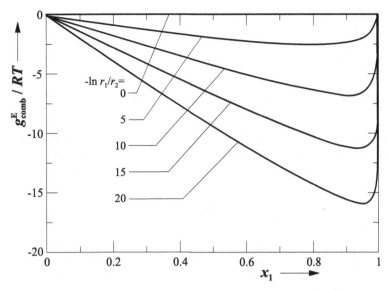

Fig. 4.20. Excess Flory–Huggins free energy for various size ratios.

The excess entropy s_{comb}^E is positive; i.e., it counteracts the negative excess entropy of the simple QUAC, i.e., (4.52). We further find that the Guggenheim–Stavermann modification reduces the absolute value of the excess entropy according to the Flory–Huggins formula. Quantitatively we find that the Flory–Huggins excess free energy amounts to $g^E = -36.6$ kJ/kmol, so that the differentiation between internal and external surfaces for this system introduces a correction of about 20%.

[1] A. Bondi. *Physical Properties of Molecular Crystals, Liquids and Glasses.* Wiley, New York, 1968.

As an illustration of size effects, we show in Figure 4.20 the effect of increasing size ratios of the molecules on the excess free energy of the mixture as calculated from the Flory–Huggins model. Clearly, the effect becomes quite significant at larger size ratios and may even dominate the excess free energy for polymers.

EXERCISE 4.3

The excess entropy in the Flory–Huggins lattice approximation is determined by the sizes of the molecules without taking internal bonds between the volume segments into account. Show that it is consistent with the excess entropy following from the repulsive part of the van der Waals equation of state.

Solution

The repulsive part of the famous van der Waals equation of state reads

$$p_{VDW}^{rep} = \frac{RT}{v - b},$$

where p, T, v are the pressure, the temperature, and the molar volume, respectively, and b is a size parameter specific to each molecule. We note that p_{VDW}^{rep} is identical to the repulsive parts of the well-known equations of Redlich and Kwong and Peng and Robinson, which have become very popular in chemical engineering.

From the equations of classical thermodynamics, cf. Section 2.1, we find for the excess entropy as calculated from an arbitrary equation of state $p(T, v, \{x_i\})$

$$s^{\mathrm{E}} = \lim_{v' \to \infty} \left[\int_{v'}^{v} \left(\frac{\partial p}{\partial T} \right)_{v,\{x_i\}} \mathrm{d}v - \sum_i x_i \int_{v'}^{v_i} \left(\frac{\partial p_i}{\partial T} \right)_{v_i} \mathrm{d}v_i \right].$$

Here v' is a reference molar volume that is set to ∞ in the ideal gas limit in which ideal mixing occurs. State quantities without an index refer to the mixture and index i refers to pure component i. For the repulsive part of the van der Waals equation of state we find

$$s_{\mathrm{VDW}}^{\mathrm{E}} = R \ln(v - b) - R \sum_i x_i \ln(v_i - b_i),$$

where we set $b = \sum x_i b_i$, in agreement with common practice. We note that the molar volumes of the mixture and of the pure components are not yet specified. We now have to make an appropriate choice ensuring consistency between the two models to be compared. Excess functions are always considered at constant temperature, but the second variable to be held constant can vary. Frequently, it is the pressure or the density; cf. (4.1). In the lattice, the derivation of the Flory–Huggins model shows that it is the packing fraction $v_i/(\frac{1}{6}\pi r_i d^3) = v/(\frac{1}{6}\pi r d^3)$ that is the second constant state quantity. In the van der Waals equation $\frac{1}{6}\pi r_i d^3 \sim b_i$, $\frac{1}{6}\pi r d^3 \sim b$ and so we postulate for a comparison of the Flory–Huggins model with the van der Waals equation of state

$$\frac{b}{v} = \frac{b_i}{v_i}.$$

This gives the excess entropy of the repulsive van der Waals term as

$$s_{\mathrm{VDW}}^{\mathrm{E}} = R \left\{ \ln \left[b \left(\frac{v}{b} - 1 \right) \right] - \sum_i x_i \ln \left[b_i \left(\frac{v_i}{b_i} - 1 \right) \right] \right\}$$

$$= -R \sum_i x_i \ln \frac{b_i}{b} = -R \sum_i x_i \ln \frac{r_i}{r}.$$

We see that the van der Waals equation of state and the Flory–Huggins model give identical results for the excess entropy due to size effects at constant packing fraction. The limitations of the van der Waals repulsive term will be discussed in the context of equation of state models; cf. Exercise 5.8. This equivalence to the Flory–Huggins model emphasizes the crude approximations in that model for taking into account size effects. We reiterate that a comparison with the Flory–Huggins model requires $s_{\mathrm{VDW}}^{\mathrm{E}}$ to be evaluated at constant packing fraction. In contrast, the repulsive van der Waals equation of state produces a density-dependent excess entropy at constant volume and a value of zero at constant pressure.

4.4.2 Surface Effects

As a second step in applying the excess free energy equation (4.52) to molecules having individual external surface areas over which the intermolecular interactions occur, we note that the interactions between the molecules are now proportional to their external surfaces instead to just the number of the molecules. We thus replace the mole fractions x_i by surface fractions θ_i, cf. (4.75), and the local mole fractions x_{ij} by local surface fractions θ_{ij} associated with the molecules i and j, such as, cf. (4.12),

$$\theta_{ij} = \frac{\theta_i}{\theta_i + \theta_j G_{ij}}, \tag{4.76}$$

where G_{ij} is a nonrandomness parameter. The conservation equations then read

$$\sum_j \theta_{ji} = 1 \quad \text{for} \quad i = 1, 2, \ldots, \tag{4.77}$$

and the excess free energy (4.52) due to attractive intermolecular contact interactions transforms via an analogous derivation into

$$
\begin{aligned}
A_{\text{att}}^{\text{E}} &= \frac{1}{4} z N q \sum_i \sum_j \theta_i \theta_{ji} \left(\omega_{ji} + 2kT \ln \frac{\theta_{ji}}{\theta_i} \right) \\
&= \frac{1}{4} z N \sum_i \sum_j x_i q_i \theta_{ji} \left(\omega_{ji} + 2kT \ln \frac{\theta_{ji}}{\theta_i} \right).
\end{aligned}
\tag{4.78}
$$

Here

$$q = \sum_k x_k q_k, \tag{4.79}$$

and $q = 1$ in the case of (4.52). The assumption of independent pair interactions of molecules now has been extended to molecular volume segments with their associated surfaces; i.e., the free segment approximation has been used. The attractive part of the excess free energy depends on the external surfaces of the molecules and the exchange energy ω_{ji} now refers to the interactions between surfaces of type i and j. Here, i.e., in the case of uniform surface properties of the molecular segments, cf. Figure 4.18, the interactions between surfaces reduce to interactions between molecules of type i and j.

We note that (4.78) is independent of the parameter r_i, i.e., of molecular volume. Together with (4.74), which introduces size and shape effects, however, it constitutes a complete model for the excess free energy of a fluid composed of complicated, i.e., structured molecules with uniform surface properties.

Various semiempirical excess free energy models of the local composition type applying to structured molecules with uniform surface properties have been proposed in the literature. One, which has become rather popular, is UNIQUAC (universal quasichemical) [18]. It attributes a size and a shape to a molecular component i based on a nondimensional volume r_i and a nondimensional external surface q_i. Size and shape effects are taken into account by the Guggenheim–Stavermann combinatorial contribution to the excess free enthalpy, which reads for a binary system, cf. (4.74),

$$\left(\frac{g^{\text{E}}}{RT} \right)_{\text{comb}} = x_1 \ln \frac{\phi_1}{x_1} + x_2 \ln \frac{\phi_2}{x_2} + \frac{1}{2} z \left[q_1 x_1 \ln \frac{\theta_1}{\phi_1} + q_2 x_2 \ln \frac{\theta_2}{\phi_2} \right], \tag{4.80}$$

where ϕ_i, θ_i are the volume and surface fractions of the molecules as defined before. The intermolecular attractive part is given for a binary system as

$$\left(\frac{g^{E}}{RT}\right)_{att} = -x_1 q_1 \ln(\theta_1 + G_{21}\theta_2) - x_2 q_2 \ln(G_{12}\theta_1 + \theta_2) \qquad (4.81)$$

with the nonrandom factors defined by

$$G_{21} = e^{-\Delta U_{21}/kT} \qquad (4.82)$$

and

$$G_{12} = e^{-\Delta U_{12}/kT}, \qquad (4.83)$$

where ΔU_{12}, ΔU_{21} are two attractive energy parameters. By the same arguments as used before for the Wilson equation, cf. Section 4.3, it can be seen that UNIQUAC is formally identical to the quasichemical approximation. However, its nonrandomness factors are not consistent with (4.78) and also the symmetry relations for the surface contacts are violated.

The interaction properties associated with the surface of a real molecule are not uniform. Any model making this assumption thus will have to absorb the associated inaccuracies into its adjustable parameters. This leads to the result that the parameters adjusted to data will not have much physical significance and so the extrapolation capacity of the model is limited. This is typically confirmed for the models discussed so far. Although a set of parameters adjusted to vapor liquid equilibrium data is quite able to correlate these and is adequate for interpolation purposes, any type of extrapolation to situations not covered by the data is subject to uncertainty. Usually, one finds that parameters obtained from vapor–liquid equilibrium data are unable to provide reliable extrapolations to the excess enthalpy and to liquid–liquid equilibria. For more fundamental modeling, in particular for predictive purposes, it is indispensable to take nonuniform surface effects into account. They will only affect the attractive part, whereas the combinatorial part can be taken from the preceding section without change.

It is straightforward to extend the attractive part of the excess free energy model (4.78) to molecules with nonuniform surface effects, as visualized in Figure 4.19. We again make use of the free segment approximation. Thus we assume that the attractive potential energy of the system derives from pair contacts between independent groups via their external surface segments and that the pair formation of these surface contacts is independent of the positions of the surface segments within the molecules. Then the attractive part of the excess free energy, i.e., the free energy of the mixture of groups minus the free energies of the pure group systems, is directly obtained from (4.52) and (4.78), respectively, as

$$
\begin{aligned}
(A^E)_{att}^{gr} &= \frac{1}{4} \sum_\alpha \sum_\beta N_{\beta\alpha} \left(\omega_{\beta\alpha} + 2kT \ln \frac{\theta_{\beta\alpha}}{\theta_\beta} \right) \\
&= \frac{1}{4} z n_{gr} \sum_\alpha \sum_\beta q_\alpha x_\alpha \theta_{\beta\alpha} \left(\omega_{\beta\alpha} + 2kT \ln \frac{\theta_{\beta\alpha}}{\theta_\beta} \right).
\end{aligned}
\tag{4.84}
$$

Here n_{gr} is the total number of groups, and $x_\alpha = n_\alpha/n_{gr}$ is the mole fraction of group α in the mixture of groups with n_α as the number of α-groups in the mixture. The local surface fraction associated with contacts between groups α and β is $\theta_{\beta\alpha} = N_{\beta\alpha}/n_\alpha q_\alpha z$, where $N_{\beta\alpha}$ is the number of contacts between surface of type β and surface of type α and (zq_α) is the number of external contacts , i.e., of external surface segments, that one single group α can have. Thus, $n_\alpha q_\alpha z$ is the total number of external contacts involving α groups in the mixture and, accordingly, $\theta_{\beta\alpha}$ is the ratio of $\beta\alpha$-surface contacts to the total number of contacts involving α. Finally, $\omega_{\beta\alpha}$ is the exchange energy related to the interactions between surface types α and β, i.e., groups α and β. As in the earlier equations, it may basically be temperature-dependent. To transform the group excess free energy in terms of the number of groups (4.84) to an excess free energy in term of the number of molecules, we use the fact that the total external surface area q_{tot} of the molecular system in nondimensional form is given either as the sum of all group contributions or as the sum of all molecule contributions, as

$$
q_{tot} = \sum_\alpha n_\alpha q_\alpha = \sum_i N_i q_i.
\tag{4.85}
$$

This can be written as

$$
\frac{n_{gr}}{N} = \frac{\sum_i x_i q_i}{\sum_\alpha x_\alpha q_\alpha}
$$

and leads to the molar excess free energy of a system of groups in a form that is now entirely analogous to the first row in (4.78),

$$
(A^E)_{att}^{gr} = \frac{1}{4} z N q \sum_\alpha \sum_\beta \theta_\alpha \theta_{\beta\alpha} \left(\omega_{\beta\alpha} + 2kT \ln \frac{\theta_{\beta\alpha}}{\theta_\beta} \right).
\tag{4.86}
$$

Here N is the number of molecules in the mixture. Further,

$$
q = \sum_i x_i q_i,
\tag{4.87}
$$

$$
q_i = \sum_\alpha n_{\alpha,i} q_\alpha,
\tag{4.88}
$$

$$
\theta_\alpha = \frac{x_\alpha q_\alpha}{\sum_\beta x_\beta q_\beta},
\tag{4.89}
$$

and

$$x_\alpha = \frac{\sum_i x_i n_{\alpha,i}}{\sum_i x_i n_i}, \tag{4.90}$$

with $n_{\alpha,i}$ the number of groups α and n_i the total number of groups in a molecule of species i. Further, the quasichemical relations (4.55) to (4.58) hold for the local surface fractions $\theta_{\beta\alpha}$ instead of the local mole fractions x_{ij}. We finally have to transform the molar excess free energy in terms of groups to one in terms of molecules. For this purpose we need the difference between the free energies of the mixture and those of the pure molecular components. This is given by the difference of the excess free energy of the group system of the mixture and the sum of excess free energies of the group systems of the pure components, i.e.,

$$(A^E)_{att} = (A^E)^{gr}_{att} - \sum_i (A^E)^{gr, 0i}_{att}. \tag{4.91}$$

We note that the basic new content of (4.84) and (4.86) is the fact that the interactions are now determined by the number of surface contacts of the groups; i.e., $N_\alpha = q_\alpha z$. The numbers of surface contacts or, equivalently, the external surface can be quite different for different groups and these are the quantities to which the free segment approximation is applied. Equation (4.86) has been applied to some classes of substances by dividing the molecules into positively charged, negatively charged, and neutral groups under the acronym GEQUAC (group explicit quasichemical) [19]; cf. Exercise 4.4.

An equivalent, but simpler, formulation of the model in terms of activity coefficients is offered by (4.59), written for surface contacts between groups in the form, cf. App. 9,

$$1/\gamma_\alpha = \sum_\beta \theta_\beta \tau_{\alpha\beta} \gamma_\beta, \tag{4.92}$$

with

$$\tau^2_{\alpha\beta} = e^{-\omega_{\alpha\beta}/kT}.$$

The attractive activity coefficient of component i is then, analogously to (4.91),

$$\ln \gamma_{i,att} = \sum_\alpha N_{\alpha,i}(\ln \gamma_\alpha - \ln \gamma^{0i}_\alpha), \tag{4.93}$$

where $N_{\alpha,i}$ is the number of surface contacts involving group α in a molecule of component i and γ^{0i}_α is the activity coefficient of group α in pure component i. In agreement with (4.84), the number of surface contacts involving one single group α is $N_\alpha = zq_\alpha$, and so $N_{\alpha,i} = n_{\alpha,i} zq_\alpha$. The expressions (4.93) and (4.92) have been derived in the context of the COSMO-RS surface interaction potential model and are known under the acronyms COSMOSPACE [20] and COSMOSAC [21].

Although the above equations have been so far discussed with respect to groups, i.e., volume segments of the molecules interacting via their surfaces, they are equally applicable to surface segments without any attachments to groups. We then switch from the lattice model to the space-filling polyhedral surface model, and the surface parameter q_α, as well as the coordination number z, becomes meaningless. The surface fraction of a segment of type α in (4.92) is then given by

$$\theta_\alpha = \frac{\sum_i x_i A_{\alpha,i}}{\sum_i x_i A_i} = \frac{\sum_i x_i N_{\alpha,i}}{\sum_i x_i \sum_\alpha (N_{\alpha,i})}, \tag{4.94}$$

where $A_{\alpha,i}$ is the surface area of segment α in molecule i and A_i is the total area of molecule i. Further, $N_{\alpha,i} = A_{\alpha,i}/A_{\mathrm{eff}}$ is the number of surface contacts of surface α in a molecule of type i, with A_{eff} as the effective surface of a thermodynamically independent contact, cf. (4.67), and $\sum_\alpha (N_{\alpha,i})$ as the total number of contacts associated with molecule i.

We note that (4.86) and (4.92) give the attractive contribution to the excess free energy in a system of groups or surface segments, i.e., the contribution that depends on the intermolecular attractive potential energies between the groups or surface segments. In the free segment approximation the effect of the intermolecular attractions on the mixing process is assumed to be independent of the actual locations of the groups or external surface segments in the molecules. The groups or surface segments are thus assumed to be decoupled from the molecules and the effect of the interaction energies between them is modeled by interactions via independently formed pairs. As noted before, the decoupling assumption is equivalent to the assumption of independent pairs in the quasichemical approximation. So (4.86), as well as (4.92), is also a formulation of the quasichemical approximation.

The QUAC model derived here on the basis of the free segment approximation cannot be an accurate representation of the mixing effects in fluid systems. So it is important to investigate its performance under various fluid system conditions. To illustrate, we show in Figure 4.21 a comparison of computer simulation results in a three-dimensional cubic lattice with the prediction of the model for a mixture of a weakly interacting component, such as an alkane, with a strongly interacting component characterized by polarity and the capacity of hydrogen bonding, referred to as component 1 [11]. Both types of molecules are considered to be cubes of equal sizes, which are represented in two-dimensional form in Figure 4.21. Here z and $-z$ denote the front and the back side, y and $-y$ the top and the bottom side, and x and $-x$ the left and the right side of a cube. The weakly interacting sites are denoted with γ and δ, the strongly interacting sites with μ and ν. The interaction energies involving weakly interacting sites are normalized to zero, whereas those between strongly interacting sites are modeled by positive charges (site μ) and negative charges (site ν). The interaction energies chosen are $\varepsilon_{\mu\nu}/R = \varepsilon_{\nu\mu}/R = -1200$ K and $\varepsilon_{\mu\mu}/R = \varepsilon_{\nu\nu}/R = 1200$ K,

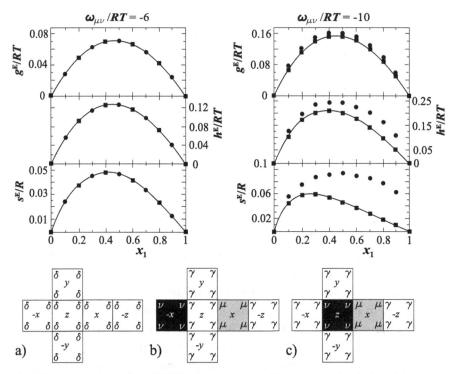

Fig. 4.21. Comparison of the GEQUAC model (—) with computer simulations; (a) weakly interacting molecule; (b) strongly interacting molecule, opposite •; (c) strongly interacting molecule, corner ■.

with all others being zero. The associated exchange energies as calculated from $\omega_{\alpha\beta}/4 = \varepsilon_{\alpha\beta} + \varepsilon_{\beta\alpha} - \varepsilon_{\alpha\alpha} - \varepsilon_{\beta\beta}$ are thus $\omega_{\mu\nu}/R = -19{,}200$ K and $\omega_{\mu\gamma}/R = \omega_{\mu\delta}/R = \omega_{\nu\gamma}/R = \omega_{\nu\delta}/R = -4800$ K, whereas all others are zero. Two significantly different arrangements of polarity on the strongly interacting component are considered, an opposite side arrangement and a corner side arrangement; cf. (b) and (c) in Figure 4.21. The opposite side arrangement will be capable of forming highly ordered molecular clusters, more so than the corner side arrangement. We therefore expect stronger deviations from the free segment approximation for the opposite side arrangement. The surface parameters of the four types of interaction sites are $q_\mu = q_\nu = 0.1666$, $q_\gamma = 0.6$, and $q_\delta = 1$. Looking at the comparison in detail we realize that at $\omega_{\mu\nu}/RT = -6$ the model reproduces the computer-simulated data essentially within their accuracy. So the difference between the two polarity arrangements is irrelevant at this interaction strength, in agreement with the free segment approximation. We note that the excess enthalpy of such a system at equimolar conditions and room temperature would be about 300 J/mol, typical for a mildly nonideal system. At $\omega_{\mu\nu}/RT = -10$, which corresponds to an excess enthalpy of about 500 J/mol, we find again a perfect agreement between the model and the computer simulation for the corner side arrangement for all three excess functions. For the opposite side arrangement we still find good agreement for the excess free energy and thus the activity coefficient, but deviations of about 25% for the excess enthalpy and a very large deviation for the excess entropy. Clearly, this points to a failure of the free segment approximation for this arrangement, which tends to favor ordered molecular clusters. Many more simulations have been performed [11], and the general result is that the free segment approximation performs surprisingly well for many molecular charge distributions. At low temperatures, there are particular charge arrangements favoring ordered molecular structures where the model induces nonnegligible errors in the excess enthalpy and excess entropy, although these tend to cancel in the excess free energy. We finally note that the model, in contrast to the simple QUAC, produces positive as well as negative excess entropies, depending on the charge distribution, in agreement with the simulated data and experimental evidence for real fluid mixtures.

EXERCISE 4.4

Calculate the attractive part of the molar excess free energy of the system 2-butanone–hexane at $x = 0.5$ and $t = 60°C$ from the models derived above in the GEQUAC parameterization. The polar molecule 2-butanone (1) is assumed to consist of a negatively charged group μ, a positively charged group ν and a neutral group γ. The alkane hexane is assumed to consist of a single neutral group of type δ; cf. Figure E 4.4.1. The relative surface areas of the groups are [1]

$$q_\mu = 0.28025, \quad q_\nu = 0.11513, \quad q_\gamma = 2.4806, \quad \text{and} \quad q_\delta = 3.8560$$

and the segment interaction exchange energies at $60°C$ are given in J/mol by [1]

$$\omega_{\mu\nu} = -37,284$$
$$\omega_{\mu\gamma} = -10,438$$
$$\omega_{\mu\delta} = -10,732$$
$$\omega_{\nu\gamma} = -23,239$$
$$\omega_{\nu\delta} = -14,603$$
$$\omega_{\gamma\delta} = +19.$$

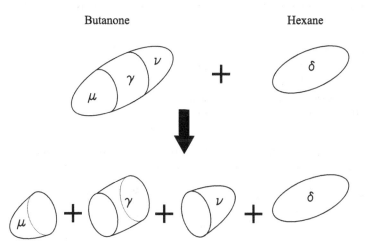

Fig. E 4.4.1. GEQUAC surface interaction model for the system 2-butanone–hexane.

Solution

The total relative surfaces of the molecules are obtained by (4.88) as

$$q_1 = 2.8760$$

and

$$q_2 = 3.8560.$$

These total values agree with those used in Exercise 4.2. We need the surface fractions

$$\theta_\alpha = \frac{x_\alpha q_\alpha}{\sum_\beta x_\beta q_\beta}$$

and thus have to calculate the mole fractions of the groups. For group μ we have the equation

$$x_\mu = \frac{\sum x_i n_{\mu,i}}{\sum_i x_i n_i} = \frac{x_1 \times 1}{x_1 \times 3 + x_2 \times 1} = \frac{0.5}{2} = 0.25$$

and also

$$x_\nu = x_\gamma = x_\delta = 0.25.$$

This gives

$$\theta_\mu = \frac{0.25 \times 0.28025}{0.25 \times 0.28025 + 0.25 \times 0.11513 + 0.25 \times 2.4806 + 0.25 \times 3.8550} = 0.04163.$$

Analogously we find

$$\theta_\nu = 0.01710,$$
$$\theta_\gamma = 0.36848,$$
$$\theta_\delta = 0.57279.$$

For the parameter q we have, according to (4.87),

$$q = x_1 q_1 + x_2 q_2 = 0.5 \cdot 2.8760 + 0.5 \cdot 3.8560 = 3.3660.$$

The local surface fractions have to be determined from the six equations

$$\frac{\theta_{\alpha\beta}\theta_{\beta\alpha}}{\theta_{\alpha\alpha}\theta_{\beta\beta}} = e^{-w_{\alpha\beta}/RT} = \tau_{\alpha\beta}^2$$

plus the four equations

$$\sum_\alpha \theta_{\alpha\beta} = 1 \quad \beta = \nu, \mu, \gamma, \delta$$

and in addition the six equations

$$\theta_\alpha \theta_{\beta\alpha} = \theta_\beta \theta_{\alpha\beta}.$$

These are in total 16 equations for the 16 unknown local surface fractions $\theta_{\nu\mu}$. Solution gives

$\theta_{\alpha\beta}$	$\alpha = \mu$	ν	γ	δ
$\beta = \mu$	0.00091	0.13715	0.03983	0.04290
ν	0.05634	0.00001	0.03002	0.00645
γ	0.35255	0.64682	0.36024	0.36663
δ	0.59020	0.21602	0.56991	0.58402

The attractive part of the molar excess free energy for the group pair interactions in the mixture is then calculated from (4.86) as ($z = 10$)

$$(g^E)_{\text{att}}^{\text{gr}} = (a^E)_{\text{att}}^{\text{gr}} = \frac{1}{4} zq \sum_\alpha \sum_\beta \theta_\alpha \theta_{\beta\alpha} \left(w_{\beta\alpha} + 2RT \ln \frac{\theta_{\beta\alpha}}{\theta_\beta} \right)$$

$$= -13,084 \text{ kJ/kmol}.$$

To transform the attractive part of the molar excess free enthalpy on the group interaction basis to the molecular interaction basis we must subtract the corresponding values for the pure components; cf. (4.91). Pure 2-butanone has the groups μ, ν, and γ. We find

$$x_\mu^{01} = \frac{n_{\mu,1}}{n_1} = 0.3333 = x_\nu^{01} = x_\gamma^{01}$$

and thus

$$\theta_\mu^{01} = \frac{0.3333 \times 0.28025}{0.3333 \times 0.28025 + 0.3333 \times 0.11513 + 0.3333 \times 2.4806} = 0.097445$$

and also

$$\theta_\nu^{01} = 0.04003$$

$$\theta_\gamma^{01} = 0.86252.$$

The local surface fractions in the pure system 2-butanone are obtained as

$\theta_{\alpha\beta}^{01}$	$\alpha = \mu$	ν	γ
$\beta = \mu$	0.002571	0.188450	0.103939
ν	0.077418	0.000008	0.037665
γ	0.920011	0.811542	0.858396

and the pure 2-butanone excess free energy on a group pair interaction basis is

$$(g^{\mathrm{E}})_{\mathrm{att}}^{\mathrm{gr},01} = -27{,}835 \text{ kJ/kmol}.$$

Because pure component 2 has only one type of interacting surface, we have

$$(g^{\mathrm{E}})_{\mathrm{att}}^{\mathrm{gr},02} = 0 \text{ kJ/kmol}.$$

Thus, the attractive part of the molar excess free enthalpy is obtained as

$$g_{\mathrm{att}}^{\mathrm{E}} = -13{,}084 - [0.5 \cdot (-27{,}835) - 0.5(0)]$$
$$= 833 \text{ kJ/kmol}.$$

The experimental value is $g^{\mathrm{E}} = 810$ kJ/kmol [2]. We note that we have to add the combinatorial contribution, which is $g_{\mathrm{comb}}^{\mathrm{E}} = -31$ kJ/kmol, as computed in Exercise 4.2, and thus contributes no more than about 5% to the total excess free energy. So the correlation of the excess free energy of this system by GEQUAC is quite accurate.

Alternatively, we can use COSMOSPACE, i.e.,

$$g_{\mathrm{att}}^{\mathrm{E}} = RT \sum_i x_i \ln \gamma_{i,\mathrm{att}},$$

with, cf. (4.93),

$$\ln \gamma_{i,\mathrm{att}} = \sum_\alpha N_{\alpha,i} q_\alpha \left(\ln \gamma_\alpha - \ln \gamma_\alpha^{0i} \right) = z \sum_\alpha n_{\alpha,i} q_\alpha \left(\ln \gamma_\alpha - \ln \gamma_\alpha^{0i} \right)$$

for the activity coefficient of component i and, cf. (4.92),

$$\ln \gamma_\alpha = -\ln \left\{ \sum_\beta \theta_\beta \tau_{\alpha\beta} \gamma_\beta \right\}$$

for the activity coefficient of an arbitrary group α. Because we have four groups μ, ν, γ, δ, the group activity coefficients are obtained from the above four equations, which have to be solved iteratively. For group μ, for example, we have

$$\ln \gamma_\mu = -\ln \left\{ \theta_\mu \tau_{\mu\mu} \gamma_\mu + \theta_\nu \tau_{\mu\nu} \gamma_\nu + \theta_\gamma \tau_{\mu\gamma} \gamma_\gamma + \theta_\delta \tau_{\mu\delta} \gamma_\delta \right\}.$$

Using the values for θ_α and $\tau_{\alpha\beta}$ as above we find

$$\ln \gamma_\mu = -1.91714$$
$$\ln \gamma_\nu = -3.62106$$
$$\ln \gamma_\gamma = -0.01131$$
$$\ln \gamma_\delta = 0.00971.$$

The analogous procedure for the group activity coefficients in the pure components gives

$$\ln \gamma_\delta^{02} = 0$$
$$\ln \gamma_\mu^{01} = -1.81735$$
$$\ln \gamma_\nu^{01} = -4.25358$$
$$\ln \gamma_\gamma^{01} = -0.00240.$$

We thus find for the activity coefficients of the components

$$\ln \gamma_{1,\mathrm{att}} = z\left[q_\mu\left(\ln \gamma_\mu - \ln \gamma_\mu^{01}\right) + q_\nu\left(\ln \gamma_\nu - \ln \gamma_\nu^{01}\right) + q_\gamma\left(\ln \gamma_\gamma - \ln \gamma_\gamma^{01}\right)\right] = 0.2270$$

and

$$\ln \gamma_{2,\mathrm{att}} = zq_\delta\left(\ln \gamma_\delta - \ln \gamma_\delta^{02}\right) = 0.3744.$$

This gives for the attractive contribution to the molar excess free energy

$$g_{\mathrm{att}}^{\mathrm{E}} = RT\left(x_1 \ln \gamma_{1,\mathrm{att}} + x_2 \ln \gamma_{2,\mathrm{att}}\right) = 833 \text{ kJ/kmol},$$

in agreement with the earlier result. The significant simplification due to the circumvention of the local surface fractions in COSMOSPACE is evident.

[1] G. H. Ehlker and A. Pfennig. *Fluid Phase Equilibria*, 203:53, 2002.
[2] D. O. Hanson and M. van Winkle. *J. Chem. Eng. Data*, 12:319, 1967.

4.4.3 Predictive Models

The excess function models presented above contain various parameters related to the structure of the molecules and to the intermolecular potential energies of interacting pairs at contact. These parameters are few for uniform surface properties. In fact, there is only one potential parameter for each binary interaction in the quasichemical approximation and there are two for the semiempirical models Wilson or UNIQUAC and others. In addition, there is one volume and one surface parameter for each molecule. On the other hand, many parameters are involved when nonuniform surface properties are considered; cf. Exercise 4.4. The procedure of adjusting these parameters to data is not straightforward in situations where there are many of them. Usually, the parameters are correlated and physically, meaningful values are not obtained. Further, in many cases of interest, data may not be available for all components and interactions in a system. So there is a strong interest in developing predictive methods for excess functions. These must be based on nonuniform surface interaction models, which are much closer to reality than the uniform approximation. Two major types of predictive models have established themselves, the more empirical group contribution models and the more physical surface charge interaction models.

A group interaction model is based on the assumption that the molecules can be structured into functional groups with particular surface sizes and interaction properties, like the one shown in Figure 4.19. The excess free energy is then obtained by introducing the group interactions into an appropriate statistical model, such as the one derived above. The values for the group interaction parameters are obtained by adjusting the model to a large number of data. Because many molecules of technical interest consist of only a limited number of functional groups, such as CH_3, CH_2, and OH, it becomes feasible to adjust those in a meaningful way. It is then possible to predict the properties of a system for which no data are available from its group contributions as determined from experimentally investigated systems with the same groups. The most popular group contribution model currently available is not based on the model derived above, but rather is UNIFAC [22], based on UNIQUAC; cf. Section 4.4.2. Written in terms of an activity coefficient model and for an arbitrary number of components, the attractive term of the UNIQUAC model reads [18]

$$\ln \gamma_{i,\text{att}} = q_i \left[1 - \ln \left(\sum_j \theta_j G_{ji} - \sum_j \frac{\theta_j G_{ij}}{\sum_k \theta_k G_{kj}} \right) \right], \tag{4.95}$$

where G_{ji} is the nonrandomness factor, given by

$$G_{ji} = e^{-\Delta U_{ji}/T}. \tag{4.96}$$

In (4.95) and (4.96) the nondimensional surface parameters q_i and the interaction energies ΔU_{ji} in K refer to molecules and molecular pairs, respectively. They are thus based on the uniform surface interaction model. However, it is possible to transform them formally into parameters for groups and group interactions. For the group-α activity coefficient one then has, instead of (4.95),

$$\ln \gamma_\alpha = q_\alpha \left[1 - \ln \left(\sum_\beta \theta_\beta G_{\beta\alpha} - \sum_\beta \frac{\theta_\beta G_{\alpha\beta}}{\sum_\delta \theta_\delta G_{\delta\beta}} \right) \right]. \tag{4.97}$$

Here q_α is the dimensionless surface parameter of group α and θ_β is the surface fraction of group β. Finally, $G_{\alpha\beta}$ is the nonrandomness factor associated with the interaction of the groups α and β; i.e.,

$$G_{\alpha\beta} = e^{-a_{\alpha\beta}/T}.$$

Values for the group surface parameters q_α and the interaction parameters $a_{\alpha\beta}$ are tabulated for many groups and group interactions, respectively [14, 22]. We note in particular that in UNIFAC $a_{\alpha\beta} \neq a_{\beta\alpha}$, unlike in the quasichemical approximation, as discussed in Section 4.4. The attractive part of the activity coefficient of component i in the mixture is then given by, cf. (4.93),

$$\ln \gamma_{i,\text{att}} = \sum_\alpha n_{\alpha,i} \left(\ln \gamma_\alpha - \ln \gamma_\alpha^{0i} \right), \tag{4.98}$$

where $n_{\alpha,i}$ is the number of groups of type α in a molecule of type i. We note that here we do not consider the number of α-surface contacts $N_{\alpha,i}$, contrary to (4.93), because UNIFAC is just an ad hoc transformation of the molecule-based model UNIQUAC to groups, whereas (4.93) is truly based on surface segment interactions.

EXERCISE 4.5

Calculate the attractive part of the molar excess free energy of the system 2-butanone–hexane at $x = 0.5$ and $t = 60°C$ from UNIFAC.

Solution

In UNIFAC the polar molecule 2-butanone(1) is assumed to consist of three functional groups, $\mu = CH_3$, $\nu = CH_2$, and $\gamma = CH_3CO$, each one appearing once. Hexane(2) is assumed to consist of two functional groups, $\mu = CH_3$ and $\nu = CH_2$, where there are two groups μ and four groups ν in the molecule. We find for the nondimensional surfaces of the groups [1, 2]

$$q_\mu = 0.848$$
$$q_\nu = 0.540$$
$$q_\gamma = 1.488,$$

leading to $q_1 = 2.8760$ and $q_2 = 3.8560$, in agreement with Exercises 4.2 and 4.4. The mole fraction of group μ is calculated from, cf. Exercise 4.1,

$$x_\mu = \frac{\sum_i x_i n_{\mu,i}}{\sum_i x_i n_i} = \frac{0.5 \times 1 + 0.5 \times 2}{0.5 \times 3 + 0.5 \times 6} = 0.3333.$$

Analogously, we find

$$x_\nu = 0.5556$$

$$x_\gamma = 0.1111.$$

This gives the surface fractions

$$\theta_\mu = \frac{x_\mu q_\mu}{x_\mu q_\mu + x_\nu q_\nu + x_\gamma q_\gamma} = \frac{0.3333 \times 0.848}{0.3333 \times 0.848 + 0.5556 \times 0.540 + 0.1111 \times 1.488}$$
$$= 0.3779$$

and also

$$\theta_\nu = 0.4011$$

$$\theta_\gamma = 0.2210.$$

The nonrandomness parameters $G_{\alpha\beta}$ at $t = 60°C$ are calculated from tabulated values [2] for the interaction parameters $a_{\alpha\beta}$ as

$$G_{\mu\mu} = 1 = G_{\nu\mu} = G_{\mu\nu} = G_{\nu\nu} = G_{\gamma\gamma}$$
$$G_{\gamma\mu} = 0.9228 = G_{\gamma\nu}$$
$$G_{\mu\gamma} = 0.2393 = G_{\nu\gamma}.$$

This gives the attractive part of the activity coefficient of group μ in the mixture, as, cf. (4.97),

$$
\begin{aligned}
\ln \gamma_\mu = 0.848 \Big[1 &- \ln(0.37791 \times 1 + 0.4011 \times 1 + 0.2210 \times 0.9228) \\
&- \frac{0.3779 \times 1}{0.3779 \times 1 + 0.4011 \times 1 + 0.2210 \times 0.9228} \\
&- \frac{0.4011 \times 1}{0.3779 \times 1 + 0.4011 \times 1 + 0.2210 \times 0.9228} \\
&- \frac{0.2210 \times 0.2393}{0.3779 \times 0.2393 + 0.4011 \times 0.2393 + 0.2210 \times 1} \Big] \\
= 0.080458
\end{aligned}
$$

and also

$$
\ln \gamma_\nu = 0.051235
$$

and

$$
\ln \gamma_\gamma = 0.92872.
$$

The corresponding group activity coefficients for the pure components are

$$
\ln \gamma_\mu^{01} = 0.29027
$$
$$
\ln \gamma_\nu^{01} = 0.18484
$$
$$
\ln \gamma_\gamma^{01} = 0.26191
$$

and

$$
\ln \gamma_\mu^{02} = 0
$$
$$
\ln \gamma_\nu^{02} = 0.
$$

The attractive parts of the activity coefficients on a molecular basis are thus found to be

$$
\begin{aligned}
\ln \gamma_{1,\text{att}} &= 1(0.080458 - 0.29027) + 1(0.051235 - 0.18484) + 1(0.92872 - 0.26191) \\
&= 0.32339
\end{aligned}
$$

and

$$
\ln \gamma_{2,\text{att}} = 2(0.080458 - 0) + 4(0.051235 - 0) = 0.365856.
$$

This gives the attractive part of the molar excess free energy as

$$
(g^E)_{\text{att}} = (0.5 \times 0.32339 + 0.5 \times 0.365856)\, RT = 955 \text{ kJ/kmol.}
$$

Adding the combinatorial contribution, cf. Exercise 4.2, leads to $g^E = 924$ kJ/kmol. This is somewhat larger than the experimental value, which is $g^E = 810$ kJ/kmol [3]. The error in g^E is 14%. In the predicted pressure this results in an error of 4%, whereas in the vapor phase composition the error is negligible due to cancellation.

[1] A. Bondi. *Physical Properties of Molecular Crystals, Liquids and Glasses*. Wiley, New York, 1968.
[2] A. Fredenslund, J. Gmehling, and P. Rasmussen. *Vapor–Liquids Equilibria Using UNIFAC*. Elsevier, Amsterdam, 1977.
[3] D. O. Hanson and M. van Winkle. *J. Chem. Eng. Data*, 12:319, 1967.

Group contribution models such as UNIFAC are predictive methods in the practical sense that they can predict the activity coefficients and thus also the excess functions of a system for which no data are available. In a more basic sense they must be considered rather as an interpolation tool, because they are

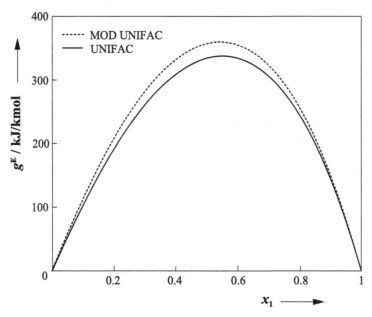

Fig. 4.22. Excess free energy of the system benzene–hexane at $T = 328.15$ K from UNIFAC and MOD UNIFAC.

based on the data of numerous systems used to establish the group parameters. Systems with groups for which no parameters are available cannot be predicted. The group classification of UNIFAC, as well as the excess function model UNIQUAC, is an approximation to the true molecular behavior in a liquid mixture. It is thus not surprising that interaction parameters obtained from vapor–liquid equilibria are usually unable to predict excess enthalpies or liquid–liquid equilibria. By using an ever-growing database including various types of data for many systems and making the interaction parameters temperature-dependent, many of these limitations have practically been overcome [3].

As an illustration, we consider predictions for the system benzene–hexane by UNIFAC and an extended version of it (MOD UNIFAC) [23], which, besides other minor refinements, uses temperature-dependent interaction parameters for better reproduction of the excess enthalpy and the liquid–liquid equilibrium. Figure 4.22 shows the excess free energy of benzene–hexane at 328.15 K for both models and Figure 4.23 the corresponding total pressure in vapor–liquid equilibrium. As can be seen, both models are rather close and in good agreement with the data. Figure 4.24 shows the predictions of both models for the excess enthalpy, from which it is clear that MOD UNIFAC is a significant improvement over the original UNIFAC and gives an almost perfect prediction for this property. As a further illustration we look at the prediction of the liquid–liquid equilibrium for the system butanone–water in Figure 4.25. UNIFAC does not predict liquid–liquid demixing for this system, which

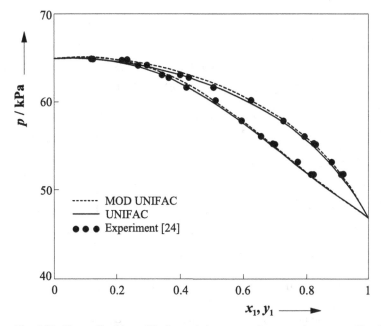

Fig. 4.23. Vapor–liquid equilibrium of the system benzene–hexane at $T = 328.15$ K from UNIFAC and MOD UNIFAC.

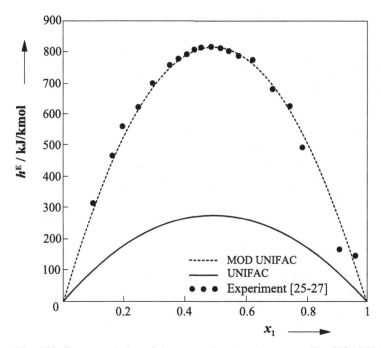

Fig. 4.24. Excess enthalpy of the system benzene–hexane at $T = 323.15$ K from UNIFAC and MOD UNIFAC.

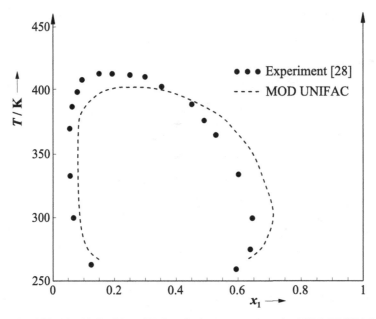

Fig. 4.25. Liquid–liquid equilibrium for butanone–water for MOD UNIFAC.

reiterates the basic limitations of this model. MOD UNIFAC, in contrast, predicts the mutual solubilities of the two components, in fair agreement with the data. We note that another rather similar modified UNIFAC model is also available [29].

Although group contribution models such as UNIFAC and MOD UNIFAC have established themselves as valuable and simple prediction tools for the excess functions of liquid mixtures, more fundamental approaches that are less dependent on data and overcome the basic limitations associated with the segmentation of a molecule into independent functional groups are desired. Such a more fundamental approach is offered by applying a surface charge interaction model, such as COSMO-RS, to an excess free energy model suitable to incorporating nonuniform surface interactions, such as that derived in this section. In the COSMO-RS method the nonuniform interaction properties of the molecules are no longer represented in terms of groups, but rather in terms of surface charges, so-called σ-profiles; cf. Section 4.2. These σ-profiles represent surface areas of a molecule associated with particular (screening) charge densities. Electrostatic interactions between these surface segments are modeled in terms of these charges and some additional parameters that are adjusted to data as discussed in Section 4.2. These parameters are relatively few and universal for all systems; i.e., they are model parameters. The liquid model adopted is that of space-filling polyhedral surface segments as visualized in Figure 4.11.

The basic equation for the attractive activity coefficient of a component i is (4.93). It is expressed in terms of the surface segment activity coefficients in the mixture and in pure component i which in turn are given by (4.92). The surface segments here arise from a proper discretization of the σ-profiles associated with the components in the mixture. The parameter of major importance in this latter equation is $\tau_{\alpha\beta}$, which is calculated from the exchange energy related to segment types α and β. If these segment types are characterized by the charge densities σ_α and σ_β, the associated interaction energy is given by, cf. (4.25),

$$\phi_{\alpha\beta} = A_{\text{eff}} \left[\frac{\alpha}{2} (\sigma_\alpha + \sigma_\beta)^2 \right],$$

and the exchange energy can be calculated from it. Here A_{eff} is a universal computational contact area for the independent segments, for which a value has been found from fitting to a large database. It is not identical to the actual surface area of a segment in the electrostatic COSMO calculation nor to that of a segment used in evaluating the activity coefficients. Rather, in combination with the external surface $A_{\sigma,i}$ associated with a discrete σ-value of the σ-profile of molecule i, it defines an effective number of independent surface contacts $N_{\sigma,i}$ by (4.67) with $A_i = A_{\sigma,i}$ and $A_{\text{ref}} = A_{\text{eff}}$, over which a surface segment of charge type σ in molecule i can interact. This number is needed in (4.93). We here note that the actual calculation of the activity coefficients is carried out with segments of various surface areas, which are found from the discretization of the σ-profile; cf. Figure 4.12. An order of magnitude of 100 segments is typical. Clearly, the surfaces associated with the various σ-values must sum up to the total external COSMO-surface of the molecule, which is known. A proper discretization and averaging of the σ-profile has consequences for the production of independent segments, which is crucial to the model. It can be executed in various different ways. Details may be found in [30, 31]. The electrostatic misfit energy coefficient α is a further universal parameter of the model to be fitted to data. Further contributions to the pair contact interactions, notably due to hydrogen bonding and dispersion, have been formulated semiempirically; cf. Section 4.2. To perform the COSMO calculations for the σ-profiles, element specific cavity radii are required. They are roughly 20% larger than the established van der Waals radii [20] for the atoms and fitted to data as further universal parameters of the model. From these enlarged atomic radii the surface of a cavity associated with a molecule may be calculated. The model is then fully predictive.

EXERCISE 4.6

Calculate the molar excess free energy of the system 2-butanone–hexane at $x = 0.5$ and $t = 60°C$ from the COSMO-RS model. The model parameters are $A_{\text{eff}} = 6.15$ Å^2, $\alpha = 6,635,000$ JÅ^2/(mol e^2), $r_{\text{H}} = 1.30$ Å, $r_{\text{C}} = 2.00$ Å, and $r_{\text{O}} = 1.72$ Å.

Solution

The σ-profiles of the two components butanone (1) and hexane (2) are shown in Figure E 4.6.1; the associated numerical values are summarized in Table E 4.6.1. These data are obtained from the COSMO files after suitable averaging with r_{av}, cf. Section 4.2. Plotted are the surface areas for the various charge densities in Å^2 over the charge densities in $e/\text{Å}^2$. The tabulated values are discretized with respect to σ in steps of 0.001 $e/\text{Å}^2$, giving a total of 31 segments. This defines the number of interactions to be considered. The sum over all discretized surfaces gives the total surface of a molecule, as obtained from a COSMO calculation. These surfaces are

$$A_1 = 120.832 \text{ Å}^2$$

and

$$A_2 = 156.895 \text{ Å}^2.$$

Table E 4.6.1. Smoothed σ-profiles for butanone (1) and hexane (2)

	$\sigma/e\text{Å}^{-2}$	$A_{\sigma,1}/\text{Å}^2$	$A_{\sigma,2}/\text{Å}^2$
	−0.010	0	0
	−0.009	0.207	0
	−0.008	1.950	0
μ :	−0.007	5.953	0
	−0.006	10.065	0.039
	−0.005	11.806	1.269
	−0.004	11.459	6.851
	−0.003	10.500	16.514
	−0.002	9.473	24.208
	−0.001	9.805	24.840
ν :	0	10.052	20.201
	0.001	7.799	16.964
	0.002	5.749	16.269
	0.003	4.165	15.274
	0.004	1.786	10.827
	0.005	0.498	3.589
	0.006	0.543	0.025
γ :	0.007	0.693	0
	0.008	0.918	0
	0.009	1.301	0
	0.010	1.247	0
	0.011	1.451	0
	0.012	2.827	0
	0.013	3.552	0
δ :	0.014	2.517	0
	0.015	1.651	0
	0.016	1.533	0
	0.017	1.007	0
	0.018	0.298	0
	0.019	0.026	0
	0.020	0	0

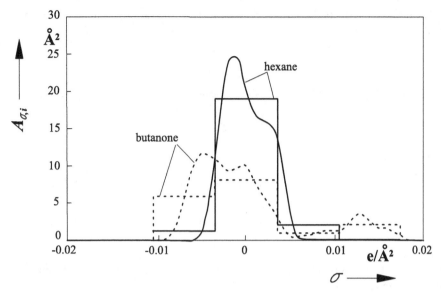

Fig. E 4.6.1. Representation of the σ-profiles for butanone and hexane.

To demonstrate the basic procedure by a hand calculation, we represent the σ-profiles by just four discrete σ-values between $\sigma = -0.01$ e/Å^2 and $\sigma = +0.017$ e/Å^2. The total of 31 tabulated values are thus structured into four crude σ-segments with the values

$$\sigma_\mu = -0.007 \text{ e/Å}^2$$
$$\sigma_\nu = 0 \text{ e/Å}^2$$
$$\sigma_\gamma = +0.007 \text{ e/Å}^2$$
$$\sigma_\delta = +0.014 \text{ e/Å}^2.$$

Clearly, neglecting the final three contributions to butanone results in a slightly smaller surface of $A_1 = 120.508$ Å^2. By summing up the surface patches in Table E 4.6.1 associated with the four segments, we get for butanone

$$A_{\mu,1} = 41.441 \text{ Å}^2$$
$$A_{\nu,1} = 57.543 \text{ Å}^2$$
$$A_{\gamma,1} = 6.986 \text{ Å}^2$$
$$A_{\delta,1} = 14.538 \text{ Å}^2$$

and for hexane

$$A_{\mu,2} = 8.159 \text{ Å}^2$$
$$A_{\nu,2} = 134.259 \text{ Å}^2$$
$$A_{\gamma,2} = 14.441 \text{ Å}^2$$
$$A_{\delta,2} = 0.$$

Again, summing up the surface areas belonging to the chosen segments gives the total surface areas of the molecules. Figure E 4.6.1 shows the crudely discretized σ-profiles along with the smoothed ones from a COSMO calculation.

The number of effective external contacts involving the various segments is obtained by dividing the surface associated with one particular segment by the effective surface $A_{\text{eff}} = 6.15 \text{ Å}^2$, i.e., the average independent contact area, to give for butanone

$$N_{\mu,1} = 6.7384$$
$$N_{\nu,1} = 9.3566$$
$$N_{\gamma,1} = 1.1359$$
$$N_{\delta,1} = 2.3639.$$

and for hexane

$$N_{\mu,2} = 1.3267$$
$$N_{\nu,2} = 21.8366$$
$$N_{\gamma,2} = 2.3481$$
$$N_{\delta,2} = 0.$$

The surface fractions or, alternatively, contact fractions of the segments in the mixture are calculated from (4.6)

$$\theta_\alpha = \frac{\sum_i x_i A_{\alpha,i}}{\sum_i x_i A_i} = \frac{\sum_i x_i N_{\alpha,i}}{\sum_i x_i (\sum_\alpha N_{\alpha,i})},$$

with the result

$$\theta_\mu = 0.1788$$
$$\theta_\nu = 0.6916$$
$$\theta_\gamma = 0.0772$$
$$\theta_\delta = 0.0524.$$

To apply the COSMOSPACE equation we need the $\tau_{\alpha\beta}$, i.e., the exchange energies for interactions between the groups α and β. The interaction energy between an α-segment and a β-segment is given by (4.25) as

$$\phi_{\alpha\beta} = A_{\text{eff}} \left[\frac{\alpha}{2} (\sigma_\alpha + \sigma_\beta)^2) \right],$$

where $\alpha = 6{,}635{,}000$ J Å^2/(mol e^2). The resulting symmetric matrix of interaction energies $\phi_{\alpha\beta}$ in J/mol is

$\phi_{\alpha\beta}$	μ	ν	γ	δ
μ	3,998.91	999.73	0	999.73
ν	999.73	0	999.73	3,998.91
γ	0	999.73	3,998.91	8,997.56
δ	999.73	3,998.91	8,997.56	15,995.66

from which the symmetric matrix of $\tau_{\alpha\beta}$ is obtained as

$\tau_{\alpha\beta}$	μ	ν	γ	δ
μ	1	1.43464	4.23618	25.74510
ν	1.43464	1	1.43464	4.23618
γ	4.23618	1.43464	1	1.43464
δ	25.74510	4.23618	1.43464	1

Substituting into the COSMOSPACE equation and solving iteratively as in Exercise 4.4 gives

$$\gamma_\mu = 0.63699$$
$$\gamma_\nu = 1.02176$$
$$\gamma_\gamma = 0.64310$$
$$\gamma_\delta = 0.16728.$$

Similar procedures for pure butanone give

$$\theta_\mu^{01} = 0.3439$$
$$\theta_\nu^{01} = 0.4775$$
$$\theta_\gamma^{01} = 0.0580$$
$$\theta_\delta^{01} = 0.1206$$
$$\gamma_\mu^{01} = 0.66688$$
$$\gamma_\nu^{01} = 1.05879$$
$$\gamma_\gamma^{01} = 0.57640$$
$$\gamma_\delta^{01} = 0.12517$$

and for pure hexane

$$\theta_\mu^{02} = 0.0520$$
$$\theta_\nu^{02} = 0.8560$$
$$\theta_\gamma^{02} = 0.0920$$
$$\theta_\delta^{02} = 0$$
$$\gamma_\mu^{02} = 0.65061$$
$$\gamma_\nu^{02} = 1.00122$$
$$\gamma_\gamma^{02} = 0.69498$$
$$\gamma_\delta^{02} = 0.21733.$$

Finally, the attractive part of the activity coefficient is obtained from (4.93). So we find that

$$\ln \gamma_{1,\mathrm{att}} = 6.7384\,[(-0.4510) - (-0.4051)] + 9.3566\,[(0.0215) - (0.0571)]$$
$$+ 1.1359\,[(-0.4415) - (-0.5510)] + 2.3639\,[(-1.7881) - (-2.0781)]$$
$$= 0.168$$

and

$$\ln \gamma_{2,\mathrm{att}} = 0.233.$$

This gives $g_{\mathrm{att}}^{\mathrm{E}} = 556$ J/mol. Using the commercially available COSMO-RS software, which apart from using more segments includes many refinements [32], gives $g_{\mathrm{att}}^{\mathrm{E}} = 613$ J/mol and thus a deviation of about 25% from the true value; cf. Exercise 4.2. Such a deviation, although significant, is not severe in view of the fact that no substance-specific parameters are involved. Although somewhat inferior to UNIFAC, which is about 14% high, cf. Exercise 4.5, this is again considered a satisfactory result for such a small number of universal parameters.

The combinatorial part is again modeled by the Guggenheim–Stavermann equation (4.74). In contrast to Exercise 4.2, where tabulated values were used for r_i and q_i, we here calculate the volume and surface fractions of the molecules consistently from the

COSMO results; i.e.,

$$A_1 = 120.832 \, \text{Å}^2, \quad A_2 = 156.895 \, \text{Å}^2$$

and

$$V_1 = 107.258 \, \text{Å}^3, \quad V_2 = 145.702 \, \text{Å}^3.$$

This leads to

$$\phi_1 = \frac{x_1 V_1}{x_1 V_1 + x_2 V_2} = 0.4240$$

$$\phi_2 = \frac{x_2 V_2}{x_1 V_1 + x_2 V_2} = 0.5760$$

$$\theta_1 = \frac{x_1 A_1}{x_1 A_1 + x_2 A_2} = 0.4351$$

$$\theta_2 = \frac{x_2 A_2}{x_1 A_1 + x_2 A_2} = 0.5649.$$

The total numbers of surface contacts associated with the molecules are

$$zq_1 = N_1 = \frac{A_1}{A_{\text{eff}}} = \frac{120.832}{6.15} = 19.595$$

$$zq_2 = N_2 = \frac{A_2}{A_{\text{eff}}} = \frac{156.895}{6.15} = 25.511.$$

This leads to

$$g_{\text{comb}}^{\text{E}} = 8.315 \times 333.15 \left[0.5 \ln \frac{0.4240}{0.5} + 0.5 \ln \frac{0.5760}{0.5} + 0.25 \times 19.595 \ln \frac{0.4351}{0.4240} \right.$$
$$\left. + 0.25 \times 25.511 \ln \frac{0.5649}{0.5760} \right]$$
$$= -32.38 + 6.90 = -25.48 \, \text{J/mol},$$

in reasonable agreement with the earlier result.

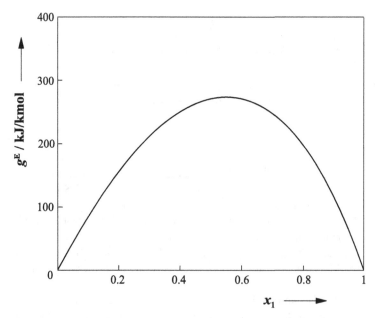

Fig. 4.26. Excess free energy of the system benzene–hexane at $T = 328.15$ K from COSMO-RS.

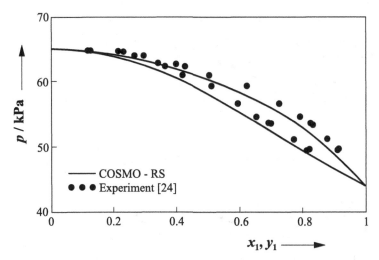

Fig. 4.27. Vapor–liquid equilibrium of the system benzene–hexane at $T = 328.15$ K from COSMO-RS.

As an illustration, Figures 4.26 and 4.27 show the excess free energy and the vapor–liquid equilibrium of the system benzene–hexane at 328.15 K, whereas Figure 4.28 shows the excess enthalpy of this system at 323.15 K, as calculated from [32]. Comparing this to Figures 4.22 to 4.24 for the same system from UNIFAC and MOD UNIFAC, we realize a similar although slightly inferior performance for g^E and the vapor–liquid equilibrium, whereas for the excess enthalpy COSMO-RS performs significantly better than UNIFAC, although

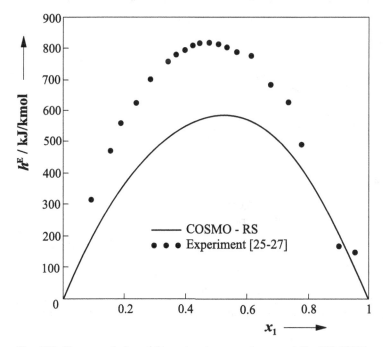

Fig. 4.28. Excess enthalpy of the system benzene–hexane at $T = 323.15$ K from COSMO-RS.

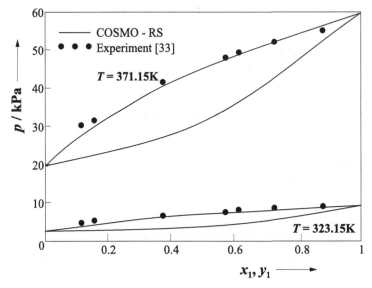

Fig. 4.29. Vapor–liquid equilibrium of the system n-butylisocyanate–n-nonane from COSMO-RS.

definitely inferior to MOD UNIFAC. We note again that only rather few universal parameters have been fitted to a data base in the COSMO-RS model, whereas many more group-specific parameters are used in UNIFAC and, in particular, in MOD UNIFAC. This points to the principal physical significance of the σ-profile as a sensible descriptor of a molecule's electrostatic interaction properties. It is also reassuring to note that a system such as benzene–toluene, when predicted from COSMO-RS, is reproduced as a practically ideal system,

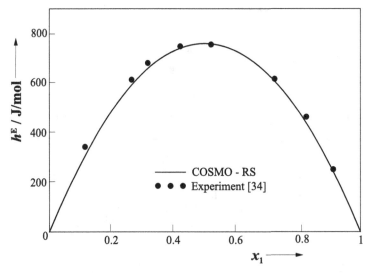

Fig. 4.30. Excess enthalpy of the system n-butylisocyanate–n-nonane from COSMO-RS at $T = 298.15$ K.

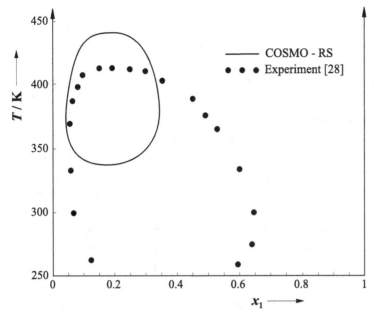

Fig. 4.31. Liquid–liquid equilibrium for butanone–water from COSMO-RS.

as anticipated from the σ-profiles of Figure 4.12. As a further illustration of the COSMO-RS model we consider a system of more complicated molecules, n-butylisocyanate and n-nonane, for which the vapor–liquid equilibrium at two temperatures is shown in Figure 4.29 and the excess enthalpy in Figure 4.30. The agreement with experimental data is rather satisfactory. We note that this system cannot be treated by UNIFAC or MOD UNIFAC, because the required group interaction parameters are not available. As a final illustration we look at the prediction of the liquid–liquid equilibrium for the system butanone–water in Figure 4.31. Although quantitatively unsatisfactory, COSMO-RS at least predicts a liquid–liquid demixing, contrary to UNIFAC, which again points to the physically sound basis of this fluid model. Clearly further refinements, notably for the dispersion part of the intermolecular potential energy model, are required to develop COSMO-RS to a generally satisfactory molecular model for fluids.

4.5 Summary

Liquid mixtures are most conveniently described in terms of excess functions or, equivalently, activity coefficients. Any type of excess free energy model should consist of a hard core term, related to size and shape effects, and an attractive term, taking account of the attractive intermolecular interactions. Also, any model must in some way introduce short-range nonrandomness, as manifested by deviations between local compositions and bulk compositions.

When the free segment approximation is made, the picture of a liquid is one in which molecular segments interact independently, i.e., decoupled from their actual connectivities in the real molecular structures. A further approximation, adequate for liquids, is that the segments are arranged in a space-filling manner; i.e., each segment has a direct neighbor in a dense liquid with which it interacts via surface contacts. The associated interaction energies can be calculated from experimental group contribution data or, more fundamentally, from quantum-chemical charge distributions. If volume segments are considered, a lattice is an adequate model for a liquid. Alternatively, the segments may be considered to be polyhedral surface patches representing the shape of a molecule. An expression for the attractive excess free energy can be derived in the free segment approximation, either from lattice theory or without the lattice framework. The free segment approximation converges to the well-known quasichemical approximation of independent pairs for simple molecules of equal sizes. The hard core contribution to the excess free energy degenerates to a combinatorial term for the excess entropy. Density-dependent free volume effects are lost. This excludes some important applications such as gas solubilities in polymers and high-pressure vapor–liquid equilibria, including critical phenomena. Different models for different views about the type of intermolecular interactions follow, including well-known semiempirical local composition models. The most fruitful developments originating from these theories are the predictive models, based either on group contributions or on surface charge distributions. It is to be expected that the latter approach will become the most important branch for progress in the future, although at present, highly parameterized group contribution methods are generally more accurate.

4.6 References to Chapter 4

[1] K. H. Lee, S. I. Sandler, and N. C. Patel. *Fluid Phase Equilibria*, 25:31, 1986.

[2] A. Klamt, and F. Eckert. *Fluid Phase Equilibria*, 172:43, 2000.

[3] J. Gmehling. *Fluid Phase Equilibria*, 144:37, 1998.

[4] A. Klamt, and G. Schüürmann. *J. Chem. Soc. Perkin Trans.*, 2:799, 1993.

[5] A. Klamt. *J. Phys. Chem.*, 99:2224, 1995.

[6] A. Klamt, V. Jonas, T. Bürger, and J. Lohrenz. *J. Phys. Chem. A*, 102:5074, 1998.

[7] E. A. Guggenheim. *Mixtures*, Oxford University Press, 1952.

[8] Y. Hu, E. G. Azevedo, and J. M. Prausnitz. *Fluid Phase Equilibria*, 13:351, 1983.

[9] B. L. Larsen, and P. Rasmussen. *Fluid Phase Equilibria*, 28:1, 1986.

[10] C. Panayiotou, and J. H. Vera. *Fluid Phase Equilibria*, 5:55, 1980.

[11] G. Pielen. Ph.D. thesis, RWTH Aachen, 2005.

[12] G. M. Wilson. *J. Am. Chem. Soc.*, 86:127, 1964.

[13] J. H. Hildebrand, J. M. Prausnitz, and R. L. Scott. *Regular and Related Solutions*. Van Nostrand Reinhold, New York, 1970.

[14] A. Bondi. *Physical Properties of Molecular Crystals, Liquids and Glasses.* Wiley, New York, 1968.

[15] P. J. Flory. *J. Chem. Phys.*, 9:660, 1941.

[16] M. L. Huggins. *J. Chem. Phys.*, 9:440, 1941.

[17] A. J. Stavermann. *Recl. Trav. Chim. Pays-Bas*, 69:163, 1950.

[18] D. S. Abrams, and J. M. Prausnitz. *AIChE J.*, 21:16, 1975.

[19] K. Egner, J. Gaube, and A. Pfennig. *Ber. Bunsenges. Phys. Chem.*, 101:209, 1997.

[20] A. Klamt, G. J. P. Krooshof, and R. Taylor. *AIChE J.*, 48:2332, 2002.

[21] S. T. Lin, and S. I. Sandler. *Ind. Eng. Chem. Res.*, 41:899, 2003.

[22] A. Fredenslund, J. Gmehling, and P. Rasmussen. *Vapor–Liquid Equilibria Using UNIFAC.* Elsevier, Amsterdam, 1977.

[23] J. Gmehling, J. Li, and M. Schiller. *Ind. Eng. Chem.*, 32:178, 1993.

[24] J. C. K. Ho, and B. C. Y. Lu. *J. Chem. Eng. Data*, 8:549, 1963.

[25] J. Gmehling. and U. Onken. *DECHEMA Chemistry Data Series.* Vol. 1. Part 1, 2a and 2b. DECHEMA, Frankfurt, 1977.

[26] M. J. Paz-Andrade. *Int. Data Series A.* Thermodynamic Research Center, A & M University Texas, 1973.

[27] M. Diaz-Penar, and C. Menduina. *J. Chem. Thermodyn.*, 6:1097, 1974.

[28] K. Ochi, M. Tada, and K. Kojima. *Fluid Phase Equilibria* , 56:341, 1990.

[29] B. L. Larsen, P. Rasmussen, and A. Fredenslund. *Ind. Eng. Chem. Res.*, 26:2274, 1987.

[30] S. Lin, J. Chang, S. Wang, W. A. Goddard, and S. I. Sandler. *J. Phys. Chem. A*, 108:7439, 2004.

[31] A. Klamt. *COSMO-RS: From Quantum Chemistry to Fluid Phase Thermodynamics.* Elsevier, Amsterdam, 2005.

[32] COSMOTHERM. Computer Package. COSMOlogic, 2003.

[33] S. Ahmad, R. Giesen, and K. Lucas. *J. Chem. Eng. Data*, 49:826, 2004.

[34] S. Ahmad. Ph.D. thesis, RWTH Aachen, 2005.

5 Equation of State Models

When fluid phase behavior over a large region of states is considered, excess function models are no longer appropriate. They are designed to address mixing effects at constant density or pressure. Effects of varying density are most conveniently treated in terms of an equation of state. The thermodynamic relations for computing fluid phase behavior from an equation of state in combination with ideal gas properties are well established; cf. Section 2.1. Although they are more demanding computationally than excess function models, there are now many well-tested computer codes available that allow the computation of fluid phase behavior from an equation of state. Basically, this approach is free from any of the restrictions associated with the use of excess function models. In principle, an equation of state model is generally applicable, including simple and complicated molecules, and, in particular, mixtures of small and large molecules, as in polymer solutions. In practice, different equation of state models are used for different applications and no general model suitable for all applications has yet emerged. The generality of the equation of state approach requires full generality of the potential energy model. A formulation in terms of contact energies, adequate for excess function models, is unsufficient. Rather, the potential energy will in principle depend on the distances between the molecular centers, on the orientations of the molecules, and, in the most general case, also on their internal coordinates. In this book we shall concentrate on equation of state models for systems composed of small molecules, such as those typically encountered in the gas industries. This allows the introduction of some simplifying approximations with respect to the intermolecular interactions, while at the same time illustrating the basic principles involved in the development of such molecular models. Equation of state models for more complex systems will be based on those for simple molecules by some kind of perturbation framework. The focus is on transparent and relatively simple approaches. A modern and more comprehensive review of equation of state theory is available in the literature [1].

5.1 General Properties

Some general information about the equation of state is available that can usefully be incorporated into the development of such molecular models.

5.1.1 The Low-Density Limit

At low densities all equation of state models must reproduce the universal limiting law

$$\lim_{n \to 0} \frac{pV}{NkT} = \lim_{n \to 0} \frac{pV}{NRT} = 1, \tag{5.1}$$

where p is the pressure, V the volume, N the number of molecules or moles, respectively, k the Boltzmann constant, R the gas constant, T the thermodynamic temperature, and $n = N/V$ the number density or molar density, respectively. A model gas whose behavior is governed by this equation of state over the whole range of temperatures and pressures is called an ideal gas; i.e.,

$$\left(\frac{pV}{NkT} \right)^{ig} = \left(\frac{pV}{NRT} \right)^{ig} = 1.$$

It was shown in Chapter 3 that this equation of state follows from the general laws of statistical mechanics when the intermolecular potential energy is zero, i.e., when there are no intermolecular forces. Because the influence of intermolecular forces becomes very small at sufficiently low densities, it is plausible that the equation of state for the ideal gas will be realistic for real gases in that limit.

5.1.2 The Low-Density Expansion

The ideal gas equation of state can be applied to real gases even at finite densities, e.g., usually at normal atmospheric pressure, with excellent accuracy. At higher pressures and densities, however, and also close to the dew point even at relatively low pressures, the real gas behavior starts to deviate from that of an ideal gas. At moderate densities the deviations are small enough to be treated as perturbations of the limiting law. In terms of the intermolecular potential energy this means that at moderate density only two or at most three molecules will be found simultaneously in the region of their mutual-interaction force field. The thermodynamic properties of such real gases can be described in terms of a universal density expansion of the compressibility factor around the zero-density limiting law, the virial equation.

The virial equation is an expansion of the compressibility factor, $Z = pV/NkT = pV/NRT$, along individual isotherms in terms of density around $n = 0$ according to

$$Z = 1 + Bn + Cn^2 + \cdots. \tag{5.2}$$

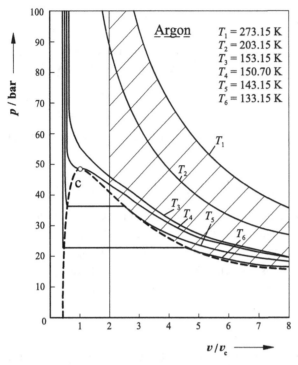

Fig. 5.1. Region of validity of the virial equation (5.2) up to the third virial coefficient for argon (dashed area).

The expansion coefficients $B, C, \ldots$, referred to as virial coefficients, are defined as

$$B = \left(\frac{\partial Z}{\partial n}\right)_{n=0} \tag{5.3}$$

$$C = \frac{1}{2!}\left(\frac{\partial^2 Z}{\partial n^2}\right)_{n=0}. \tag{5.4}$$

They are properties of the gas at $n = 0$. and thus do not depend on density but only on temperature and, in mixtures, on composition. The virial equation is the preferred form for the equation of state of real gases at moderate densities. For temperatures below the critical it is valid up to the saturation line; for supercritical temperatures the region of convergence is not entirely clear. Because only the second virial coefficient and, with less accuracy, the third virial coefficient can be determined experimentally or theoretically, the application of the virial equation is limited practically to densities up to about half the critical density; cf. Exercise 5.1. This is a large region of states in the p–v–T diagram, as demonstrated by the dashed area in Figure 5.1 for argon. So the virial equation is a practically rather useful equation of state model. Besides, it is theoretically important for model development. Because any equation of state can be cast into a virial form, the resulting expression can be checked against experimental

or theoretical results for the virial coefficients. This provides a valuable test of
the model.

EXERCISE 5.1

Expand the equation of state of Redlich and Kwong into the form of a virial equation
and investigate its convergence up to the third virial coefficient.

Solution

The Redlich–Kwong [1] equation is given by

$$p_{RK} = \frac{NkT}{V - b} - \frac{a}{T^{0.5}V(V + b)}.$$

Here V is the volume of the gas, T its temperature, and p its pressure. The quantities a
and b are parameters that are specific for each substance.

For the compressibility factor we have

$$Z_{RK} = \frac{1}{1 - b/V} - \frac{a/(Nk)}{T^{3/2}V(1 + b/V)} = \frac{V^*}{V^* - 1} - \frac{1}{T^*(V^* + 1)}$$

$$= \frac{V_c^*}{V_c^* - 1/V_r} - \frac{1/V_r}{T^*(V_c^* + 1/V_r)},$$

with

$$T^* = T^{3/2}\frac{(Nk)b}{a},$$

$$V^* = V/b, \quad \text{and} \quad V_r = V/V_c,$$

where V_c is the volume at the critical point,

Expansion into the virial form leads to

$$Z_{RK} = 1 + \left(\frac{\partial Z_{RK}}{\partial(1/V_r)}\right)_{1/V_r=0}\left(\frac{1}{V_r}\right) + \left(\frac{1}{2!}\right)\left(\frac{\partial^2 Z_{RK}}{\partial(1/V_r)^2}\right)_{1/V_r=0}\left(\frac{1}{V_r}\right)^2 + \cdots$$

$$= 1 + \frac{(1 - 1/T^*)}{V_c^*}\left(\frac{1}{V_r}\right) + \frac{1}{2!}\frac{(2 + 2/T^*)}{(V_c^*)^2}\left(\frac{1}{V_r}\right)^2 + \cdots$$

$$= 1 + \left(\frac{NB_{RK}}{V_c}\right)\left(\frac{1}{V_r}\right) + \left(\frac{N^2C_{RK}}{V_c^2}\right)\left(\frac{1}{V_r}\right)^2 + \cdots.$$

From the conditions of the critical point,

$$\left(\frac{\partial p}{\partial V}\right)_T = \left(\frac{\partial^2 p}{\partial V^2}\right)_T = 0,$$

we find from the Redlich–Kwong equation

$$V_c^* = 3.847$$

and

$$T_c^* = 0.203.$$

We can now draw the state diagram $Z_{RK} = f(T^*, 1/V_r)$ according to the full Redlich–
Kwong equation and compare it to its virial expansion truncated after the second or
the third virial coefficient. From Figure E 5.1.1 we note that the truncation after the
second virial coefficient is valid up to about $0.2v_c/v$ and that generally the third virial
coefficient must be included to get convergence up to half the critical density. The
permissible density range increases for increasing temperature. The dew point line, too,

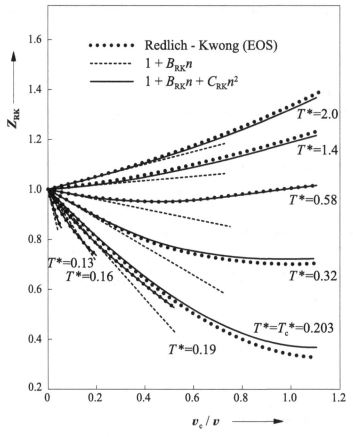

Fig. E 5.1.1. The convergence of the virial equation for the Redlich–Kwong fluid.

can generally only be described with sufficient accuracy when the third virial coefficient is taken into account.

[1] O. Redlich and Y. N. S. Kwong. *Chem. Rev.*, 44:233, 1949.

Virial coefficients can be calculated from p–v–T measurements. At normal temperatures the second virial coefficient is known with satisfactory accuracy; i.e., its uncertainty is usually no more than a few cm^3/mol. Much higher uncertainties must be allowed for the third virial coefficient. The temperature dependence of the second virial coefficient is displayed schematically in Figure 5.2. At low temperatures the second virial coefficient is negative. In this region of temperature, intermolecular attractive forces are dominant. Molecular pairs are formed over nonnegligible periods of time and lead to a reduction of pressure relative to that calculated from the ideal gas law. At high temperatures the attractive forces lose their influence due to the high kinetic energy of the molecules. Instead, the repulsive forces dominate due to the finite volumes of the molecules. This leads to a decrease of the volume in which the gas molecules can move about, with a corresponding increase of pressure relative to the ideal gas value,

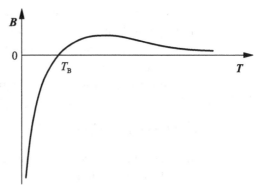

Fig. 5.2. Temperature dependence of the second virial coefficient (schematic).

i.e., to a positive second virial coefficient. At very high temperature the positive second virial coefficient decreases, because the molecular diameter decreases due to the soft nature of the intermolecular repulsive forces. The temperature at which the second virial coefficient passes through zero is called the Boyle temperature, T_B.

If the virial coefficients of a gas are known as a function of temperature and, for mixtures, of composition, all thermodynamic functions depending on the equation of state may be calculated from them. As an example, one finds for the residual internal energy, i.e., the part of the internal energy that is controlled by the intermolecular forces, cf. Section 2.1,

$$U^{\text{res}}(T, V, \{N_j\}) = U(T, V, \{N_j\}) - U^{\text{ig}}(T, \{N_j\})$$
$$= -NkT^2 \left[n \left(\frac{\partial B}{\partial T} \right)_{\{x_j\}} + \frac{n^2}{2} \left(\frac{\partial C}{\partial T} \right)_{\{x_j\}} + \cdots \right], \quad (5.5)$$

where N_j is the number of molecules of component j and the composition dependence is in the virial coefficients. We immediately note that

$$(\partial u^{\text{res}}/\partial v)_{T,\{x_j\}} = (kT^2/v^2)(\partial B/\partial T)_{\{x_j\}} + \cdots$$

goes to zero for $v \to \infty$, in agreement with Joule's experiments and the behavior of an ideal gas. We also note, however, that

$$(\partial u^{\text{res}}/\partial n)_{T,\{x_j\}} = kT^2(\partial B/\partial T)_{\{x_j\}} + \cdots$$

remains finite at $n \to 0$, in contrast to ideal gas behavior. This underlines the statement made in Chapter 3 that a real gas in the limit of zero density is not identical with an ideal gas. A particular thermodynamic function that is related to the second virial coefficient is the isenthalpic throttling coefficient or Joule–Thomson coefficient extrapolated to zero pressure:

$$\mu^{\circ}(T) = \lim_{p \to 0} \left(\frac{\partial T}{\partial p} \right)_{h,\{x_j\}} = \frac{1}{c_p^{\text{ig}}} \left(T \left(\frac{\partial B}{\partial T} \right)_{\{x_j\}} - B \right). \quad (5.6)$$

This coefficient can be measured with good accuracy and yields valuable direct information about the temperature dependence of the second virial coefficient and thus the potential energy model; cf. Section 5.3. We note that the isenthalpic Joule–Thomson coefficient μ° is identically zero for the ideal gas, whereas it is generally nonzero for real gases in the zero-pressure limit.

5.1.3 The Hard Body Limit

Although at low densities all equation of state models must approach the universal low-density limit, i.e., the ideal gas equation of state, there is another limit at high densities and high temperatures. As we have noted in (4.7), the thermodynamic functions can generally be split into a repulsive and an attractive contribution. Similarly to the excess function models, we thus also expect that equation of state models can be represented as a sum of a repulsive and attractive part; i.e.,

$$p^{\mathrm{res}} = p^{\mathrm{res}}_{\mathrm{rep}} + p^{\mathrm{res}}_{\mathrm{att}}, \tag{5.7}$$

where the suffix res stands again for residual and points to that contribution to the equation of state that is controlled by intermolecular interactions. At high temperatures the attractive part of the intermolecular potential energy becomes negligible and the equation of state is controlled by the repulsive part. Then, at liquid-like densities, fluid phase behavior can be represented quite accurately by modeling the molecules as hard bodies. Because analytical equation of state models for hard bodies of various shapes are available, it is very useful to incorporate this limit into any molecular model for the equation of state. Most modern equations of state make use of this separation (5.7) in one way or another.

5.2 Intermolecular Potential Energy

The intermolecular potential energy for equation of state models must reflect the density dependence of this energy contribution, i.e., the dependence on the distances between the molecules. Contact values for pair interactions, adequate for excess function models, are no longer sufficient. As a consequence, more complicated formulations of the intermolecular potential energy are needed, taking into account distance- and orientation-dependent repulsive and attractive interactions. Some basic aspects of the distance dependence are well known. Gases with large average distances between the molecules condense to liquids when the temperature and thus the kinetic energy of the molecules is lowered. This points to the existence of attractive forces at large distances between the molecules. Liquids with short average distances between the molecules show a strong resistance to compression, from which the existence of short-range repulsive forces can be concluded.

When studying the intermolecular interactions on the basis of classical electrostatics in Section 2.4 and quantum mechanical perturbation theory in Section 2.5, we found that there are three types of forces at large intermolecular distances, i.e., at distances far beyond the molecular diameter. First, there are the electrostatic interactions that arise from rigid nonsymmetric charge distributions, as represented by the multipoles of a molecule. They are obtained either from classical electrostatics or from first-order quantum mechanical perturbation theory. Additional contributions originate from the fact that the charge distributions in molecules are not rigid. They appear in the higher orders of the perturbation expansion. So, in an interaction of one molecule with permanent multipoles and another molecule with or without permanent multipoles, the electrical field of the polar molecule induces a redistribution of the charges in the second molecule, an effect referred to as polarization, and described by a particular property of the molecules referred to as polarizability. By way of polarization, multipole moments are induced in the second molecule, which interact with the permanent multipoles of the first molecule. The resulting intermolecular forces are called induction forces. Finally, there obviously exist long-range attractive forces in such cases, where none of the considered molecules has permanent multipole moments, as in the interaction between two argon atoms. The electrons in an atom or molecule are in continual motion; i.e., the electron density fluctuates, while the molecule runs through its excited states. So, in all molecules, even those with a spherical symmetrical charge distribution on the average, instantaneous asymmetries of the charge distribution will arise, which can be considered as instantaneous multipole moments. An instantaneous multipole in one molecule induces an instantaneous multipole in another molecule. The associated intermolecular forces are referred to as dispersion forces. They are present in all molecular interactions and in many applications are much more important than the multipole and induction forces. All these long-range forces, for which closed formulae are available on the basis of quantum mechanical perturbation theory, are essentially attractive. They have significant influence on the condensation behavior of vapors, in particular the vapor–liquid equilibria in mixtures. We note that some fluids are further strongly influenced by a particular type of attractive interaction, referred to as hydrogen bonding; cf. Section 4.2.

Short-range forces, in contrast, i.e., forces arising at intermolecular distances of about the molecular diameter and below, are essentially repulsive. They arise when two atoms approach each other so closely that their electron clouds overlap and the electrons repel each other. In this spirit they are frequently referred to as overlap forces. Their origin is due to the Pauli principle, which restricts the number of electrons in the overlapping region and modifies the shapes of the electron clouds in such a way that repulsive interactions result. These strong repulsive forces depend significantly on the shape of the molecules. They tend to determine the structure of liquids and thus also have a strong influence on the macroscopic properties. Unfortunately, and unlike the long-range forces, there are not yet analytical formulations for them available that are based on quantum mechanics and can be practically applied. However, reasonable semiempirical models, based on the shape of the molecules, can be formulated.

5.2.1 The Pairwise Additivity Approximation

In a typical thermodynamic system, e.g., a liquid or a dense gas, very many molecules interact simultaneously, as visualized in Figure 5.3 for a system of four molecules. To a first approximation the intermolecular potential energy in a system of spherical molecules is then given by the summation over all pair energies, i.e.,

$$U = \sum \sum_{i < j} \phi(r_{ij}),$$
(5.8)

where $\phi(r_{ij})$ is the pair potential, depending on the distance between the molecular centers of the molecules i and j. This is the pairwise additivity approximation. It cannot be an accurate representation of the intermolecular energy function of the system, because the presence of a third molecule will generally disturb the interaction between two molecules because of polarization effects; cf. Section 2.4. However, it is surprisingly good at all densities. For accurate modeling it is possible to correct for nonadditivity effects, e.g., by adding non-additive three-body and higher order forces; cf. App. 5. We then have the general equation

$$U = \sum \sum_{i < j} \phi(r_{ij}) + \sum \sum \sum_{i < j < k} \Delta\phi(r_{ij}r_{ik}r_{jk}) + \cdots.$$
(5.9)

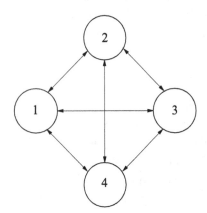

Fig. 5.3. The pairwise additivity approximation.

$U(r_1, r_2, r_3, r_4) = \phi_{12} + \phi_{13} + \phi_{14} + \phi_{23} + \phi_{24} + \phi_{34}$

The term involving the triple sum is the *additional* contribution to the sum of pair potentials made by the presence of a third molecule. It represents the first-order nonadditivity correction. The idea behind writing the intermolecular energy function in the form of (5.9) is the expectation that the summation will rapidly converge. Clearly, the pairwise additivity approximation will become exact at sufficiently low density and progressively worse at increasing density. However, not all properties are very sensitive to the associated inaccuracies. Due to some cancellation in the higher-order terms, the pairwise additivity approximation gives better results at liquid densities than could otherwise be expected. In most molecular models to be discussed in this book we use the pairwise additivity approximation. We note that the multipolar forces are strictly pairwise additive, because they reflect single-molecule charge distributions in the ground state, which are independent of neighbor molecules.

5.2.2 The Rigid Molecule Approximation

In this book, we focus on equation of state models for relatively simple molecules. In small and many medium-sized molecules, such as those shown in Figure 1.13, the internal motions of the molecules are rapid and of small scale. They thus depend only on the *intra*molecular forces that are active between various locations within a molecule, such as atoms or groups of atoms, but do not depend on the external coordinates of the other molecules in the system. The internal motions are then fully kinetic, i.e., do not have a potential contribution, because they are properties of the single molecules and do not depend on the molecular environment. By the same token, the intermolecular potential energy is independent of the internal coordinates of the molecules and just depends on the locations of the molecular centers and the external orientation coordinates. We refer to such small molecules as rigid molecules and the approximation is called the rigid molecule approximation. Also, for small molecules, it is plausible to assume that the various contributions to the kinetic energy of a molecule are independent of each other. So we assume in particular that the kinetic translational motion will not be coupled to the other types of kinetic motion and further that these latter will also be independent of each other. In this spirit we write for the Hamilton function of the molecular system in the rigid molecule approximation

$$H = H_{tr}^{kin} + H_r^{kin} + H_v^{kin} + H_{ir}^{kin} + U. \tag{5.10}$$

On the basis of (5.10) we thus can treat the kinetic external rotational energy, say, as independent of the potential energy, although, clearly, the total external rotational motion of a molecule, just like translation, may well be influenced by neighbor molecules. This is taken into account by the external rotational coordinates of the intermolecular potential energy U. We thus have a factorization of the canonical partition function as

$$Q = Q_{tr}^{kin} Q_r^{kin} Q_v^{kin} Q_{ir}^{kin} Q^C, \qquad (5.11)$$

where Q_{tr}^{kin} contains the correction for translational indistinguishability and the translational part of the uncertainty correction. Further quantum mechanical corrections due to indistinguishability and uncertainty arise in the semiclassical treatment of the other molecular degrees of freedom and have been discussed in Chapter 3. Finally, Q^C is the configurational partition function, also referred to as the configurational integral. The contributions to the thermodynamic functions due to the configurational partition function are referred to as configurational contributions and receive a suffix C, as before. The configurational partition function is an integral over all the configurational coordinates of the molecules on which the intermolecular potential energy depends. In the rigid molecule approximation the intermolecular energy function contains only translational and external rotational coordinates, i.e., those related to the locations of the molecular centers and the associated orientations. We thus have

$$Q^C = \int\limits_{r^N} \int\limits_{\omega^N} e^{-U(r^N\omega^N)/kT} \mathrm{d}r^N \mathrm{d}\omega^N. \qquad (5.12)$$

We note that the translational coordinates of the configurational integral obviously make the configurational partition function depend on volume, in contrast to the kinetic contributions to the partition function, which thus only depend on temperature. In particular, for rigid molecules this implies that the contributions of internal molecular motions are density-independent. This, together with the relatively simple form of the intermolecular potential energy $U(r^N\omega^N)$, simplifies the equation of state considerably.

5.2.3 Spherical Interaction Models

We consider a system of two argon atoms, 1 and 2, in vacuo. They each consist of a positively charged nucleus that is surrounded by a negatively charged spherically symmetric electron cloud. Clearly, the force field is spherically symmetric, i.e., just depends on the distance between them. When the two molecules are infinitely far apart ($r_{12} \to \infty$), there is no interaction between them and the total energy of the system is equal to the sum of the individual kinetic energies of the molecules. When the two atoms approach each other and reach a finite distance r_{12}, intermolecular forces between them become important. The total energy of the system is then no longer equal to the sum of the individual energies but rather contains an additional term $\phi(r_{12})$, to which we refer as

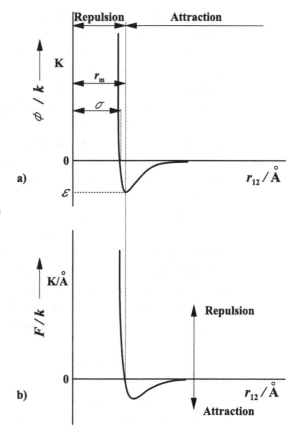

Fig. 5.4. (a) Pair potential ϕ and **(b)** Force
F between two atoms.

the intermolecular potential energy, in particular the pair potential. The force
exerted by atom i on atom j is calculated from ϕ via

$$\boldsymbol{F}_j = -\frac{d\phi}{d\boldsymbol{r}_j} = -\frac{d\phi}{d\boldsymbol{r}_{ij}} = -\boldsymbol{e}_{ij}\frac{d\phi(r_{ij})}{dr_{ij}}, \qquad (5.13)$$

with $\boldsymbol{r}_{ij} = \boldsymbol{r}_j - \boldsymbol{r}_i$ as the distance vector between the two atoms pointing from
i to j and $\boldsymbol{e}_{ij}$ as the unit vector in this direction. We thus have

$$\phi(r_{12}) = -\int_{\infty}^{r_{12}} \boldsymbol{e}_{12} \cdot \boldsymbol{F}_2 dr'_{12} = -\int_{\infty}^{r_{12}} F dr'_{12}, \qquad (5.14)$$

as discussed more specifically for the interaction between single charges in
Section 2.4. The negative sign in (5.13) and (5.14) ensures that a force is counted
as positive if it is repulsive, because diminishing the distance between the two
molecules will increase the potential energy between them at short distances.
A negative force is, therefore, attractive. Figure 5.4 shows a typical plot of the
intermolecular interaction energy between two atoms and the associated force
F. The essential parameters of the pair potential are the distances σ or r_m
where the function passes through zero or has its minimum, respectively, and
the depth of the potential well, which is referred to as ε. A strong repulsive force

is shown at short distances, whereas a far weaker attractive force is effective at large distances. The units chosen are Å for the distance and K for the energy, which is measured here in multiples of the Boltzmann constant. Molecule 1 is located at the origin of the coordinate system. We note that for $\sigma < r_{12} < r_m$ we have a repulsive force although the potential energy is negative, which conflicts with the common identification of negative interaction energies with attractive forces. The reference state of the intermolecular potential energy, here in particular of the pair potential, is the ideal gas, i.e., zero energy. This is obvious from the formal definition of $\phi(r_{12})$ in (5.14) and is consistent with the standard reference state for equations of state.

We note that the interaction between two argon atoms can be represented in terms of a function that depends on nothing else than the distance between the nuclei. This is a consequence of the Born–Oppenheimer approximation, according to which the motions of the nuclei and the electrons can be separated; cf. Section 2.3 and 2.5. Due to the very large mass ratio between the atomic nucleus and the electrons, the motion of the electrons is very much faster than that of the nuclei. At each fixed distance r_{12} between the nuclei in motion there is thus a sufficient number of electronic fluctuations to build up the force field; i.e., the nuclei move in a force field that is due to the rapid fluctuations of the electron configurations and that has a fixed value at each value of the distance r_{12}. In the following we shall omit the index 12 at the distance between the molecular centers for notational simplicity.

A pair potential such as the one shown in Figure 5.4 can be represented analytically. A well-known empirical function with only two adjustable parameters ε and σ is the Lennard–Jones (12–6) potential, i.e.,

$$\phi(r) = 4\varepsilon \left[\left(\frac{r}{\sigma}\right)^{-12} - \left(\frac{r}{\sigma}\right)^{-6} \right], \tag{5.15}$$

where the r^{-6}-term follows from quantum mechanical perturbation theory, cf. App. 5, whereas the r^{-12}-term is empirical. Far more complicated functions with many parameters to be fitted to experimental data have been used to describe the interaction between two atoms [2]. A sensible compromise between simplicity and accuracy is provided by the MSK potential [3],

$$\phi(r) = \frac{6}{n-6} \, \varepsilon \left[\left(\frac{r_m - d}{r - d}\right)^n - \frac{n}{6} \left(\frac{r_m - d}{r - d}\right)^6 \right], \tag{5.16}$$

with

$$n = 12 + 5 \left(\frac{r}{r_m} - \frac{d}{r_m} - 1 \right), \tag{5.17}$$

containing three adjustable parameters, i.e., ε, r_m, and d. A large variety of potential functions for spherical molecules have been suggested [4].

For fundamental theoretical investigations, highly simplified pair potential functions are useful, which, while allowing analytical solutions of the statistical

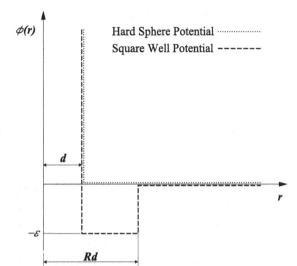

Fig. 5.5. Model potentials of the hard-sphere type.

mechanical equations, still contain the basic features of realistic intermolecular interactions. The two most common models of this kind are the hard-sphere and the square-well potential. The hard-sphere potential is defined by

$$\phi(r) = \infty \quad \text{for } r \leq d_-, \tag{5.18}$$

$$\phi(r) = 0 \quad \text{for } r \geq d_+. \tag{5.19}$$

Here d_- denotes an approach from $r < d$, whereas d_+ denotes an approach from $r > d$. This potential describes the interaction between two hard spheres of diameter d and thus cannot model the attractive forces between molecules. It is plotted as the dotted line in Figure 5.5. The square-well potential is defined by

$$\phi(r) = \infty \qquad \text{for } r \leq d_-, \tag{5.20}$$

$$\phi(r) = -\varepsilon \qquad \text{for } (Rd)_- \geq r \geq d_+, \tag{5.21}$$

$$\phi(r) = 0 \qquad \text{for } r \geq (Rd)_+. \tag{5.22}$$

Here ε is the depth and R a measure for the width of the potential well. This model represents a rather simple extension of the hard-sphere model to incorporate attractive energies and is plotted as the dashed line in Figure 5.5.

In mixtures, combining rules for the potential parameters are needed. As shown in Section 2.5, the approximate London formula for the dispersion interaction between spherical molecules on the dipole polarizability level leads to

$$\varepsilon_{\alpha\beta}\sigma_{\alpha\beta}^6 = \frac{2(\varepsilon_{\alpha\alpha}\sigma_{\alpha\alpha}^6)(\varepsilon_{\beta\beta}\sigma_{\beta\beta}^6)}{\varepsilon_{\alpha\alpha}\sigma_{\alpha\alpha}^6\alpha_\beta^2 + \varepsilon_{\beta\beta}\sigma_{\beta\beta}^6\alpha_\alpha^2}\alpha_\alpha\alpha_\beta \tag{5.23}$$

for the ε-parameters, whereas, by analogy to hard spheres, it is plausible to assume that

$$\sigma_{\alpha\beta} = \frac{1}{2}(\sigma_{\alpha\alpha} + \sigma_{\beta\beta}), \qquad (5.24)$$

with analogous rules for d and r_m. The rule (5.23) is frequently replaced by a simple geometric mean, cf. (2.255), and (2.255) and (5.24) are then referred to as the Lorentz–Berthelot combining rules. Although (5.23) is in most cases an improvement over the geometric mean for spherical interactions, it is by no means exact. So, although predictions for mixtures from pure components using the above combining rules are frequently at least reasonable approximations, there remains some uncertainty about the modeling of the dispersion forces, which precludes reliable predictions of mixture behavior from that of the pure components. For accurate representation of data, therefore, it has become common practice to introduce a binary interaction parameter to correct the geometric mean rule. We shall not use this, however, because we wish to demonstrate the progress achieved in using molecular models without empirical corrections based on data.

5.2.4 Nonspherical Interaction Models

Although the intermolecular potential energy in systems of spherical molecules can be modeled in terms of just the position coordinates of the molecular centers, this is no more than a crude approximation for polyatomic molecules. The force field between two polyatomic molecules is no longer spherically symmetric. At a fixed distance between the molecular centers, the pair potential depends not only on the distance coordinate but also on the orientations of the two molecules. As a consequence, the rotation of the molecules, like the translation, contains a potential part in addition to the kinetic energy. Further, the pair potential will in principle also depend on the vibrational and internal rotational motions of the molecule. Simple molecules, however, can be treated in the rigid molecule approximation. Their vibrational frequencies are high and essentially identical in the gaseous and liquid states. The vibrational contribution thus does not depend on the molecular environment and configuration, and thus there are no vibrational coordinates in the expression for the intermolecular energy function. Vibration is then taken care of entirely by the ideal gas contribution; cf. Chapter 3. Analogous conclusions are valid for internal rotations and for the electronic energy. Thus, the intermolecular energy function of rigid molecules can be evaluated for the intramolecular ground state. We further refer by simple molecules to those that are small enough so that they have a genuine center. As a result, the pair potential between two polyatomic molecules, 1 and 2, can be expressed in terms of the distance r

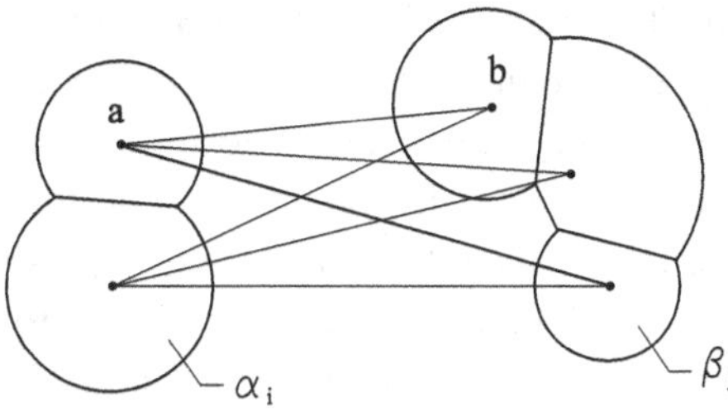

Fig. 5.6. The "site–site" potential.

between the molecular centers and the orientations of the molecules ω_1 and ω_2; i.e.,

$$\phi = \phi(r\omega_1\omega_2). \tag{5.25}$$

Here ω is a shorthand for the three Euler angles (θ, ϕ, χ) defining the orientation of a three-dimensional rigid structure in a space fixed coordinate system; cf. Figure 2.7.

There are various different origins for the orientational dependence of the pair potential between two simple nonspherical molecules. In Section 2.4 and 2.5 we treated various types of long-range forces. In particular, we derived results for the multipolar forces in the form of (5.25). In App. 5 general formulae for induction and dispersion pair potentials in the form of (5.25) are summarized. Of particular relevance, and not treated until now, are the orientational effects at short range due to the nonspherical shape of a molecule. There is as yet no analytically evaluable theory based on quantum mechanics for this contribution to the intermolecular potential energy. A straightforward semiempirical extension of the spherical model to the orientation-dependent pair potential of simple nonspherical molecules due to nonspherical shape is to assume spherical repulsive interactions between strategic sites on the molecules. This model is referred to as the "site–site" repulsion potential model (SSR); cf. Figure 5.6. The sites can be located at the atomic centers. Frequently, a more realistic model results when their distances from the molecular centers are adjusted by a proportionality factor on the basis of experimental or quantum-chemical data. In any case, the quantum-chemically defined shapes of the molecules are thus incorporated into the model, for linear as well as for more complicated molecular shapes. Obviously, such a model yields nonspherical potential contours, i.e., nonspherical lines of constant potential, as required for nonspherical molecules. The particular repulsive law between the sites is not prescribed theoretically. A convenient assumption is that it follows an r^{-12} dependence, as in the Lennard–Jones (12–6) potential (5.15); i.e.,

$$\phi_{\alpha_i\beta_j}^{SSR}(r_{\alpha_i\beta_j}\omega_{\alpha_i}\omega_{\beta_j}) = \sum_{a,b} 4\varepsilon_{ab}\sigma_{ab}^{12}r_{ab}^{-12}. \tag{5.26}$$

Here r_{ab} is the distance between site a in one molecule and site b in another molecule. For dimensional reasons this distance is reduced by a parameter σ_{ab}

and the associated energy parameter is ε_{ab}. Both parameters are empirical and have to be introduced, as in the Lennard–Jones potential between molecular centers, because there are no appropriate molecular constants in terms of which the repulsive pair potential can be formulated. According to (5.26), we need for each pair a, b of interacting repulsive sites in different molecules a value for the associated parameter $\varepsilon_{ab}\sigma_{ab}^{12}$. Because the site–site repulsive potential must converge to the Lennard–Jones r^{-12} repulsive potential between molecular centers for zero eccentricity, there will be a general relation between the parameters,

$$\sum_{a,b} \varepsilon_{ab}^* \sigma_{ab}^{*12} = 1, \tag{5.27}$$

where $\varepsilon_{ab}^* = \varepsilon_{ab}/\varepsilon$ and $\sigma_{ab}^* = \sigma_{ab}/\sigma$, with ε, σ as the parameters of the spherical center-to-center Lennard–Jones (12–6) potential. For homonuclear molecules such as oxygen there is only one type of interaction and the associated parameter of the site–site repulsive potential is, therefore, fixed as

$$\varepsilon_{aa}^* \sigma_{aa}^{*12} = 0.25.$$

For molecules made up of different atoms, (5.27) still holds but is insufficient to determine the parameters of the site–site repulsive potential. Different atoms will experience different repulsive forces at the same distances. It is plausible that the parameter $\varepsilon\sigma^{12}$ will increase with increasing diameter of the atom. An ad hoc rule based on this tendency is

$$\varepsilon_{aa}\sigma_{aa}^{12} = \left(\frac{m_a}{m_b}\right)^n \varepsilon_{bb}\sigma_{bb}^{12}, \tag{5.28}$$

where $n = 2$ has been found adequate but is empirical. A square mass-ratio law for $\varepsilon\sigma^{12}$ is essentially supported by the Lennard–Jones parameters of the heavy noble gases. Further, an additional rule for $\varepsilon_{ab}\sigma_{ab}^{12}$ is needed. A simple combination rule can be developed from a mechanical spring model [5] and gives

$$\varepsilon_{ab}\sigma_{ab}^{12} = \left[\frac{\left(\varepsilon_{aa}\sigma_{aa}^{12}\right)^{1/13} + \left(\varepsilon_{bb}\sigma_{bb}^{12}\right)^{1/13}}{2}\right]^{13}. \tag{5.29}$$

In a mixture, ε is replaced by $\varepsilon_{\alpha\beta}$ and σ by $\sigma_{\alpha\beta}$. The combination rule (5.29) then transforms into

$$\frac{\varepsilon_{a\alpha b\beta}}{\varepsilon_{\alpha\beta}}\left(\frac{\sigma_{a\alpha b\beta}}{\sigma_{\alpha\beta}}\right)^{12} = \left[\frac{\left(\frac{\varepsilon_{\alpha\alpha}\sigma_{\alpha\alpha}^{12}}{\varepsilon_{\alpha\beta}\sigma_{\alpha\beta}^{12}}\varepsilon_{a_\alpha a_\alpha}^* \sigma_{a_\alpha a_\alpha}^{*12}\right)^{1/13} + \left(\frac{\varepsilon_{\beta\beta}\sigma_{\beta\beta}^{12}}{\varepsilon_{\alpha\beta}\sigma_{\alpha\beta}^{12}}\varepsilon_{b_\beta b_\beta}^* \sigma_{b_\beta b_\beta}^{*12}\right)^{1/13}}{2}\right]^{13}. \tag{5.30}$$

By using these rules the parameters of the site–site repulsive potential can be calculated a priori, i.e., without any fitting to data. Clearly, except for (5.27), these rules are semiempirical, without a firm quantum mechanical basis, as is the site–site model itself.

EXERCISE 5.2

Consider a diatomic molecule AB with the center-to-center Lennard–Jones(12–6) parameters $\varepsilon/k = 600$ K and $\sigma = 3.700$ Å. The interatomic distance is $d = 1.400$ Å; the molar masses of the atoms are $M_A = 12$ g/mol and $M_B = 10$ g/mol. Repulsive sites are located at the centers of the atoms.

  (a) Sketch qualitatively the potential contours for the SSR potential
  (b) Evaluate the interaction energy between two AB molecules for the configuration
      $\theta_1 = \theta_2 = 90°$ and $\phi_{12} = \phi_2 - \phi_1 = 0$ and a center-to-center distance of 3.700 Å.

Solution

The site–site potential parameters can be calculated from the equations (5.27), (5.28), and (5.29). We have

$$\varepsilon_{AA}\sigma_{AA}^{12} + 2\varepsilon_{AB}\sigma_{AB}^{12} + \varepsilon_{BB}\sigma_{BB}^{12} = \varepsilon\sigma^{12} = 5.4533 \cdot 10^{-14} \text{ J Å}^{12}$$

$$\varepsilon_{AA}\sigma_{AA}^{12} = \left(\frac{M_A}{M_B}\right)^2 \varepsilon_{BB}\sigma_{BB}^{12}$$

$$\varepsilon_{AB}\sigma_{AB}^{12} = \left[\frac{(\varepsilon_{AA}\sigma_{AA}^{12})^{1/13} + (\varepsilon_{BB}\sigma_{BB}^{12})^{1/13}}{2}\right]^{13} .$$

This gives

$$\varepsilon_{AA}\sigma_{AA}^{12} = 1.1260 \times 10^{-14} \text{ J Å}^{12}$$

$$\varepsilon_{BB}\sigma_{BB}^{12} = 1.6214 \times 10^{-14} \text{ J Å}^{12}$$

$$\varepsilon_{AB}\sigma_{AB}^{12} = \varepsilon_{BA}\sigma_{BA}^{12} = 1.3529 \times 10^{-14} \text{ J Å}^{12} .$$

  (a) The potential contours are obtained from the spherically symmetric force fields associated with both atoms and are shown schematically in Figure E 5.2.1
  (b) The configuration $\theta_1 = \theta_2 = 90°$, $\phi_{12} = 0$ is sketched in Figure E 5.2.2. The distance between sites A and B is found to be $r_{AB} = 3.9560$ Å, leading to

$$\phi^{SSR} = 4\left(\varepsilon_{AA}\sigma_{AA}^{12}r_{AA}^{-12} + 2\varepsilon_{AB}\sigma_{AB}^{12}r_{AB}^{-12} + \varepsilon_{BB}\sigma_{BB}^{12}r_{BB}^{-12}\right)$$
$$= 2.4061 \cdot 10^{-20} \text{ J}$$

    or

$$\phi^{SSR}/k = 1742.7 \text{ K} .$$

Fig. E 5.2.1. The potential contours (schematic).

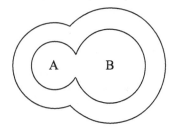

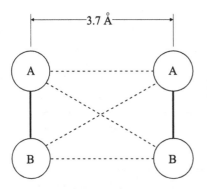

Fig. E 5.2.2. The considered configuration.

The dependence of the site–site repulsion potential on the center-to-center distance r and the orientational angles ω_1 and ω_2 is not explicit in (5.26). So we have not yet formulated the short-range forces in the form of (5.25) analytically. If required, such an explicit formulation can be worked out by suitable coordinate transformations [6]. However, in most numerical work, the potential energy for the various configurations can simply be evaluated along the lines of Exercise 5.2. Also, the SSR potential can be represented in the form of a spherical harmonic expansion. This representation brings (5.26) into a form analogous to that of the long-range interactions (2.125) and (2.126). Although the convergence of this expansion can be shown to be poor [6], it is a good approximation for small excentricities and is useful in perturbation models; cf. Section 5.5.

We can combine the SSR potential for the anisotropic short-range repulsive forces with the long-range potential as obtained from quantum mechanical perturbation theory in the multipole approximation. We then arrive at a comprehensive model for the intermolecular pair interactions of simple polyatomic molecules, referred to as the SSR-MPA model. As an illustration, Figure 5.7 shows the r-dependence of a SSR-MPA pair potential, fitted to experimental data of CO_2 for various orientations. The adjusted parameters are the spherical interaction parameters ε, σ and the excentricity $r_a = r_b$, where r_a, r_b are the locations of the two repulsive sites with respect to the molecular center of CO_2. CO_2 being a linear molecule, two angles θ, ϕ fully define the orientation of one molecule in space. Depending on the orientation, the r-dependence may become qualitatively different from that of a monatomic substance. It is clear then that the parameters ε, σ here do not have the same simple significance as for the monatomic fluid. Some basic features of the pair potential in Figure 5.7 are quite evident. When both θ-angles are 90°, the two molecules can approach each other quite closely before repulsive interactions become relevant. The opposite is true when both angles are zero, because then the electron clouds of two atoms belonging to different molecules overlap already at much greater distances between the molecular centers. Clearly, this will also affect the attractive part of

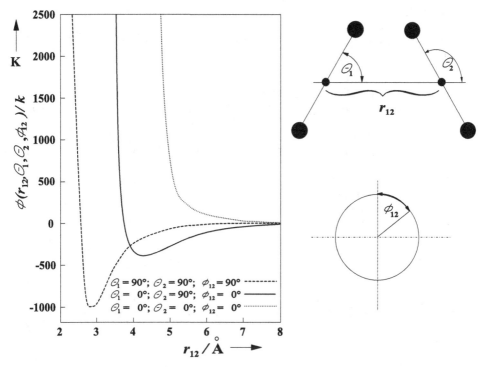

Fig. 5.7. Pair potential of two linear molecules for different orientations (fitted to data of CO_2).

the potential energy. Similar effects can be attributed to a variation of the angle ϕ_{12}. To demonstrate the quality of this pair potential, Table 5.1 shows a comparison of computer-simulated data for CO_2 with experimental data over a large region of states [7]. The precision of the Monte Carlo simulations was checked by performing successive calculations with an increasing number of configurations and by varying the number of molecules; cf. Section 2.6. Generally, 256 molecules were found to be sufficient here, in the sense that check calculations with 500 molecules did not alter the calculated pressure significantly. There are, however, a few state points close to the saturation line and/or in regions of high compressibility where one must rely on calculations with 500 molecules. Such state points are identified in the table. The number of configurations was between 2.5 and 10 million. Typically, a statistical error of ±5 bar in these calculations has to be taken into account for the pressure, whereas the configurational internal energy has a precision of ±10 J/mol. When looking at the comparison between experimental data and Monte Carlo data of CO_2, we see that the differences for the pressure at most state points are larger than those due to simulation imprecision. We thus have a deficiency due to the model of the intermolecular forces, as could be expected. Still, the overall agreement between simulation and experiment is quite satisfactory. We note a particularly close agreement

Table 5.1. Monte Carlo data for CO_2 compared with experimental data

T K	v cm^3/mol	$-p_{MC}^{res}$ bar	$-p_{exp}^{res}$ bar	$-u_{MC}^{C}$ kJ/mol	$-u_{exp}^{C}$ kJ/mol
230	38.941	449	481	13.19	13.24
	38.202	386	401	13.43	13.49
	36.974	181	217	13.87	13.92
	36.082	5	30	14.18	14.25
250	42.051	476	476	12.01	12.05
	40.864	389	409	12.35	12.40
	37.663	39	52	13.36	13.43
	35.690	−395	−418	14.03	14.16
270	45.906	457	439	10.83	10.86
	44.485	418	405	11.14	11.18
	39.515	76	68	12.50	12.53
	36.919	−388	−392	13.30	13.41
300	190.650	66	66	3.33	3.41
	54.871	364*	355	8.969*	8.92
	48.567	307	314	9.982	9.924
	42.755	84	83	11.25	11.21
	40.145	−169	−179	11.96	11.93
	38.964	−354	−360	12.31	12.29
350	490.44	9	9	1.178	1.188
	192.26	51	51	2.838	2.845
	71.54	220*	207	6.661*	6.623
	49.700	95	86	9.322	9.218
	42.799	−335	−320	10.78	10.68
400	115.54	94	88	4.105	4.089
	65.19	118	108	6.910	6.803
	58.979	76	64	7.604	7.490
	50.161	−121	−137	8.888	8.767
	47.159	−295	−295	9.446	9.323
500	187.16	22	22	2.339	2.326
	82.218	14	6	5.108	5.038
	73.180	−24	−32	5.737	5.628
	57.042	−271	−271	7.337	7.186
600	244.00	5	4	1.670	1.616
	106.02	−26	−29	3.740	3.624
	76.627	−139	−149	5.163	4.989
700	294.92	−4	−3	1.298	1.226
	128.22	−46	−46	2.948	2.790
	78.489	−245	−259	4.778	4.576

Note: Calculations with *were performed with 500 molecules, all others with 256.

between the simulated and experimental data for the configurational part of the internal energy. It is typically better than 1%, except for high temperatures, where the accurately known ideal gas contribution will dominate the final values. The less impressive agreement for the pressure, notably in the typical liquid region, results from the high sensitivity of the virial function to intermolecular configuration in that range. The computer-generated data can be used to fit an empirical equation of state from which then all thermodynamic properties of CO_2 in the fluid state can be predicted with good technical accuracy [7].

As an alternative to the SSR-MPA model, the site–site potential model can be applied to the dispersion part of the interactions between the sites as well. We then arrive at the site–site interaction model, which also takes the anisotropy of the dispersion interaction into account in an empirical manner. It can be combined with point charges to include electrostatic effects and then is referred to as the SS-PC model. Also, we can combine the site–site interaction model with point multipoles at the molecular centers, as in the SSR-MPA, and then refer to it as the SS-PM model. Various other representations of the pair potential between nonspherical molecules have been suggested in the literature. To apply them to mixtures, we need combination rules for $\varepsilon_{\alpha\beta}$ and $\sigma_{\alpha\beta}$. Because these refer to spherical interactions, the London formula can be used. Although this is frequently a good approximation, it must be realized that it is not exact. Also, the pure fluid parameters are fitted to data and thus contain contributions beyond the spherical dispersion interactions, which may disturb the predictive power of the London formula. For a reliable representation of binary interactions, the simple geometrical mean (2.255) can be used and corrected by an empirical interaction parameter. Predictions of multicomponent interactions are then usually satisfactory. As explained above, this will not be done here, in order to demonstrate the predictive capacity of molecular models.

The pair potential of simple nonspherical molecules depends on the mutual orientations of the two molecules, in addition to the dependence on the distance between the molecular centers. In numerous applications, it is desired to avoid any explicit consideration of the angular dependence and to consider the intermolecular interaction as spherically symmetric even for nonsymmetric molecules. This can properly be done by replacing the Boltzmann factor associated with the angle-dependent pair energy $\phi(r\omega_1\omega_2)$ by a Boltzmann factor associated with a spherical, temperature-dependent pair energy $\widetilde{\phi}(r; T)$ defined as [8]

$$e^{-\widetilde{\phi}(r;T)/kT} = \frac{\iint e^{-\phi(r\omega_1\omega_2)/kT} d\omega_1 d\omega_2}{\iint d\omega_1 d\omega_2}. \tag{5.31}$$

By performing the temperature derivative of (5.31), it is easily shown that

$$\widetilde{\phi}(r; T) = \overline{\phi}(r; T) + T\frac{d\widetilde{\phi}(r; T)}{dT},$$

where

$$\overline{\phi}(r; T) = \frac{\iint \phi(r\omega_1\omega_2)e^{-\phi(r\omega_1\omega_2)/kT} d\omega_1 d\omega_2}{\iint e^{-\phi(r\omega_1\omega_2)/kT} d\omega_1 d\omega_2}. \tag{5.32}$$

By an appeal to classical thermodynamics, cf. (2.2), it thus becomes clear that $\widetilde{\phi}(r; T)$ is an intermolecular free energy, in contrast to $\overline{\phi}(r; T)$, which is just the usual intermolecular energy of the pair interaction averaged over the angles. In agreement with this, (5.31) immediately gives

$$\tilde{\phi}(r; T) = -kT \ln \left[\frac{\iint e^{-\phi(r\omega_1\omega_2)/kT} d\omega_1 d\omega_2}{\iint d\omega_1 d\omega_2} \right], \tag{5.33}$$

where the term in brackets represents the canonical partition function of the system under consideration. We note that we can thus replace the angle-dependent multipole interaction energy with a spherical interaction energy that depends on temperature and actually has the significance of an intermolecular free energy.

EXERCISE 5.3

Derive an expression for the orientational average of the pair potential between two dipoles.

Solution

To compute the orientational average of the Stockmayer potential, numerical values for many orientations must be generated and associated with proper weights. The simple average $\iint \phi d\omega_1 d\omega_2$, giving equal weights to all possible configurations, will yield a value of zero. We know, however, that the statistical weights of specific mutual orientations of the two molecules are determined by their Boltzmann factors, cf. Section 2.2, according to

$$\overline{\phi^{\mu\mu}} = \frac{\int \phi e^{-\phi/kT} d\omega_1 d\omega_2}{\int e^{-\phi/kT} d\omega_1 d\omega_2}.$$

By expanding the exponentials in the numerator and the denominator we find for the canonical average pair potential due to dipole–dipole interactions

$$\overline{\phi^{\mu\mu}} = \frac{+\mu^2}{r^3} \int [\sin\theta_1 \sin\theta_2 \cos\phi_{12} - 2\cos\theta_1 \cos\theta_2]$$

$$\cdot \frac{\left[1 - \frac{\mu^2/kT}{r^3} (\sin\theta_1 \sin\theta_2 \cos\phi_{12} - 2\cos\theta_1 \cos\theta_2) + \cdots \right] d\omega_1 d\omega_2}{\int \left[1 - \frac{\mu^2/kT}{r^3} (\sin\theta_1 \sin\theta_2 \cos\phi_{12} - 2\cos\theta_1 \cos\theta_2) + \cdots \right] d\omega_1 d\omega_2}.$$

For the individual integrals we have

$$\int_0^\pi \int_0^\pi \int_0^{2\pi} \sin\theta_1 \sin\theta_2 d\theta_1 d\theta_2 d\phi_{12} = 8\pi,$$

$$\int_0^\pi \int_0^\pi \int_0^{2\pi} (\sin\theta_1 \sin\theta_2 \cos\phi_{12} - 2\cos\theta_1 \cos\theta_2) \sin\theta_1 \sin\theta_2 d\theta_1 d\theta_2 d\phi_{12}$$

$$= \int_0^{2\pi} \cos\phi_{12} d\phi_{12} \left(\int_0^\pi \sin^2\theta d\theta \right)^2 - 2 \times 2\pi \left(\int_0^\pi \cos\theta \sin\theta d\theta \right)^2$$

$$= [\sin\phi_{12}]_0^{2\pi} \left\{ \left[-\frac{\sin\theta \cos\theta}{2} \right]_0^\pi + \frac{1}{2} \int_0^\pi \sin^0\theta d\theta \right\}^2 - 4\pi \left\{ \left[\frac{1}{2} \sin^2\theta \right]_0^\pi \right\}^2 = 0,$$

and

$$\int_0^\pi \int_0^\pi \int_0^{2\pi} [\sin^2\theta_1 \sin^2\theta_2 \cos^2\phi_{12} - 4\sin\theta_1\cos\theta_1\sin\theta_2\cos\theta_2\cos\phi_{12} + 4\cos^2\theta_1\cos^2\theta_2]$$

$$\cdot \sin\theta_1\sin\theta_2 d\theta_1 d\theta_2 d\phi_{12}$$

$$= \left(\int_0^\pi \sin^3\theta d\theta\right)^2 \int_0^{2\pi} \cos^2\phi_{12}\, d\phi_{12} - 4\left(\int_0^\pi \sin^2\theta\cos\theta d\theta\right)^2 \int_0^{2\pi} \cos\phi_{12}\, d\phi_{12}$$

$$+ 8\pi\left(\int_0^\pi \cos^2\theta\sin\theta d\theta\right)^2 = \frac{16\pi}{3}.$$

This leads to

$$\overline{\phi^{\mu\mu}} = -\frac{2}{3}\frac{\mu^4}{r^6 kT} + 0\left(\frac{1}{T^2}\right)$$

and

$$\overline{F_2^{\mu\mu}} = -e_{12}\frac{d\overline{\phi^{\mu\mu}_{12}}}{dr} = -e_{12}\frac{4}{r^7}\frac{\mu^4}{kT} + 0\left(\frac{1}{T^2}\right).$$

We learn from Exercise 5.3 that the canonically averaged dipole–dipole interaction leads to an attractive force between the two molecules that depends on temperature. At high temperatures the high kinetic energy of the molecules tends to toss them around more randomly and the average effect of polarity dies off. Evaluating the averaged pair potential according to (5.31) gives [9]

$$\tilde{\phi}^{\mu\mu}_{12} = -\frac{1}{3}\frac{\mu^4}{r^6 kT} + O\left(\frac{1}{T^2}\right).$$

5.3 The Statistical Virial Equation

In Section 5.1 we introduced the virial equation as a general form for the equation of state at moderate density. Although the virial coefficients there were identified as coefficients of a formal Taylor expansion around zero density, we will now show that they can be calculated from the intermolecular potential energy.

5.3.1 Pure Gases

Particularly straightforward access to the virial form of the equation state by statistical mechanics is provided by exploiting the properties of the grand canonical ensemble. The statistical analog of the expression pV, i.e., the grand canonical potential, is given according to (2.69) by

$$pV(T, V, \mu) = kT \ln \Xi. \tag{5.34}$$

Here, Ξ is the grand canonical partition function, which for pure fluids reads, cf. (2.53),

$$\Xi = \sum_{N \geq 0} Q_N(T, V, N) e^{N\mu/kT}, \tag{5.35}$$

with Q_N as the canonical partition function of an N-molecule system and μ as the chemical potential per molecule of the fluid. The grand canonical partition function has the important property of being representable as a series with individual terms containing subsystems of $1, 2, \ldots, N$ molecules in an otherwise empty system. It is thus a most convenient starting point for the derivation of the virial equation within the framework of statistical mechanics [10,11], although the virial equation was originally derived in the canonical ensemble [12].

For simplicity we go through the derivation for monatomic molecules and generalize later. In the special case of monatomic molecules the canonical partition function contains only translational contributions and (5.35) becomes

$$\Xi = \sum_{N \geq 0} \frac{1}{N!} \Lambda^{-3N} Q_N^C e^{N\mu/kT} = \sum_{N \geq 0} \frac{Q_N^C}{N!} a^N \tag{5.36}$$

with, cf. (2.75),

$$Q_N^C = \int \cdots \int e^{-U(r_1, r_2, \ldots, r_N)/kT} dr^N \tag{5.37}$$

and

$$a = \Lambda^{-3} e^{\mu/kT} = \left(\frac{2\pi mkT}{h^2} \right)^{3/2} e^{\mu/kT}. \tag{5.38}$$

Q_N^C is the configuration integral or the configurational partition function of a pure system consisting of N monatomic molecules in the volume V. It is that part of the phase integral that is determined by the configuration of the molecules and thus by the intermolecular potential energy U. The quantity a is referred to as the absolute activity and is related to the thermodynamic activity; cf. Section 2.1.

The grand canonical ensemble thus yields an expansion for the expression $e^{pV/kT}$ in terms of the absolute activity. Successive terms are evaluated for clusters of molecules of successively higher order in an otherwise empty system. Although formally different, we note the analogy in spirit to the virial equation. Because we wish to derive an expansion for the compressibility factor $Z = pV/NkT$ in terms of density, we first reformulate the expansion of the function $e^{pV/kT}$ into an expansion of pV/kT in terms of a and finally eliminate the absolute activity in favor of the density. We assume the following expansion of pV/kT in terms of a,

$$\frac{pV}{kT} = V \sum_{i \geq 1} b_i a^i, \tag{5.39}$$

where the coefficients b_i are to be determined in such a way that the assumed expansion (5.39) and the theoretical statistical expansion (5.36) are consistent, and $i = 1, 2, \ldots$ stands for the number of molecules in the system.

Equation (5.39) is consistent with the equations for an ideal gas. Because, according to (3.47),

$$\left(\frac{G}{N}\right)^{\mathrm{ig}} = \mu^{\mathrm{ig}} = kT \ln \frac{N\Lambda^3}{V}, \tag{5.40}$$

we find for the absolute activity of an ideal gas

$$a^{\mathrm{ig}} = e^{\ln N/V} = n. \tag{5.41}$$

In the ideal gas, all interactions disappear and only single molecules have to be considered in (5.39); i.e., the expansion is truncated at $i = 1$ and (5.39) degenerates to

$$\left(\frac{pV}{kT}\right)^{\mathrm{ig}} = Vb_1 a^{\mathrm{ig}}, \tag{5.42}$$

which for $b_1 = 1$, cf. (5.44), gives $(pV)^{\mathrm{ig}}/NkT = 1$. This demonstrates that (5.39) is consistent with the virial expansion for $n \to 0$. Apart from that, however, (5.39) is an assumption, and it is by no means proven that this expansion generally exists.

To determine the b-coefficients in such a way that the expansions (5.39) and (5.36) are consistent, we first write (5.39) as

$$e^{pV/kT} = e^{V\sum_{i\geq 1} b_i a^i} = \prod_{i\geq 1} e^{Vb_i a^i} = \prod_{i\geq 1} \sum_{m_i\geq 0} \frac{1}{m_i!}(Vb_i a^i)^{m_i}. \tag{5.43}$$

Here the exponential function is expressed in terms of its series expansion. We now compare the coefficients of the statistical thermodynamical expansion (5.36) for equal numbers of molecules, i.e., equal power of the activity in the grand canonical partition function, with those of the expansion (5.43) and find

$$\begin{aligned} N = 0: \quad & 1 = 1 \\ N = 1: \quad & Q_1^C = Vb_1, \text{ i.e., } b_1 = Q_1^C/V = 1 \\ N = 2: \quad & \frac{1}{2!}Q_2^C = \frac{1}{2!}(Vb_1)^2 + \frac{1}{1!}Vb_2 \\ N = 3: \quad & \frac{1}{3!}Q_3^C = \frac{1}{3!}(Vb_1)^3 + \frac{1}{1!}(Vb_1)\frac{1}{1!}(Vb_2) + \frac{1}{1!}(Vb_3) \\ & \cdots \cdots \end{aligned} \tag{5.44}$$

This makes the assumed expansion (5.39) consistent with the expansion for the grand canonical partition function. The expansion coefficients b_i can thus be expressed in terms of the configuration integrals of from 1 up to at most N molecules. So b_2 contains the interaction of two molecules, b_3 that of three molecules, etc., where it is understood that the two, three, etc. molecules are alone in the total volume of the system, i.e., at zero density. Such arrangements are referred to as molecular clusters and the b_i are therefore called cluster integrals. Because the virial coefficients can, as we shall see, be expressed in terms of the cluster integrals, they are properties at zero density too, in agreement with their mathematical background as coefficients of a Taylor expansion of the compressibility factor around zero density.

Equation (5.39) can be cast into the form of the virial equation if the absolute activity a can be eliminated in favor of the number density n. For this purpose we first show the general relationship between n and a. For the number of molecules we have in the grand canonical ensemble, according to (2.65),

$$N = kT \left(\frac{\partial \ln \Xi}{\partial \mu} \right)_{T,V} = \left(\frac{\partial \ln \Xi}{\partial \ln a} \right)_{T,V}. \tag{5.45}$$

With (5.34) and (5.39) we thus find the following formal relationship between n and a:

$$\frac{N}{V} = n = \frac{a}{V} \left(\frac{\partial (pV/kT)}{\partial a} \right)_{T,V} = \Sigma_{i \geq 1} i b_i a^i. \tag{5.46}$$

We invert this expansion; i.e., we formulate the absolute activity in terms of the number density as

$$a = \Sigma_{j \geq 1} c_j n^j = c_1 n + c_2 n^2 + c_3 n^3 + \dots. \tag{5.47}$$

This expansion contains (5.41) as a limiting case, because $c_1 = 1$ and $c_2, c_3, \dots = 0$ for an ideal gas; cf. (5.48) and (5.44). Introducing (5.47) into (5.46) and comparing coefficients reveals relations between the c_i and the b_i:

$$n : c_1 = 1/b_1$$
$$n^2 : c_2 = -2b_2/b_1^3$$
$$n^3 : c_3 = -3b_3/b_1^4 + 8b_2^2/b_1^5$$
$$\dots \tag{5.48}$$

If we now introduce the density expansion (5.47) for a into the assumed expansion for the compressibility factor (5.39), we find a density expansion for the compressibility factor, i.e., the virial equation, as

$$Z = \frac{pV}{NkT}$$

$$= \frac{V}{N} \sum_{i \geq 1} b_i a^i = \frac{V}{N} \left[b_1 \left\{ \frac{1}{b_1} n - \frac{2b_2}{b_1^3} n^2 + \left(\frac{8b_2^2}{b_1^5} - \frac{3b_3}{b_1^4} \right) n^3 + \cdots \right\} \right.$$

$$\left. + b_2 \left\{ \frac{1}{b_1} n - \frac{2b_2}{b_1^3} n^2 + \cdots \right\}^2 + b_3 \left\{ \frac{1}{b_1} n - \frac{2b_2}{b_1^3} n^2 + \cdots \right\}^3 + \cdots \right]$$

$$= 1 + (b_2/b_1 - 2b_2/b_1^2)n + (8b_2^2/b_1^4 - 3b_3/b_1^3 - 4b_2^2/b_1^4 + b_3/b_1^3)n^2 + \cdots. \tag{5.49}$$

Comparison with (5.2) yields expressions for the virial coefficients in terms of the $b_1, b_2, b_3, \ldots$. For the second virial coefficient of monatomic gases in the semiclassical approximation we find

$$
\begin{aligned}
B = -b_2 &= -\frac{1}{2V} Q_2^C + \frac{1}{2} V = -\frac{1}{2V} \iint e^{-U(r_1, r_2)/kT} dr_1 dr_2 + \frac{1}{2} V \\
&= -\frac{1}{2V} \iint (e^{-U(r_1, r_2)/kT} - 1) dr_1 dr_2 \\
&= -2\pi \int_0^\infty (e^{-\phi(r)/kT} - 1) r^2 dr.
\end{aligned}
\tag{5.50}
$$

Here $\phi(r)$ is the pair potential between two monatomic molecules, which depends only on the intermolecular distance r. The integration strictly extends only over the volume of the system. However, because $\phi(r) \to 0$ at large intermolecular distances, which are still very small as compared to the macroscopic dimensions of the system, we can also integrate from 0 to ∞.

For the third virial coefficient we find in an analogous manner

$$
\begin{aligned}
C = 4b_2^2 - 2b_3 &= -\frac{1}{3V} \left(Q_3^C - \frac{3}{V} (Q_2^C)^2 + 3V Q_2^C - V^3 \right) \\
&= -\frac{1}{3V} \left[\iiint e^{-U(r_1, r_2, r_3)/kT} dr_1 dr_2 dr_3 \right. \\
&\quad - \frac{3}{V} \iint e^{-U(r_1, r_2)/kT} dr_1 dr_2 \iint e^{-U(r_1, r_3)/kT} dr_1 dr_3 \\
&\quad \left. + 3V \iint e^{-U(r_1, r_2)/kT} dr_1 dr_2 - \iiint dr_1 dr_2 dr_3 \right] \\
&= \left[\frac{1}{V} \iint (e^{-U(r_1, r_2)/kT} - 1) dr_1 dr_2 \right] \\
&\quad \cdot \left[\frac{1}{V} \iint (e^{-U(r_1, r_3)/kT} - 1) dr_1 dr_3 \right] \\
&\quad - \frac{1}{3V} \iiint \left(e^{-U(r_1, r_2, r_3)/kT} - e^{-U(r_1, r_2)/kT} - e^{-U(r_1, r_3)/kT} \right. \\
&\quad \left. - e^{-U(r_2, r_3)/kT} + 2 \right) dr_1 dr_2 dr_3.
\end{aligned}
\tag{5.51}
$$

For the intermolecular potential energy of a system consisting of three monatomic molecules we write, cf. Section 5.2,

$$
U(r_1, r_2, r_3) = U(r_1, r_2) + U(r_2, r_3) + U(r_1, r_3) + \Delta U(r_1, r_2, r_3).
\tag{5.52}
$$

Here, $\Delta U(r_1, r_2, r_3)$ represents the nonadditive three-body interaction. With this relation, we find for the third virial coefficient

$$
\begin{aligned}
C = &\left[\frac{1}{V} \iint (e^{-U(r_1, r_2)/kT} - 1) dr_1 dr_2 \right] \\
&\cdot \left[\frac{1}{V} \iint (e^{-U(r_1, r_3)/kT} - 1) dr_1 dr_3 \right]
\end{aligned}
$$

$$-\frac{1}{3V}\iiint\left\{e^{-U(r_1,r_2)/kT}e^{-U(r_2,r_3)/kT}e^{-U(r_1,r_3)/kT}\right.$$

$$\left.\cdot\left[e^{\Delta U(r_1,r_2,r_3)/kT}-1\right]\right\}d\mathbf{r}_1d\mathbf{r}_2d\mathbf{r}_3$$

$$-\frac{1}{3V}\iiint\left\{e^{-U(r_1,r_2)/kT}e^{-U(r_2,r_3)/kT}e^{-U(r_1,r_3)/kT}\right.$$

$$\left.-e^{-U(r_1,r_2)/kT}-e^{-U(r_2,r_3)/kT}-e^{-U(r_1,r_3)/kT}+2\right\}d\mathbf{r}_1d\mathbf{r}_2d\mathbf{r}_3$$

$$=-\frac{1}{3V}\iiint\left\{e^{-U(r_1,r_2)/kT}e^{-U(r_2,r_3)/kT}e^{-U(r_1,r_3)/kT}\right.$$

$$\left.-3e^{-U(r_1,r_2)/kT}+2+6e^{-U(r_1,r_2)/kT}-3\right\}d\mathbf{r}_1d\mathbf{r}_2d\mathbf{r}_3$$

$$+\frac{1}{V^2}\iint e^{-U(r_1,r_2)/kT}d\mathbf{r}_1d\mathbf{r}_2\iint e^{-U(r_1,r_3)/kT}d\mathbf{r}_1d\mathbf{r}_3$$

$$-\frac{1}{3V}\iiint\left[e^{-\Delta U(r_1,r_2,r_3)/kT}-1\right]$$

$$\cdot\left[e^{-[U(r_1,r_2)+U(r_1,r_3)+U(r_2,r_3)]/kT}\right]d\mathbf{r}_1d\mathbf{r}_2d\mathbf{r}_3$$

$$=-\frac{1}{3V}\iiint\left[e^{-U(r_1,r_2)/kT}-1\right]\left[e^{-U(r_1,r_3)/kT}-1\right]$$

$$\cdot\left[e^{-U(r_2,r_3)/kT}-1\right]d\mathbf{r}_1d\mathbf{r}_2d\mathbf{r}_3$$

$$-\frac{1}{3V}\iiint\left[e^{-\Delta U(r_1,r_2,r_3)/kT}-1\right]$$

$$\cdot\left[e^{-[U(r_1,r_2)+U(r_1,r_3)+U(r_2,r_3)]/kT}\right]d\mathbf{r}_1d\mathbf{r}_2d\mathbf{r}_3, \tag{5.53}$$

where the transformation (A 11.4) has been used; cf. App. 11. Practical integration variables are introduced via (A 11.1) and we find

$$C(T)=-\frac{8}{3}\pi^2\int_0^\infty\int_0^\infty\int_{-1}^1\left[e^{-\phi(r_{12})/kT}-1\right]\left[e^{-\phi(r_{13})/kT}-1\right]$$

$$\cdot\left[e^{-\phi(r_{23})/kT}-1\right]r_{12}^2r_{13}^2dr_{12}dr_{13}d(\cos\alpha)$$

$$-\frac{8}{3}\pi^2\int_0^\infty\int_0^\infty\int_{-1}^1\left[e^{-\frac{\Delta\phi_{123}(r_{12}r_{13}r_{23})}{kT}}-1\right]$$

$$\cdot e^{-[\phi(r_{12})+\phi(r_{13})+\phi(r_{23})]/kT}r_{12}^2r_{13}^2dr_{12}dr_{13}d(\cos\alpha)$$

$$=C_{\text{add}}+C_{\text{nonadd}}, \tag{5.54}$$

where the integration for $(\cos\alpha)$ runs from -1 to $+1$ here and ϕ, $\Delta\phi_{123}$ are the pair potential and the nonadditive three-body potential, respectively. C_{add} is the contribution to the third virial coefficient that remains when $\Delta\phi_{123}=0$, i.e., under the assumption of pairwise additive interaction potentials. When the integrations in the above equations are evaluated numerically, r_{23} in ϕ and in $\Delta\phi_{123}$ has to be expressed in terms of r_{12}, r_{13}, and $\cos\alpha$; cf. (A 11.2).

Equations (5.50) and (5.54) remain unchanged when polyatomic molecules with isotropic interactions are considered. The absolute activity then contains additional contributions due to internal degrees of freedom. The relation to the number density, however, remains unchanged, and so does the virial expansion.

EXERCISE 5.4

(a) Calculate the second virial coefficient for a hard-sphere gas.
(b) Calculate the second virial coefficient for a square-well gas.

Solution (a)

The hard-sphere potential is defined by (5.18) and (5.19) as

$$\phi(r) = \infty \quad \text{for } r \leq d_-$$
$$\phi(r) = 0 \quad \text{for } r \geq d_+,$$

where d_- and d_+ represent the limiting value $r \to d$ in an approach from $r < d$ and $r > d$, respectively. According to (5.50), we find for the second virial coefficient in cm^3/mol

$$B = -2\pi N_A \int_0^\infty \left[e^{-\phi(r)/kT} - 1 \right] r^2 \mathrm{d}r$$

$$= -2\pi N_A \left[\int_0^d \left[e^{-\infty} - 1 \right] r^2 \mathrm{d}r + \int_d^\infty \left[e^{-0} - 1 \right] r^2 \mathrm{d}r \right] = \frac{2}{3}\pi N_A d^3.$$

The second virial coefficient of a hard-sphere gas does not depend on temperature, contrary to experimental observation. The hard-sphere potential is thus no realistic model for intermolecular interactions at low density.

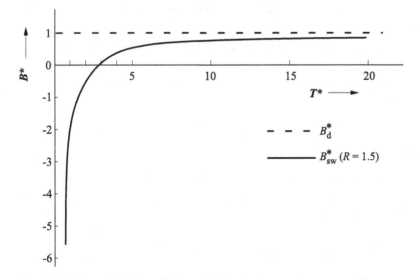

Fig. E 5.4.1. Second virial coefficients of a hard-sphere and a square-well gas.

Solution (b)

The square-well potential is defined by (5.20) through (5.22) as

$$\phi(r) = \infty \quad \text{for } r \leq d_-$$
$$\phi(r) = -\varepsilon \quad \text{for } d_+ \leq r \leq (Rd)_-$$
$$\phi(r) = 0 \quad \text{for } r \geq (Rd)_+.$$

According to (5.50), we find for the second virial coefficient

$$B = -2\pi N_A \left\{ \int_0^d [e^{\infty} - 1] r^2 dr + \int_d^{Rd} [e^{\varepsilon/kT} - 1] r^2 dr + \int_{Rd}^{\infty} [e^{-0} - 1] r^2 dr \right\}$$

$$= \frac{2}{3}\pi N_A d^3 \left[1 - (R^3 - 1)(e^{\varepsilon/kT} - 1) \right].$$

These results are shown graphically in Figure E 5.4.1, with $B^* = B/(2\pi/3 N_A d^3)$ and $T^* = kT/\varepsilon$. We note that the square-well potential gives a qualitatively correct temperature dependence at low and medium temperatures; cf. Figure 5.2. At high temperatures the influence of the attractive forces is lost and the second virial coefficient yields to that of a hard-sphere gas. The observed temperature dependence at high temperatures is reproduced when the decrease of the hard sphere diameter with increasing temperature is incorporated.

As an illustration, we show the representation of the thermodynamic data of gaseous argon in Figures 5.8 and 5.9 [13]. The intermolecular pair potential energy of this monatomic fluid is modeled by the MSK potential, cf. (5.16). The three adjustable parameters of this pair potential have been found by a simultaneous fit to the second virial and Joule–Thomson coefficient data [14]. In Figure 5.8 we show the correlation of these data [15–17]. The dashed areas represent the estimated experimental precision of the second virial coefficient and a 1% deviation for the Joule–Thomson coefficient, respectively. In Figure 5.9 it is shown that we are able to predict the third virial coefficient within experimental error [16] from this pair potential when the nonadditive three-body forces modeled by the Axilrod–Teller term are included [13]. Also, by performing computer simulations, good results for the thermodynamic properties of dense gaseous and liquid argon are obtained from this model of the intermolecular forces [13].

5.3.2 Gas Mixtures

The grand canonical partition function of a binary mixture of N_A monatomic molecules of component A and N_B monatomic molecules of component B reads, according to (2.54),

$$\Xi = \sum_{N_A, N_B \geq 0} Q_{N_A N_B}^C (T, V, N_A, N_B) e^{N_A \mu_A/kT} e^{N_B \mu_B/kT}$$

$$= \sum_{N_A, N_B \geq 0} \frac{Q_{N_A N_B}^C}{N_A! N_B!} a_A^{N_A} a_B^{N_B}, \tag{5.55}$$

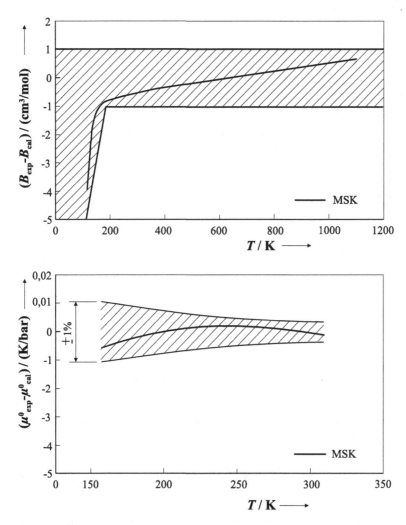

Fig. 5.8. Correlation of second viral and Joule–Thomson coefficients for argon.

with

$$Q^C_{N_A N_B} = \int \cdots \int \exp\left(-U\left(\boldsymbol{r}_{A_1}, \boldsymbol{r}_{B_1}, \ldots \boldsymbol{r}_{A_{N_A}}, \boldsymbol{r}_{B_{N_B}}\right) / kT\right)$$
$$\cdot \, \mathrm{d}\boldsymbol{r}_A^{\,N_A} \mathrm{d}\boldsymbol{r}_B^{\,N_B} \tag{5.56}$$

and

$$a_A = \Lambda_A^{-3} e^{\mu_A / kT}, \tag{5.57}$$

with an analogous expression for a_B.

Statistical mechanics thus again yields a particular expansion of the grand canonical partition function in terms of a_A and a_B. For the component A in a mixture at volume V, the ideal gas chemical potential reads, cf. Chapter 3,

$$\mu_A^{\mathrm{ig}} = kT \ln \frac{N_A \Lambda_A^3}{V}, \tag{5.58}$$

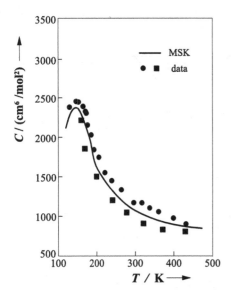

Fig. 5.9. Prediction of the third virial coefficient for argon.

and thus we have

$$a_A^{ig} = n_A, \qquad (5.59)$$

where $n_A = N_A/V$ is the number density of component A.

Analogously to the pure gas case, we expand the expression pV/kT in terms of the absolute activities. An expansion that is consistent with the statistical mechanical result of (5.55) is

$$\frac{pV}{kT} = V \sum_{i,j \geq 0} b_{ij} a_A^i a_B^j, \qquad (5.60)$$

where i and j must not be zero simultaneously. The cluster integrals b_{ij} can be expressed in terms of configuration integrals by comparing the coefficients of (5.55) with the following form of (5.60):

$$e^{pV/kT} = \prod_{i,j \geq 0} e^{V b_{ij} a_A^i a_B^j} = \prod_{i,j \geq 0} \sum_{m_{i,j} \geq 0} \frac{1}{m_{ij}!} \left(V b_{ij} a_A^i a_B^j \right)^{m_{ij}}. \qquad (5.61)$$

We thus find for the lowest cluster integrals

$$N_A = 0, N_B = 0: \qquad 1 = 1$$

$$N_A = 1, N_B = 0: \frac{1}{1!} Q_{10}^C = \frac{1}{1!} V b_{10} = V, \text{ i.e., } b_{10} = 1$$

$$N_A = 0, N_B = 1: \frac{1}{1!} Q_{01}^C = \frac{1}{1!} V b_{01} = V, \text{ i.e., } b_{01} = 1$$

$$N_A = 1, N_B = 1: \frac{1}{1!} Q_{11}^C = \frac{1}{1!} \frac{1}{1!} V^2 b_{01} b_{10} + \frac{1}{1!} V b_{11}$$

$$N_A = 2, N_B = 0 : \frac{1}{2!} Q_{20}^C = \frac{1}{2!}(Vb_{10})^2 + \frac{1}{1!}Vb_{20}$$

$$N_A = 0, N_B = 2 : \frac{1}{2!} Q_{02}^C = \frac{1}{2!}(Vb_{01})^2 + \frac{1}{1!}Vb_{02}$$

$$\cdots \cdots \tag{5.62}$$

To arrive at the virial equation for mixtures we have to eliminate the absolute activity in terms of the number density. For the number density of component A we find in the grand canonical ensemble

$$\frac{N_A}{V} = \frac{kT}{V}\left(\frac{\partial \ln \Xi}{\partial \mu_A}\right)_{T,V,\mu_A^*} = \frac{1}{V}\left(\frac{\partial \ln \Xi}{\partial \ln a_A}\right)_{T,V,a_A^*}$$

$$= \frac{a_A}{V}\left(\frac{\partial(pV/kT)}{\partial a_A}\right)_{T,V,a_A^*} = b_{10}a_A + 2b_{20}a_A^2 + b_{11}a_A a_B + \cdots . \tag{5.63}$$

Analogously, for component B, we have

$$\frac{N_B}{V} = b_{01}a_B + 2b_{02}a_B^2 + b_{11}a_A a_B + \cdots . \tag{5.64}$$

We eliminate again the quantities a_A and a_B by a consistent inversion of the expansion (5.63) and (5.64). This gives for a_A

$$a_A = c_{A1}n_A + c_{A2}n_A^2 + c_{A3}n_A n_B \ldots , \tag{5.65}$$

and an analogous equation results for a_B. Insertion of (5.65) and the corresponding equation for a_B into (5.63) and (5.64), respectively, and comparing coefficients gives for the c_{Ai} or c_{Bi}

$$c_{A1} = 1/b_{10}; \qquad c_{B1} = 1/b_{01}$$

$$c_{A2} = -2b_{20}/b_{10}^3; \qquad c_{B2} = -2b_{02}/b_{01}^3$$

$$c_{A3} = -b_{11}/(b_{10}^2 b_{01}); \quad c_{B3} = -b_{11}/(b_{01}^2 b_{10})$$

$$\cdots .$$

We thus have

$$a_A = \frac{1}{b_{10}}n_A - \frac{2b_{20}}{b_{10}^3}n_A^2 - \frac{b_{11}}{b_{10}^2 b_{01}}n_A n_B + \cdots \tag{5.67}$$

and

$$a_B = \frac{1}{b_{01}}n_B - \frac{2b_{02}}{b_{01}^3}n_B^2 - \frac{b_{11}}{b_{01}^2 b_{10}}n_A n_B + \cdots . \tag{5.68}$$

From (5.60) we thus arrive at the following virial representation of the compressibility factor for a binary real gas mixture:

$$Z = \frac{pV}{NkT} = \frac{V}{N}\left[b_{10}a_A + b_{01}a_B + b_{20}a_A^2 + b_{02}a_B^2 + b_{11}a_A a_B + \cdots \right]$$

$$= \frac{V}{N}\left\{ b_{10}\left[\frac{1}{b_{10}}n_A - \frac{2b_{20}}{b_{10}^3}n_A^2 - \frac{b_{11}}{b_{10}^2 b_{01}}n_A n_B + \cdots \right]\right.$$

$$+ b_{01}\left[\frac{1}{b_{01}}n_B - \frac{2b_{02}}{b_{01}^3}n_B^2 - \frac{b_{11}}{b_{01}^2 b_{10}}n_A n_B + \cdots \right]$$

$$+ b_{20}\left[\frac{1}{b_{10}}n_A - \frac{2b_{20}}{b_{10}^3}n_A^2 - \frac{b_{11}}{b_{10}^2 b_{01}}n_A n_B + \cdots \right]^2$$

$$+ b_{02}\left[\frac{1}{b_{01}}n_B - \frac{2b_{02}}{b_{01}^3}n_B^2 - \frac{b_{11}}{b_{01}^2 b_{10}}n_A n_B + \cdots \right]^2$$

$$+ b_{11}\left[\frac{1}{b_{10}}n_A - \frac{2b_{20}}{b_{10}^3}n_A^2 - \frac{b_{11}}{b_{10}^2 b_{01}}n_A n_B + \cdots \right]$$

$$\cdot \left[\frac{1}{b_{01}}n_B - \frac{2b_{02}}{b_{01}^3}n_B^2 - \frac{b_{11}}{b_{01}^2 b_{10}}n_A n_B + \cdots \right] + \cdots \right\}$$

$$= 1\frac{n_A}{n} + 1\frac{n_B}{n} + \left(\frac{b_{20}}{b_{10}^2} - \frac{2b_{20}}{b_{10}^2} \right)\frac{n_A^2}{n} + \left(\frac{b_{02}}{b_{01}^2} - \frac{2b_{02}}{b_{01}^2} \right)\frac{n_B^2}{n}$$

$$+ \left(\frac{b_{11}}{(b_{10}b_{01})} - \frac{b_{11}}{(b_{01}b_{10})} - \frac{b_{11}}{(b_{10}b_{01})} \right)\frac{n_A n_B}{n} + \cdots$$

$$= 1 + \left[\left(\frac{b_{20}}{b_{10}^2} - \frac{2b_{20}}{b_{10}^2} \right)x_A^2 + \left(\frac{b_{02}}{b_{01}^2} - \frac{2b_{02}}{b_{01}^2} \right)x_B^2\right.$$

$$+ \left(\frac{b_{11}}{(b_{10}b_{01})} - \frac{b_{11}}{(b_{01}b_{10})} - \frac{b_{11}}{(b_{10}b_{01})} \right)x_A x_B \right]n + \cdots . \tag{5.69}$$

Here we have introduced

$$x_A = \frac{N_A}{N} = \frac{N_A/V}{N/V} = \frac{n_A}{n} \tag{5.70}$$

as the mole fraction of component A and, correspondingly, x_B as the mole fraction of component B.

From the comparison with the empirical virial equation we find for the second virial coefficient with (5.69) and (5.62)

$$B(T, x_A) = -\left(b_{20}x_A^2 + b_{02}x_B^2 + b_{11}x_A x_B \right), \tag{5.71}$$

i.e., a quadratic mixing rule in the mole fraction, where

$$b_{20} = \frac{1}{2V}Q_{20}^C - \frac{1}{2}V = -B_{AA},$$

$$b_{02} = \frac{1}{2V}Q_{02}^C - \frac{1}{2}V = -B_{BB},$$

and

$$b_{11} = \frac{Q_{11}^C}{V} - V = 2\left(\frac{Q_{11}^C}{2V} - \frac{V}{2}\right) = -2B_{AB}. \tag{5.72}$$

Here B_{AB} is a term that is to be evaluated according to the equation for the second virial coefficient of a pure gas, but using the pair potential between a molecule of component A and a molecule of component B. It is frequently referred to as the cross interaction second virial coefficient. For the second virial coefficient of a binary gas mixture of components A and B, we thus have

$$B(T, x_A) = x_A^2 B_{AA} + x_B^2 B_{BB} + 2x_A x_B B_{AB}. \tag{5.73}$$

For multicomponent mixtures this generalizes to

$$B(T, \{x_i\}) = \sum_i \sum_j x_i x_j B_{ij}. \tag{5.74}$$

If the statistical mechanical density expansion of the compressibility factor is carried out up to the quadratic term, a cubic mixing rule is found for the third virial coefficient,

$$C(T, \{x_i\}) = \sum_i \sum_j \sum_k x_i x_j x_k C_{ijk}, \tag{5.75}$$

where C_{ijk} is the third virial coefficient to be evaluated from (5.54) for an interaction of the molecular triplet i, j, and k. This theoretically exact composition dependence of the virial coefficients, which determines the chemical potentials in a gaseous mixture, is one of the reasons for the practical significance of the virial equation.

5.3.3 Nonspherical Interactions

For nonspherical interactions we have to include integrations over the orientational angles. We thus get for the second virial coefficient in the rigid molecule approximation

$$B = -2\pi \int_0^\infty \left\langle \left(e^{-\phi(r\omega_1\omega_2)/kT} - 1 \right) \right\rangle_{\omega_1\omega_2} r^2 dr, \tag{5.76}$$

where ω_1, ω_2 are the relative orientational angles of molecules 1 and 2 referring to an axis connecting the molecular centers. Here we use the notation

$$\frac{1}{\int d\omega_1 d\omega_2} \int A d\omega_1 d\omega_2 = \langle A \rangle_{\omega_1\omega_2}. \tag{5.77}$$

The normalization of the angular averaging $\int d\omega_1 d\omega_2$ leads to a factor of $(1/4\pi)^2$ for linear and $(1/8\pi^2)^2$ for nonlinear molecules, because

$$\int d\omega_1 = \int_0^\pi \sin\vartheta_1 d\vartheta_1 \int_0^{2\pi} d\chi_1 \int_0^{2\pi} d\phi_1 = 8\pi^2$$

and $\chi = 0$ for a linear molecule. An analogous procedure yields the expression for the third virial coefficient of a gas with nonspherical interactions in the rigid molecule approximation:

$$
\begin{aligned}
C(T) = {}&-\frac{8}{3}\pi^2 \int_0^\infty\!\!\!\int \int_{-1}^1 \left\langle \left(e^{-\phi(r_{12}\omega_1\omega_2)/kT} - 1\right)\left(e^{-\phi(r_{13}\omega_1\omega_3)/kT} - 1\right)\right. \\
&\left.\cdot\left(e^{-\phi(r_{23}\omega_2\omega_3)/kT} - 1\right)\right\rangle_{\omega_1\omega_2\omega_3} r_{12}^2 r_{13}^2 dr_{12}dr_{13}d(\cos\alpha) \\
&-\frac{8}{3}\pi^2 \int_0^\infty\!\!\!\int \int_{-1}^1 \left\langle \left(e^{-\Delta\phi(r_{12}r_{13}r_{23}\omega_1\omega_2\omega_3)/kT} - 1\right)\right. \\
&\left.\cdot\left(e^{-\phi(r_{12}\omega_1\omega_2)/kT}e^{-\phi(r_{13}\omega_1\omega_3)/kT}e^{-\phi(r_{23}\omega_2\omega_3)/kT}\right)\right\rangle_{\omega_1\omega_2\omega_3} r_{12}^2 r_{13}^2 \\
&\cdot dr_{12}dr_{13}d(\cos\alpha) \\
= {}&C_{\text{add}}(T) + C_{\text{nadd}}(T).
\end{aligned}
\tag{5.78}
$$

For linear molecules the second virial coefficient thus requires integration over four variables; for nonlinear molecules the number of integration variables is six. For the third virial coefficient a ninefold integration is required for linear molecules, whereas the number of integration variables rises to 12 for a nonlinear molecule. For integrations over several variables the nonproduct method has proven to be effective [18].

As a typical illustration, we demonstrate that the thermodynamic properties of gas mixtures can rather confidently be predicted from the potential parameters fitted to pure gas data and using the combining rules for the unlike potential parameters discussed in Sections 2.5 and 5.2, notably based on the London approximation for the dispersion forces. This is shown in Figure 5.10 for the second virial coefficients in the system R12–R22–R23 [19]. The molecular models for the intermolecular forces are based on the SSR-MPA pair potential, cf. Sections 5.2, which contains at long range multipole moments and polarizabilities, taken from independent sources [20–23]. At short range, repulsive interactions of the r^{-12}-type are assumed between interaction sites in the molecules. The positions of these are proportional to the sites of the atoms but scaled with a single proportionality factor p, which is also fitted to pure component data; cf. Section 5.2. The potential model takes full account of the nonlinear nature of the considered molecules by the location of the sites as well as by the tensor elements of the multipole moments and the polarizability. It has three adjustable pure component parameters, ε, σ, and p. The pure gas second virial coefficients for R12, R22, and R23 are reproduced within experimental error and no data for them are shown. The experimental cross-interaction virial coefficients [24,25] are far from being close to a simple average of the pure gas data. They are predicted essentially within experimental accuracy. The major goal of molecular models of fluids, i.e., predicting the properties of mixtures from data

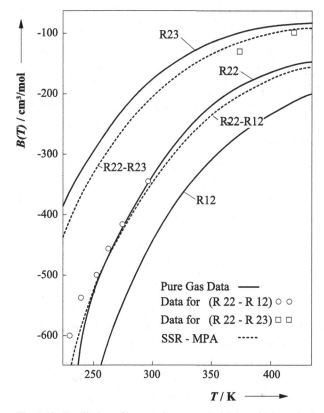

Fig. 5.10. Prediction of interaction second virial coefficients in the system R12–R22–R23 [19].

of the pure components, is thus achieved for the second virial coefficient of this system in an impressive way. Although typical for many systems, the good performance in predicting the virial coefficient of mixtures is not guaranteed for all systems. It must be realized that, although the statistical mechanics including the mixing rule is exact, this is not so for the intermolecular potential model. In particular, the model for the dispersion forces, including the combination rule for $\varepsilon_{\alpha\beta}$, is far from rigorous and may well lead to wrong predictions in unfortunate applications. Still, compared to standard empirical models, which usually predict the cross virial coefficient to be a simple average of the pure component virial coefficients, the progress is impressive.

5.4 Conformal Potential Models

As applications in fluid systems proceed to higher densities, including liquids, the virial equation of state is no longer adequate. At high density many molecules interact simultaneously, and the picture of two or three molecules forming a cluster in a vacuum becomes inappropriate. The derivation of a molecular model for an equation of state that is valid over the whole density range is not straightforward. A general approach for all fluids, comparable with the ideal gas equation of state or the virial equation, is not known. Instead, statistical mechanics can be used as a general framework from which, by introducing suitable approximations, equation of state models for various types of applications may be derived.

A particularly important class of models emerges when pairwise additivity, along with conformal pair potentials, is assumed. Then the intermolecular potential energy is approximated by

$$U(\boldsymbol{r}_1, \boldsymbol{r}_2, \ldots, \boldsymbol{r}_N) = \sum \sum_{i < j} \phi(r_{ij}),$$

(5.79)

where the pair potentials are assumed to be conformal, i.e., defined by

$$\phi = \varepsilon f_u (r/\sigma)$$

(5.80)

for pure fluids and

$$\phi_{\alpha\beta} = \varepsilon_{\alpha\beta} f_u \left(r_{\alpha_i \beta_j} / \sigma_{\alpha\beta} \right)$$

(5.81)

for mixtures. In these equations, f_u is a function that is universal for all interactions, regardless of the components considered. The individuality of a particular interaction between two molecules is taken care of exclusively by the associated potential parameters ε and σ. An example of a conformal pair potential is the Lennard–Jones (6,12) potential, cf. (5.15), or any other two-parameter pair potential. The particular function f_u does not have to be specified. Strictly, the condition of conformality as formulated in (5.80) and (5.81) can only apply to monatomic fluids, because more complex molecules have interactions characterized by a larger number of specific potential parameters than just ε and σ. It turns out, however, that many simple systems can be approximately modeled on this basis.

To introduce the assumption of conformal pair potentials, we reformulate the general statistical mechanical equations in a suitable way. For this purpose, we first note, as before in Section 4.1, that any model must take account of the fact that a fluid is characterized by a microscopic structure. So the molecules are not evenly distributed in the system as determined by the bulk density and bulk composition, but instead show a density and composition structure on the molecular scale. This structure is a consequence of the intermolecular forces and has a strong influence on the thermodynamic functions. Clearly, such a structure is also present to a lesser extent in real gases at moderate density and is fully taken care of there by the virial expansion. In the general case, molecular structure can be formulated explicitly in terms of so-called correlation functions. Under the pairwise additivity approximation these correlation functions can be simplified to take into account the structure due to only pairs of molecules. In the context of conformal potential models they are finally represented in a way adequate for molecules interacting via conformal pair potentials.

5.4.1 Correlation Functions

The assumption of pairwise additivity, along with the assumption of conformal pair potentials, is the key to a tremendous simplification with regard to the atomistic structure of the liquid and the resulting thermodynamic functions. First, instead of taking the configuration of all molecules into account simultaneously, as is done in the general formulation of the configuration integral, only the configurations of molecular pairs need to be considered under this assumption. Specific formulations for the thermodynamic functions in terms of so-called pair correlation functions can be derived on this basis. Then, by assuming conformal pair potentials, these formulations can be cast into simple and practical models.

To define correlation functions in a system of molecules interacting with spherically symmetrical forces, we investigate the probability of finding the system in a particular configurational state characterized by r^N, i.e., the position coordinates of all molecules. In particular, we assume distinguishable molecules and examine the probability of finding molecule 1 in volume element dr_1 at r_1, while molecule 2 is in dr_2 at r_2, with analogous specifications for all other molecules. We thus look for the probability of a specific configurational microstate, for which we have for the canonical ensemble and in the classical approximation, according to (2.50) and (2.74),

$$P(r^N) = \frac{e^{-U(r^N)/kT}}{\int \cdots \int e^{-U(r^N)/kT} dr^N} = \frac{e^{-U(r^N)/kT}}{Q^C}, \qquad (5.82)$$

with Q^C as the classical configurational partition function. This probability is related exclusively to the configurational microstate. Correspondingly, the exponent of the exponential function contains only that part of the Hamilton function of the system that is determined by molecular configuration, i.e., the intermolecular potential energy $U(r^N)$. The probability defined by (5.82) satisfies the normalization condition

$$\int \cdots \int P(r^N) dr^N = 1, \qquad (5.83)$$

which expresses the plausible fact that the system must certainly be in any one configurational microstate.

The molecules of a pure substance are indistinguishable. We introduce this fact and denote the probability that some molecule is in configuration 1, some other molecule in configuration 2, etc., as a distribution function. The N-molecule distribution function then is defined as

$$f_N(r^N) = N! P(r^N). \qquad (5.84)$$

Here the combinatorial factor $N!$ takes into account the number of configurational arrangements of N molecules representing the same microstate when the molecules are considered to be indistinguishable; cf. Section 2.2. A particular microstate with indistinguishable molecules is found more frequently by a factor of $N!$ than in the case of distinguishable molecules. The N-molecule distribution function is normalized according to

$$\int \cdots \int f_N\left(r^N\right) dr^N = N!. \tag{5.85}$$

We now make use of the simplifications due to pairwise additivity. Generally, in order to describe the simultaneous interaction of molecular pairs, molecular triplets, etc., it is convenient to define lower-order distribution functions. If we have a system of N distinguishable molecules and select from it a subsystem of h molecules we find for the probability that molecule 1 is in dr_1 around r_1, molecule 2 in dr_2 around r_2, ..., and molecule h in dr_h around r_h, whereas the configuration of all other molecules $h+1, h+2, \ldots, N$ is unspecified,

$$P_h\left(r^h\right) = \int \cdots \int P\left(r^N\right) dr^{N-h}$$
$$= \frac{\int \cdots \int e^{-U(r^N)/kT} dr^{N-h}}{Q^C}. \tag{5.86}$$

For indistinguishable molecules the corresponding normalized distribution function of order h is

$$f_h\left(r^h\right) = \frac{N!}{(N-h)!} P_h\left(r^h\right), \tag{5.87}$$

where the combinatorial factor here derives from

$$N(N-1)(N-2)\cdots(N-h+1) = \frac{N(N-1)(N-2)\cdots 3 \cdot 2 \cdot 1}{(N-h)(N-h-1)\cdots 3 \cdot 2 \cdot 1}$$
$$= \frac{N!}{(N-h)!}, \tag{5.88}$$

which clearly reduces to (5.85) for $h = N$, as it must.

In an ideal gas there are no intermolecular interactions. Thus all molecules are totally uncorrelated, and there is no structure. The configurational coordinates of the individual molecules in an ideal gas are thus independent of each other and the distribution function of order h factors as

$$f_h^{ig}\left(r^h\right) = f_1\left(r_1\right) f_1\left(r_2\right) \cdots f_1\left(r_h\right), \tag{5.89}$$

according to the calculus of probability. In describing a liquid, however, it is convenient to define a correlation function g_h of order h according to

$$f_h\left(r^h\right) = f_1\left(r_1\right) f_1\left(r_2\right) \cdots f_1\left(r_h\right) g_h\left(r^h\right). \tag{5.90}$$

For ideal gases we have

$$g_h^{ig}\left(r^h\right) = 1 \tag{5.91}$$

Any deviation of $g_h\left(r^h\right)$ from unity indicates a molecular structure, i.e., a correlation of the positions of the individual molecules due to intermolecular interactions. In an ideal gas the probability of finding a molecule i at r_i, irrespective of the coordinates of all other molecules, is independent of r_i. From the normalization condition we find

$$\int f_1\left(r_i\right) dr_i = N = f_1 \int dr_i = V f_1, \qquad (5.92)$$

and we thus have

$$f_1 = \frac{N}{V} = n, \qquad (5.93)$$

as is immediately plausible. For the relation between distribution functions and correlation functions we thus find

$$f_h\left(r^h\right) = n^h g_h\left(r^h\right). \qquad (5.94)$$

In particular, under the assumption of pairwise additivity, only distribution functions and correlation functions of molecular pairs in terms of the relative distance between molecular centers are relevant, and we have

$$f_2\left(r_1 r_2\right) = f_2\left(r_{12}\right) = f_2\left(r\right) \qquad (5.95)$$

and

$$g_2\left(r_1 r_2\right) = g_2\left(r_{12}\right) = g_2\left(r\right), \qquad (5.96)$$

where r_{12} is the distance vector and r the scalar distance between the molecular centers. We refer to $g_2(r)$ as the radial pair correlation function.

5.4.2 Thermodynamic Functions

Under the assumption of pairwise additivity, the thermodynamic functions can be expressed rigorously in terms of the pair correlation function.

According to (2.55), the internal energy is related to the canonical partition function by

$$U = kT^2 \left(\frac{\partial \ln Q}{\partial T}\right)_{V,N}, \qquad (5.97)$$

with

$$Q = N! \Lambda^{-3N} Q^C \qquad (5.98)$$

for a monatomic fluid. If we restrict our attention to the configurational part of the internal energy U^C, i.e., that part that is determined by the intermolecular forces, the canonical partition function reduces to the configuration integral and we find

$$
U^C = kT^2 \left(\frac{\partial \ln Q^C}{\partial T} \right)_{V,N} = kT^2 \frac{1}{Q^C} \left(\frac{\partial Q^C}{\partial T} \right)_{V,N}
$$

$$
= \frac{kT^2}{Q^C} \frac{\partial}{\partial T} \int \cdots \int e^{-U(r^N)/kT} d\mathbf{r}^N
$$

$$
= \frac{kT^2}{Q^C} \int \cdots \int \frac{1}{kT^2} U(r^N) e^{-U(r^N)/kT} d\mathbf{r}^N
$$

$$
= \frac{N(N-1)}{2!} \frac{1}{Q^C} \int \cdots \int \phi(r_{12}) e^{-U(r^N)/kT} d\mathbf{r}^N
$$

$$
= \frac{1}{2} N(N-1)
$$

$$
\cdot \iint \frac{\int \cdots \int e^{-U(r^N)/kT} d\mathbf{r}^{N-2}}{Q^C} \phi(r_{12}) d\mathbf{r}_1 d\mathbf{r}_2
$$

$$
= \frac{1}{2} n^2 \iint g_2(\mathbf{r}_1, \mathbf{r}_2) \phi(r_{12}) d\mathbf{r}_1 d\mathbf{r}_2
$$

$$
= 2\pi n^2 V \int_0^\infty g(r) \phi(r) r^2 dr. \tag{5.99}
$$

For the total internal energy we get

$$
U - U^0 = U^{ig} - U^0 + U^C, \tag{5.100}
$$

where $U^{ig} - U^0$ is to be computed from the equations for the ideal gas in Chapter 3.

According to (2.61), we have the following expression for the configurational part of the pressure:

$$
p^C = kT \left(\frac{\partial \ln Q^C}{\partial V} \right)_{T,N} = kT \frac{1}{Q^C} \left(\frac{\partial Q^C}{\partial V} \right)_{T,N}. \tag{5.101}
$$

Because, for monatomic molecules, all kinetic contributions to the partition function are independent of density, this is also the equation for the total pressure and, contrary to the internal energy, not only the part determined by the intermolecular forces. In Section 2.6.1 we related the residual pressure to the virial; cf. (2.263). Introducing there the pairwise additivity and the ideal gas law, we get for the pressure

$$
p = \frac{NkT}{V} - \frac{1}{2} N(N-1) \frac{1}{Q^C} \frac{1}{3V} \int \cdots \int r_{12} \frac{d\phi(r_{12})}{dr_{12}} e^{-U(r^N)/kT} d\mathbf{r}^N
$$

$$
= \frac{NkT}{V} - \frac{1}{6V} n^2 \iint r_{12} \frac{d\phi(r_{12})}{dr_{12}} g_2(\mathbf{r}_1, \mathbf{r}_2) d\mathbf{r}_1 d\mathbf{r}_2
$$

$$
= \frac{NkT}{V} - \frac{2}{3} \pi n^2 \int_0^\infty \frac{d\phi}{dr} g(r) r^3 dr. \tag{5.102}
$$

This is a general expression for the equation of state in the pairwise additivity approximation for a pure monatomic fluid.

Extension of these equations to nonspherical interactions, although not needed in this section on conformal potential models, is straightforward under the condition of the rigid molecule approximation, because then again all kinetic contributions are independent of density, and is performed by additional integrations over the orientational angles; cf. Section 5.3.3. Extension to mixtures requires summing over all pair interactions, i.e., $\sum \sum x_\alpha x_\beta$, where x_α, x_β denote the mole fractions of the components.

5.4.3 The Pair Correlation Function

To evaluate the thermodynamic functions from the above formulations based on the pairwise additivity approximation, we need the pair correlation function. The radial pair correlation function represents the correction of the probability of finding a second molecule at a distance r from a molecule at the origin with respect to that given by the average density. To simplify notation, the index 2 at g_2 is frequently omitted in what follows. Clearly, as can be seen from (5.86), a direct calculation of the pair correlation function from its definition for a given pair potential will generally not be possible, due to the high dimensionality of the integrals involved. However, various indirect experimental and theoretical approaches are available that give insight into the appearance of $g(r)$ as a function of density and temperature. So the general shape of the radial pair correlation function is known over the full region of density.

Figure 5.11 shows $g(r)$ schematically for various densities. In the ideal gas $g(r) = 1$, as discussed above; cf. Figure 5.11a. We consider in particular

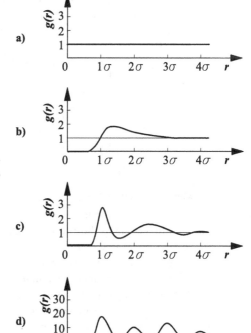

Fig. 5.11. Schematic representation of the pair correlation function: **(a)** ideal gas; **(b)** real gas at zero density; **(c)** liquid; **(d)** solid.

Figure 5.11b, which represents the limit of a real gas at zero density; cf. Exercise 5.5. At a distance of about the minimum of the pair potential, a second molecule is most likely to be situated, because attractive and repulsive forces are in equilibrium there. At very small distances, a second molecule will not be found because it is repelled by the repulsive forces of the molecule at the origin. At very large distances, when the intermolecular pair potential has fallen to essentially zero, there is no correlation between the two molecules any longer, and the probability of finding a second molecule corresponds to that given by the average density of the fluid; i.e., $g = 1$.

EXERCISE 5.5

Derive an equation for the pair correlation function in the limit of zero density.

Solution

We are looking for the function $g(0) = \lim_{n \to 0} g(r; n, T)$. By making use of the pairwise additivity approximation, the Boltzmann factor can be written as

$$e^{-U/kT} = \prod_{N \geq i > j \geq 1} e^{-\phi(r_{ij})/kT} = \prod_{i \geq j} (1 + f_{ij})$$

$$= 1 + f_{21} + f_{31} + f_{32} + \cdots + f_{21} f_{31} + f_{21} f_{32} + \cdots$$
$$+ f_{31} f_{32} + \cdots + f_{21} f_{31} f_{32} + \cdots ,$$

with

$$f_{ij} = e^{-\phi(r_{ij})/kT} - 1$$

as the so-called Mayer function. Introducing this into the zero-order density expansion coefficient of the pair correlation function, cf. (5.86), yields

$$g(0) = \lim_{n \to 0} \left\{ \frac{N(N-1)}{Q^C} \frac{1}{n^2} \int \cdots \int e^{-U(r^N)/kT} dr_3 \cdots dr_N \right\}$$

$$= \lim_{n \to 0} \left\{ \frac{N(N-1)}{Q^C} \frac{1}{n^2} \int \cdots \int (1 + f_{21} + f_{31} + f_{32} + \cdots \right.$$

$$\left. + f_{21} f_{31} + f_{21} f_{32} + \cdots + f_{31} f_{32} + \cdots + f_{21} f_{31} f_{32} + \cdots) dr_3 \cdots dr_N \right\}$$

$$= \lim_{n \to 0} \left\{ \frac{1}{Q^C} V^2 \left(V^{N-2} + V^{N-2} f_{21} \right) + 0(n) \right\}$$

$$= 1 + f_{21} = e^{-\phi(r)/kT},$$

where we made use of $N \gg 2, 1$ and $\lim_{n \to 0} Q^C = Q^{C, \text{ig}} = V^N$. It turns out that the pair correlation function in the limit of zero density is not identical to that of the ideal gas but rather contains the intermolecular energy function of isolated intermolecular pairs. So, in a gas at low density, there is structure, contrary to the picture of an ideal gas, and the associated pair correlation function can be calculated accurately from the pair potential. Further, since the limit $n \to 0$ obviously covers the density range in which the second virial coefficient describes the fluid phase behavior of a gas, cf. Section 5.3, we conclude that $g(0)$ covers the density range up to about one-fifth of the critical and that the pair correlation function does not depend on density in this range.

The structural effects become more prominent with increasing density because the intermolecular forces then play a more significant role. So a strong density dependence is to be expected of the pair correlation function at higher densities. This is confirmed by experiment in the laboratory as well as on the computer. In a regular crystal the probability of finding a molecule is highest at the lattice sites but also generally nonzero between the shells due to the small vibrations around them. Thus the pair correlation function is sharply peaked and periodic, as demonstrated in Figure 5.11d.

EXERCISE 5.6

Consider an ideal two-dimensional crystal lattice of spacing a with point particles fixed at lattice sites and construct the pair correlation function up to a distance of $2a$ from a selected particle.

Solution

In this simple case the pair correlation function can be found from trivial trigonometric considerations. Figure E 5.6.1 shows the lattice along with the coordination shells. Clearly, the first shell will be at a distance of $c_1 = \sqrt{a^2/4 + a^2/4} = a/\sqrt{2}$ from the center molecule, and the next two shells will be at $c_2 = a$ and $c_3 = \sqrt{2}a$. In the idealized situation considered there will be no vibrations around the lattice sites, and so the pair correlation function is an array of infinite sharp lines at the appropriate locations, as shown in Figure 5.6.2.

The pair correlation function of a liquid is in between those of a gas and a solid; i.e., it reveals short-range structural order, cf. Figure 5.11c, which dies off at long range, i.e., at distances of more than a few σ. This behavior of the pair correlation function can be taken to characterize the liquid state on a molecular level. Besides the density dependence, $g(r)$ also depends on temperature. Because high temperatures lead to high kinetic energies of the molecules with

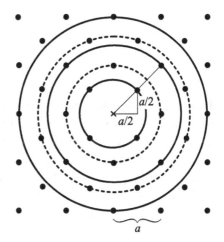

Fig. E 5.6.1. Two-dimensional lattice for an ideal crystal.

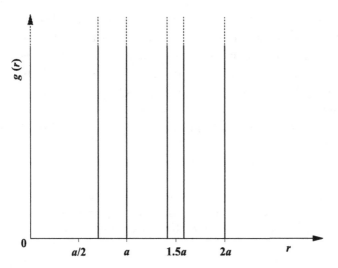

Fig. E 5.6.2. Pair correlation function of an idealized two-dimensional lattice.

an associated destruction of structure, the peaks in $g(r)$ will become smaller with increasing temperature.

5.4.4 Conformal Potentials

Although the basic features of the pair correlation function are well understood, we do not have a simple theory for it in the general case. However, by introducing the assumption of a conformal potential into (5.102) for the equation of state, we can make progress. First we make all terms in the thermodynamic functions dimensionless, by multiplying by σ^3/ε,

$$\frac{p\sigma^3}{\varepsilon} = \frac{N(kT/\varepsilon)}{V/\sigma^3} - \frac{2}{3}\pi n^2 \sigma^6 \int_0^\infty \frac{d(\phi/\varepsilon)}{d(r/\sigma)} g(r/\sigma)(r/\sigma)^3 \, d(r/\sigma),$$

which gives

$$p^* = n^*T^* - \frac{2}{3}\pi n^{*2} \int_0^\infty \frac{d\phi^*}{dr^*} g(r^*)r^{*3} dr^*$$

$$= n^*T^* + p_u^{*,\mathrm{res}}(T^*, n^*) = f_u(T^*, n^*). \tag{5.103}$$

Here we have defined the nondimensional state quantities $p^* = p\sigma^3/\varepsilon$, $n^* = n\sigma^3$, and $T^* = kT/\varepsilon$. We now use the conformal potential assumption. For conformal potentials the integral in (5.103) is universal for all fluids, and thus $p_u^{*,\mathrm{res}}$ is a universal residual pressure function. We thus find that pure substances with conformal potentials follow a universal equation of state in dimensionless variables. This is a fact of considerable practical importance and is referred to as the theorem of corresponding states or the corresponding states principle. If one can determine the function $p^*(T^*, n^*)$ from data of one particular fluid, the equation of state of any other fluid with a conformal potential is known as soon as the two potential parameters ε and σ have been determined. In principle, only two data points are necessary for this, for which frequently the critical data are chosen. These data are generally available for simple fluids.

EXERCISE 5.7

Check the corresponding states principle by calculating the critical compressibility factors and the ratios of the triple point values of temperature and pressure to the corresponding critical values for Ar, Kr, Xe, CH_4, and O_2.

Solution

The parameters ε and σ of the intermolecular pair potential may be expressed in terms of the critical data. At the critical point we have, with v as the molar volume,

$$\left(\frac{\partial p}{\partial v}\right)_T = \left(\frac{\partial^2 p}{\partial v^2}\right)_T = 0,$$

which gives from (5.103),

$$\left(\frac{\partial f_u(T^*, v^*)}{\partial v^*}\right)_{T^*} = \left(\frac{\partial^2 f_u(T^*, v^*)}{\partial v^{*2}}\right)_{T^*} = 0.$$

This leads to the following universal relationships between the potential parameters and the critical data:

$$T_c^* = C_1 = kT_c/\varepsilon$$

and

$$v_c^* = C_2 = v_c/N_A\sigma^3$$

and, consequently,

$$p_c^* = C_3 = p_c\sigma^3/\varepsilon.$$

In terms of the critical constants the corresponding states principle can thus be written as

$$p_r = f_u(T_r, v_r),$$

where

$$p_r = p/p_c, \ T_r = T/T_c, \text{ and } v_r = v/v_c.$$

Table E 5.7.1 shows the results. If the corresponding states principle is valid, all dimensionless variables with reference to critical properties should be universal numbers. For the noble gases argon, krypton, and xenon, the critical compressibility factor and the ratio of the triple point temperature to the critical temperature are roughly equal within the limits of experimental accuracy. Slight differences become evident for the ratio of the triple point pressure to the critical pressure, which is a very sensitive quantity. For methane and, particularly, for oxygen there are strong deviations for the ratios of triple point to critical point state quantities, although the critical compressibility factor essentially follows the universal pattern. The corresponding states principle has become

Table E 5.7.1. Test of the corresponding states principle

	Data	T_c K	p_c bar	v_c cm³/mol	T_t K	p_t bar	Z_c	T_t/T_c	p_t/p_c
Ar	[2]	150.9	49.0	74.6	83.8	0.689	0.291	0.556	0.0141
Kr	[2]	209.4	55.0	92.2	116.0	0.730	0.291	0.554	0.0133
Xe	[2]	289.8	58.8	118.8	161.3	0.815	0.290	0.557	0.0139
CH₄	[2]	190.6	46.0	99.0	90.68	0.1172	0.288	0.476	0.0025
O₂	[3]	154.6	50.45	73.4	54.36	0.00146	0.288	0.351	0.00003

the basis of many successful prediction methods for the fluid phase behavior of simple pure fluids, albeit with various types of empirical extensions [1].

[1] B. E. Poling, J. M. Prausnitz, and J. P. O'Connell. *The Properties of Gases and Liquids*, 5th ed., McGraw-Hill, New York, 2001.
[2] J. S. Rowlinson and F. L. Swinton. *Liquids and Mixtures*. Butterworths, London, 1982.
[3] W. Wagner, J. Ewers, and W. Pentermann. *J. Chem. Thermodyn.*, 8:1094, 1976.

For mixtures, too, the assumption of conformal potentials leads to universal results for the thermodynamic functions. Mixtures in which the components obey conformal potentials are referred to as conformal solutions. According to (5.102), the equation of state of a solution can be expressed as

$$p = nkT - \frac{2}{3}\pi n^2 \sum_\alpha \sum_\beta x_\alpha x_\beta \int_0^\infty \frac{d\phi_{\alpha\beta}}{dr_{\alpha_1\beta_2}}$$

$$\cdot g_{\alpha\beta}(r_{\alpha_1\beta_2}; T, n, \{x_\gamma\}; \{\varepsilon_{\gamma\delta}\}; \{\sigma_{\gamma\delta}\})r_{\alpha_1\beta_2}^3 \, dr_{\alpha_1\beta_2}, \qquad (5.104)$$

where $\{x_\gamma\}$ stands for all mole fractions, $\{\varepsilon_{\gamma\delta}\}$ for all energy parameters, and $\{\sigma_{\gamma\delta}\}$ for all distance parameters of all possible pair interactions. The $g_{\alpha\beta}$ are the pair correlation functions of the various pairwise interacting components in the mixture. With $r_{\alpha_1\beta_2}^* = r_{\alpha_1\beta_2}/\sigma_{\alpha\beta}$ and $\phi^* = \phi_{\alpha\beta}/\varepsilon_{\alpha\beta}$, the integral term in (5.104) can be brought into a dimensionless form and we get

$$p = nkT - \frac{2}{3}\pi n^2 \sum_\alpha \sum_\beta x_\alpha x_\beta \varepsilon_{\alpha\beta} \sigma_{\alpha\beta}^3 \int_0^\infty \frac{d\phi^*}{dr_{\alpha_1\beta_2}^*}$$

$$\cdot g_{\alpha\beta}(r_{\alpha_1\beta_2}^*; T, n, \{x_\gamma\}; \{\varepsilon_{\gamma\delta}\}; \{\sigma_{\gamma\delta}\})r_{\alpha_1\beta_2}^{*3} \, dr_{\alpha_1\beta_2}^*.$$

We can cast this equation into the form of a one-fluid model, if we now introduce one composition-dependent energy parameter ε_x and one composition-dependent distance parameter σ_x and further postulate that due to the conformity of all pair interactions the correlation effects in the mixture are adequately modeled by a pure fluid pair correlation function with these composition-dependent potential parameters. We then arrive at

$$p = nkT - \frac{2}{3}\pi n^2 \varepsilon_x \sigma_x^3 \int_0^\infty \frac{d\phi^*}{dr^*} g_0(r^*; T^*, n^*)r^{*3} dr^*, \qquad (5.105)$$

where $g_0(r^*)$ is the pair correlation function of a hypothetical pure fluid at the temperature $T^* = kT/\varepsilon_x$ and the density $n^* = n\sigma_x^3$ with $r^* = r/\sigma_x$. By comparison with (5.104) we have

$$\varepsilon_x \sigma_x^3 = \sum_\alpha \sum_\beta x_\alpha x_\beta \varepsilon_{\alpha\beta} \sigma_{\alpha\beta}^3. \tag{5.106}$$

An additional equation is required for either σ_x or ε_x in order to apply the theory. This can be derived for σ_x by making use of the fact that the structure of a fluid at high density does not depend very much on the attractive forces. With respect to the correlation function we may then use (5.106) for the particular case of universal energy parameters $\varepsilon_{\alpha\beta} = \varepsilon$ and find

$$\sigma_x^3 = \sum_\alpha \sum_\beta x_\alpha x_\beta \sigma_{\alpha\beta}^3 \tag{5.107}$$

as an independent equation for σ_x. We then know the composition-dependent potential parameters of the hypothetical pure fluid [26]. This model is referred to as the van der Waals 1 approximation (VDW1), because it yields the established one-fluid mixing rules for the constants $a \mathrel{\hat{=}} \varepsilon\sigma^3$ and $b \mathrel{\hat{=}} \sigma^3$ in this famous equation of state. Obviously the VDW1 theory becomes a better approximation as the potential parameters of the components become more similar. On the basis of the VDW1 model, if we have an equation of state for a pure fluid as $p_0^* = f(T^*, n^*)$, it can be extended to mixtures by setting

$$(p\sigma_x^3/\varepsilon_x)_{\text{VDW1}} = p_0^*(kT/\varepsilon_x, n\sigma_x^3). \tag{5.108}$$

Accurate empirical equations of state are known for a number of real pure fluids. From such equations the VDW1 model permits the calculation of the fluid phase behavior of mixtures composed of components that can be characterized by conformal pair potentials.

Conformal solution theory in terms of the VDW1 model is not an exact statistical mechanical solution to the problem of predicting mixture properties from those of the pure components, not even for strictly conformal potentials. Comparisons with computer simulations reveal, however, that it is a good approximation as long as the ratio of potential parameters remains moderate, say $\varepsilon_{22}/\varepsilon_{11} \le 1.5$ and $\sigma_{22}/\sigma_{11} \le 1.1$ [6]. This is a plausible result. In application to real fluid mixtures, it must be noted that the combining rules for the spherical potential parameters ε and σ on the basis of the London approximation are no more than approximations. So the performance of the model is not easily predicted and must be investigated by comparison with data on real fluid mixtures. To illustrate, we predict the vapor–liquid equilibrium and the excess volume for the krypton–argon mixture and all excess functions for the ethane–carbon dioxide mixture and compare to data. We determine the potential parameters ε and σ of the pure fluids argon, krypton, ethane, and carbon dioxide by fitting an equation of state for argon [27], formulated as $p_0^* = f(T^*, n^*)$, to pure fluid saturation data for these components. Because an equation of state for a real fluid is used, we take nonadditive three-body interactions implicitly into account.

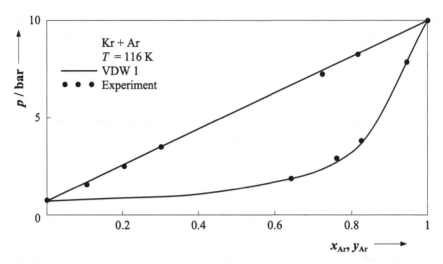

Fig. 5.12. Vapor–liquid equilibrium in the krypton–argon system from an equation of state for pure argon evaluated with VDW1 mixing rules.

We then use the combining rules (5.23) and (5.24) applied to a Lennard–Jones fluid along with the composition-dependent potential parameters according to the VDW1 model. Data are from [28–30]. Figures 5.12 and 5.13 display the vapor–liquid equilibrium and the excess volume of the krypton–argon mixture. The ratios of potential parameters are

$$\sigma_{Kr}/\sigma_{Ar} = 1.07$$

and

$$\varepsilon_{Kr}/\varepsilon_{Ar} = 1.39$$

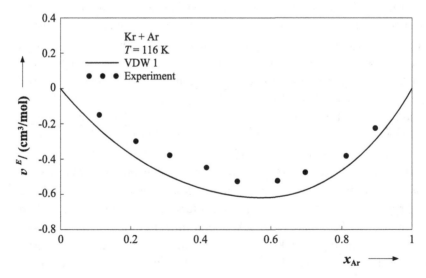

Fig. 5.13. Excess volume in the krypton–argon system from an equation of state for pure argon evaluated with VDW1 mixing rules.

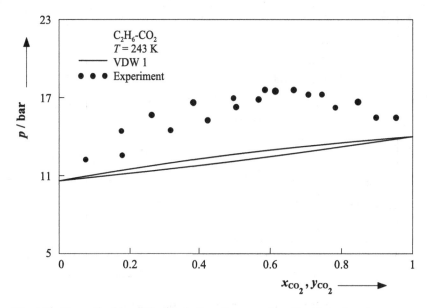

Fig. 5.14. Vapor–liquid equilibrium in the ethane–carbon dioxide system from an equation of state for pure argon evaluated with VDW1 mixing rules.

and are thus well within the range where the VDW1 rules should be valid. We find a satisfactory prediction of the mixture properties, a significant improvement over the simplest possible model, i.e., an ideal solution. This places confidence on the VDW1 as well as on the combination rules. The results for the ethane–carbon dioxide system are shown in Figures 5.14 to 5.16. Although the conformal solution model shows some superiority over the ideal solution

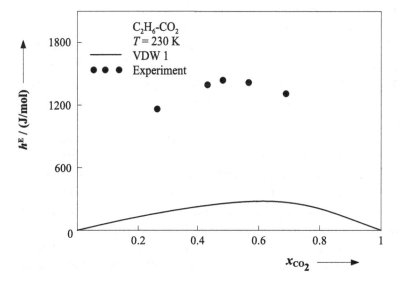

Fig. 5.15. Excess enthalpy in the ethane–carbon dioxide system from an equation of state for pure argon evaluated with VDW1 mixing rules.

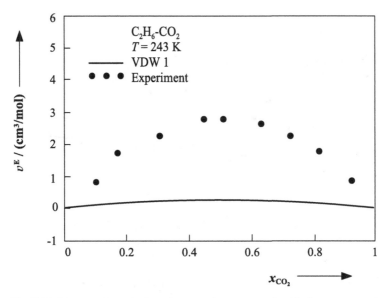

Fig. 5.16. Excess volume in the ethane–carbon system dioxide from an equation of state for pure argon evaluated with VDW1 mixing rules

model, its performance is entirely unacceptable. The ratios of potential parameters for this system are

$$\sigma_{C_2H_6}/\sigma_{CO_2} = 1.16$$

and

$$\varepsilon_{C_2H_6}/\varepsilon_{CO_2} = 0.97.$$

Although the ratio of size parameters is slightly larger than for krypton–argon, this cannot explain the large deviations with respect to the data. The failure is not primarily due to a deficiency of the VDW1 or the combining rules. Rather, it is a misapplication of the model, because the rather nonideal behavior of this system is a consequence of nonconformal interactions in the ethane–carbon dioxide system due to the differences in the charge distributions of these two molecules.

In spite of their general inadequacy, the conformal potential models have found wide application in chemical engineering. Many empirical equations of state make use of the VDW1 rules. Typically, these models are used for fluids of simple molecules as encountered in the gas industries or the oil-based industry, usually modified by some empirical correction in the combining rules for ε and σ to ensure a good representation of the data. This approach is then no longer predictive for binary systems, but it has turned out that empirical binary interaction parameters are sufficient to predict the fluid phase behavior of multicomponent systems.

5.5 Perturbation Models

Generally speaking, the assumption of conformal potentials becomes unrealistic for all but the simplest systems. Instead, the behavior of fluid phases is controlled by the specific, nonconformal interactions between the various components. For predictive purposes it is thus essential to introduce specific information about the intermolecular interactions into the general statistical mechanical equations.

One approach that has been successfully applied over the last decades and continues to produce innovative and successful equations of state is based on the fact that there are different and to some degree separable contributions to the intermolecular pair potential. Various types of separations have been investigated. To demonstrate the basic theory we here discuss two simple versions, notably that between the isotropic repulsive and attractive forces and that between the conformal part of the potential and the nonconformal contributions. Mathematically, such separations can be used as a basis for a perturbation expansion. Here, the potential is split into a reference part and a perturbation part and the perturbation contributions to the thermodynamic functions are evaluated by averaging the statistical mechanical equations over the reference system, the properties of which are assumed to be known. In this sense a perturbation expansion becomes mathematically equivalent to a Taylor series expansion. A more empirical offspring of this approach, based on a separation of repulsive and attractive forces, is generalized van der Waals theory. It leads to rather simple models that are generalizable in a semiempirical manner to more complicated systems and has been the basis for many successful equation of state models in recent years.

5.5.1 The λ-Expansion

We consider a fluid for which the intermolecular potential energy can be formulated in the rigid molecule approximation, i.e., as $U(r^N \omega^N)$, where the r's are the position coordinates of the molecular centers and the ω's are the orientational angles of the molecules. The intermolecular potential energy of a system under consideration is written as the sum of a reference part and a perturbation part [31],

$$U^\lambda = U^{\text{ref}} + \lambda U^{\text{p}}, \tag{5.109}$$

where λ is a continuous variable between 0 and 1. For $\lambda = 0$ we have $U^\lambda = U^{\text{ref}}$ and for $\lambda = 1$ we have the complete intermolecular energy function of the system.

For the difference between the residual free energies of a system with U^λ and a reference system with U^{ref} we find from (5.109)

$$A^{\text{res},\lambda} - A^{\text{res,ref}} = A^{C,\lambda} - A^{C,\,\text{ref}} = -kT \ln \frac{Q^{C,\lambda}}{Q^{C,\text{ref}}}$$

$$= -kT \ln \left\{ \frac{1}{Q^{C,\text{ref}}} \int e^{-U^\lambda(r^N\omega^N)/kT} d\mathbf{r}^N d\omega^N \right\}$$

$$= -kT \ln \langle e^{-\frac{\lambda U^{\text{p}}(r^N\omega^N)}{kT}} \rangle_{\text{ref}}. \tag{5.110}$$

Here, $\langle \ldots \rangle_{\text{ref}}$ denotes canonical ensemble averaging over the reference system and (5.110) expresses the fact that the difference in the residual free energies is given by the average of the Boltzmann factor of the perturbation in the reference system. The expansion of the exponential function yields

$$A^{\text{res},\lambda} - A^{\text{res,ref}} = -kT \ln \left[1 - \lambda \left\langle \frac{U^{\text{p}}}{kT} \right\rangle_{\text{ref}} + \frac{1}{2}\lambda^2 \left\langle \left(\frac{U^{\text{p}}}{kT}\right)^2 \right\rangle_{\text{ref}} \right.$$

$$\left. - \frac{1}{6}\lambda^3 \left\langle \left(\frac{U^{\text{p}}}{kT}\right)^3 \right\rangle_{\text{ref}} + \cdots \right].$$

We now use the expansion of the function $\ln(1 + x)$ around $x = 0$, i.e.,

$$\ln(1 + x) = x - \frac{1}{2}x^2 + \frac{1}{3}x^3 \ldots.$$

This expression converges for

$$|x| = \left| -\lambda \left\langle \frac{U^{\text{p}}}{kT} \right\rangle_{\text{ref}} + \frac{1}{2}\lambda^2 \left\langle \left(\frac{U^{\text{p}}}{kT}\right)^2 \right\rangle_{\text{ref}} - \frac{1}{6}\lambda^3 \left\langle \left(\frac{U^{\text{p}}}{kT}\right)^3 \right\rangle_{\text{ref}} + \cdots \right| \leq 1,$$

which can always be fulfilled for sufficiently small λ. Because the formal structure of a Taylor expansion is independent of convergence restrictions, we may set $\lambda = 1$ and find

$$A^{\text{res}} - A^{\text{res,ref}} = \langle U^{\text{p}} \rangle_{\text{ref}} - \frac{1}{2}\left(\frac{1}{kT}\right) \langle [U^{\text{p}} - \langle U^{\text{p}} \rangle_{\text{ref}}]^2 \rangle_{\text{ref}}$$

$$+ \frac{1}{6}\left(\frac{1}{kT}\right)^2 \langle [U^{\text{p}} - \langle U^{\text{p}} \rangle_{\text{ref}}]^3 \rangle_{\text{ref}} + \cdots$$

$$= A^\lambda + A^{\lambda\lambda} + A^{\lambda\lambda\lambda} + \cdots. \tag{5.111}$$

This is referred to as the λ-expansion. The first-order perturbation term is determined by the reference average of the perturbing potential. The higher-order perturbation terms decrease in significance with increasing temperature. It is in this sense that we also refer to (5.111) as a high-temperature expansion. The perturbation terms are determined by fluctuations of the perturbation potential around its average and thus decrease for perturbations that do not depend strongly on configuration. We thus expect this expansion to be particularly suited for isotropic and anisotropic long-range forces, for which the configuration dependence is much weaker than at short range. The expansion (5.111) is not yet restricted to a particular separation of the potential energy into a reference part and a perturbation part in (5.109).

5.5.2 The Hard Body Reference

One type of application of the λ-expansion aims at introducing the attractive forces as a perturbation around a hard body reference system, the properties of which are assumed to be known. The associated separation of the pair potential is into a repulsive and an attractive term. Many versions have appeared, but here we restrict consideration to the simplest, i.e., that for isotropic interactions. Thus, the reference system is the hard sphere fluid and the spherical attractive part of the potential is the perturbation. In this case the first-order perturbation term to the free energy in the λ-expansion reads, for a pure fluid,

$$\frac{A^\lambda}{NkT} = \frac{1}{NkT}\langle U^{\mathrm{p}}\rangle_{\mathrm{d}} = \frac{N(N-1)}{2NkT}\langle \phi^{\mathrm{p}}(r)\rangle_{\mathrm{d}}$$

$$= 2\pi n \frac{1}{kT} \int_0^\infty \phi^{\mathrm{p}}(r) g_{\mathrm{d}}(r) r^2 \mathrm{d}r, \tag{5.112}$$

where the index d refers to the hard sphere fluid and the pairwise additivity of pair potentials has been used; cf. Section 5.4. This allows formulating the peturbation in terms of the hard sphere pair correlation function. Extension to mixtures is simply done by replacing (5.112) by a double sum over all components, as usual. The free energy of the system is thus given to first order in the λ-expansion by

$$\frac{A^{\mathrm{res}}}{NkT} = \frac{A_{\mathrm{d}}^{\mathrm{res}}}{NkT} + \frac{A^\lambda}{NkT} \tag{5.113}$$

with

$$\frac{A_{\mathrm{d}}^{\mathrm{res}}}{NkT} = -\ln\left(\frac{Q_{\mathrm{d}}}{Q_{\mathrm{d}}^{\mathrm{ig}}}\right)^{1/N} = -\ln\left(\frac{V_{\mathrm{f}}}{V}\right)_{\mathrm{d}}. \tag{5.114}$$

When the hard sphere volume is zero we have $Q_{\mathrm{d}=0} = Q_{\mathrm{d}}^{\mathrm{ig}} = (1/N!)\Lambda^{-3N}V^N$, as shown in Section 3.1 for the ideal gas. The total system volume is then accessible to all molecules. Molecules of finite diameter, however, reduce the volume that is available to other molecules. In this sense we write

$$Q_{\mathrm{d}} = \frac{1}{N!}\Lambda^{-3N}(V_{\mathrm{f}})_{\mathrm{d}}^N, \tag{5.115}$$

where $(V_{\mathrm{f}})_{\mathrm{d}}$ is the free volume of a hard sphere system, i.e., the volume available to the center of mass of a hard sphere as it moves about in a container of volume V, with $V_{\mathrm{f}} < V$ at finite densities. This simple first-order perturbation model is adequate for fluids at high temperatures and high densities. To extend it to typical liquid applications the second-order perturbation term should be included.

EXERCISE 5.8

Derive the van der Waals equation of state

$$p_{\mathrm{VDW}} = \frac{RT}{v-b} - \frac{a}{v^2}$$

from the first-order λ-expansion by assuming the molecules to be hard spheres with a constant attractive intermolecular energy for $r > d$ as a perturbation and neglecting the density dependence of the hard sphere pair correlation function.

Solution

In a system of hard spheres the center of one hard sphere can approach that of a second hard sphere only up to the hard sphere diameter d. Thus the volume excluded for one hard sphere due to the presence of a second hard sphere is a sphere of radius d and thus $\frac{4}{3}\pi d^3$; cf. Fig. E 5.8.1a. In a system of N hard spheres there are $(N-1)$ such terms. Neglecting all overlap effects they add up to a total excluded volume for a selected hard sphere. Because the excluded volume for each pair can be associated in equal parts with each molecule, we introduce a factor of $\frac{1}{2}$ and find, with $N \gg 1$, for the total volume excluded to any single molecule

$$\frac{N}{2}\frac{4}{3}\pi d^3 = N\frac{2}{3}\pi d^3 = \frac{N}{N_{\mathrm{A}}}b, \ \text{where} \ b = N_{\mathrm{A}}\frac{2}{3}\pi d^3.$$

Thus we have for the free volume

$$V_{\mathrm{f,VDW}} = V - \frac{N}{N_{\mathrm{A}}}b.$$

Clearly this approximation will only be valid in the limit of low densities. At higher densities the excluded volumes associated with the $(N-1)$ hard spheres overlap, as shown in Fig. E 5.8.1b, and

$$V_{\mathrm{f}} > \left(V - \frac{N}{N_{\mathrm{A}}}b\right).$$

To obtain the van der Waals expression for A^λ we assume that the hard-sphere pair correlation function is independent of density and temperature. Although indeed the independence of temperature is correct, the independence of density is a crude approximation. When it is further introduced that the attractive potential is modeled by a uniform negative energy value, the integral over r in (5.112) is a constant depending only on the constant depth of the attractive potential energy, which, in the notation of van der Waals, is written as

$$-a/2\pi N_{\mathrm{A}}^2$$

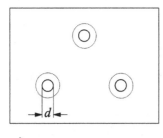

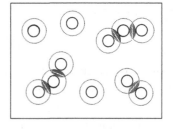

a)   b)

Fig. E 5.8.1. Excluded volume (dashed lines) at **(a)** low and **(b)** high density.

with "a" as a positive constant reflecting the attractive intermolecular potential energy ϕ^p. Then we get

$$\frac{A^\lambda}{NkT} = 2\pi n \frac{1}{kT}\left(-\frac{a}{2\pi N_A^2}\right).$$

If we introduce the van der Waals approximations for V_f and A^λ into the λ-expansion, we get the van der Waals equation of state by taking the derivative with respect to volume as

$$p_{VDW} = \frac{RT}{v-b} - \frac{a}{v^2},$$

where v is the molar volume to replace V for $N = N_A$.

A perturbation expansion about the hard-sphere reference depends crucially on an adequate representation of the hard-sphere thermodynamic and structural properties. Although based on the λ-expansion with the hard-sphere reference, the van der Waals equation of state does not make proper use of such data. The hard-sphere free volume is approximated only crudely by a term valid only at low density and the rather strong density dependence of the hard-sphere pair correlation function is neglected altogether. Systematic improvement is possible using accurate formulations of the pair correlation function and the equation of state for hard spheres.

The pair correlation function for a pure hard-sphere fluid has been determined from computer simulation [32]. It is reproduced for some densities in Figure 5.17, where $\eta = \frac{1}{6}\pi nd^3 = \frac{1}{6}\pi n^*$ is the packing fraction; cf. Exercise 4.3.

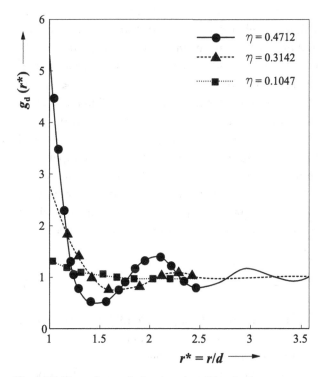

Fig. 5.17. The pair correlation function of hard spheres.

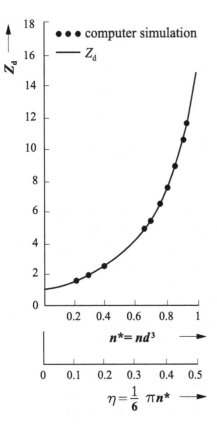

Fig. 5.18. The equation of state for hard spheres.

By comparison with the schematic representation of Figure 5.11 it becomes clear that the hard-sphere fluid at high density is a good model for the structure in fluids at high density and high temperature, although not at low density. In particular, it should be noted that, in contrast to a real fluid, the pair correlation function of hard spheres is discontinuous at $r^* = 1$, with $g(r^* < 1) = 0$. Although this is severe at low densities it is less significant at liquid densities. There is no temperature effect on the pair correlation function of a hard-sphere fluid, which is consistent with the interpretation that a real fluid of spherical molecules converges to the hard-sphere fluid in the limit of high temperature and density. An efficient numerical procedure for calculating hard-sphere correlation functions in pure fluids as well as in mixtures in good agreement with computer simulations is available [33,34]. The equation of state of a pure fluid of hard spheres has been established from a combination of statistical mechanics and computer simulation as

$$Z_d = \left(\frac{p}{nkT}\right)_d = \frac{1 + \eta + \eta^2 - \eta^3}{(1 - \eta)^3}. \qquad (5.116)$$

It is referred to as the Carnahan–Starling equation [35] and shown in Figure 5.18, along with results of computer simulations that are in good agreement with (5.116). The equation of state for hard spheres has the typical appearance of that of a high-temperature gas in the Z_d, n^*-diagram. It has been extended to mixtures [36]. For a constant hard-sphere diameter there is no temperature dependence. The residual free energy of a pure hard-sphere fluid is given by

$$
\left(\frac{A^{\text{res}}}{NkT}\right)_d = -\frac{1}{N}\ln\left(\frac{Q}{Q^{\text{ig}}}\right)_d = -\ln\left(\frac{V_f}{V}\right)_d
$$
$$
= -\int_\infty^V \left[\left(\frac{pV}{NkT}\right)_d - 1\right]\frac{dV}{V} = \int_0^\eta \frac{4\eta - 2\eta^2}{(1-\eta)^3}\frac{d\eta}{\eta} = \frac{4\eta - 3\eta^2}{(1-\eta)^2},
$$

(5.117)

which leads to the nondimensional free volume as

$$
\left(\frac{V_f}{V}\right)_d = e^{-\frac{4\eta - 3\eta^2}{(1-\eta)^2}}.
$$

(5.118)

It is evidently different from that of the van der Waals equation of state, cf. Exercise 5.8, but converges to it at low densities.

To derive a successful equation of state for real fluids with isotropic interactions from this perturbation model, the finite steepness of a real pair potential at short range, in contrast to the hard-sphere potential, has to be taken into account. This can be done by adding a short-range perturbation expansion around the hard-sphere model, referred to as the blib expansion [37]. It can be shown that its effect is adequately incorporated by making the hard-sphere diameter temperature-dependent by a well-defined theoretical prescription. By using the accurate hard-sphere properties for pure fluids, successful calculations for the phase behavior of pure fluids and fluid mixtures at high temperatures and densities on the basis of this perturbation expansion for a chosen isotropic potential model may be performed [38]. For mixtures, the accurately known mixing rules for the hard-sphere system and for the perturbation term lead to an improvement over the conformal solution's VDW1 mixing rules. Again the usual combining rules for the unlike interaction potential parameters must be applied and then give a reasonable prediction of mixture behavior from that of the pure components.

Although well established, this simple perturbation expansion is only of limited utility. Most fluids of practical interest cannot be modeled adequately by spherical potential functions. So anisotropic contributions at short range as well as at long range must be considered. In principle, the separation of the pair potential into a hard repulsive part and an attractive part is not restricted to the simple case of spherical interactions discussed above. Rather, various versions of perturbation models have been developed with both parts of the potential being anisotropic. However, because these models tend to become quite complex and thus impractical, we shall not consider them here, but rather turn to an analogous but simpler and more empirical approach in Section 5.5.4.

5.5.3 The Conformal Potential Reference

Another type of application of (5.111) is to perform a perturbation expansion around the properties of a conformal reference system, e.g., the spherical Lennard–Jones fluid. This has the advantage of using a universal reference system whose properties are well understood through computer simulation. The resulting model is a systematic correction of the conformal potential models of Section 5.4.4. Although the expansion (5.111) does not converge well for strong long range anisotropic contributions such as strong multipoles, a reasonable estimation of their effects can be achieved by using a Padé approximation, i.e. [20,39],

$$A^{\text{res}} = A^{\text{res, ref}} + A^{\lambda} + A^{\lambda\lambda}\left(\frac{1}{1 - A^{\lambda\lambda\lambda}/A^{\lambda\lambda}}\right). \tag{5.119}$$

Good agreement between the Padé and simulation data has been found up to reduced dipole and quadrupole moments of $\mu^* = \mu/\sqrt{\varepsilon\sigma^3} \approx 1$ and $\theta^* = \theta/\sqrt{\varepsilon\sigma^5} \approx 1$, respectively, which may be considered as realistic for many polar molecules. Much stronger multipole moments cannot be treated by (5.119). The model is strictly not suitable for strong shape effects due to the associated dramatic configuration dependence, as confirmed by a comparison with computer simulations [6]. However, it can also be applied effectively to mildly nonspherical molecules, as demonstrated below. Any deficiencies are, as usual, absorbed in the adjustable parameters, which here are the Lennard–Jones parameters ε, σ and a shape parameter, such as the repulsive site distance from the molecular center in the SSR-potential model; cf. Section 5.2. We assume pairwise additivity, as usual. The complete intermolecular pair potential between molecule 1 and molecule 2 is written as the sum of the spherical reference part $\phi^{\text{ref}}(r)$ and the orientation-dependent perturbation term $\phi^{\text{P}}(r\omega_1\omega_2)$, which represents all types of anisotropic forces, according to [40,41]

$$\phi(r\omega_1\omega_2) = \phi^{\text{ref}}(r) + \lambda\phi^{\text{P}}(r\omega_1\omega_2). \tag{5.120}$$

Here r is the center-to-center distance of the two molecules and ω_1, ω_2 describe their orientations, which can be written in terms of the Euler angles $(\phi\theta\chi)$ for each molecule; cf. Section 5.2. The reference part of the potential is chosen to be the unweighted orientational average of the complete pair potential, i.e.,

$$\phi^{\text{ref}}(r) \equiv \langle\phi(r\omega_1\omega_2)\rangle_{\omega_1\omega_2}, \tag{5.121}$$

where

$$\langle\cdots\rangle_{\omega_1\omega_2} = \frac{1}{\int d\omega_1 d\omega_2}\int(\cdots)d\omega_1 d\omega_2. \tag{5.122}$$

If we perform the unweighted average on (5.120) we get

$$\langle\phi(r\omega_1\omega_2)\rangle_{\omega_1\omega_2} = \phi^{\text{ref}}(r) + \langle\phi^{\text{P}}(r\omega_1\omega_2)\rangle_{\omega_1\omega_2},$$

wherefrom, by definition of ϕ^{ref}, we find

$$\langle \phi^{\text{p}}(r\omega_1\omega_2)\rangle_{\omega_1\omega_2} = 0. \tag{5.123}$$

The unweighted angular average of the perturbation part of the potential thus vanishes identically in this expansion. Equation (5.123) leads to a significant simplification in evaluating the λ-expansion. We then find for the expansion terms of the residual free energy A^{res} up to third order for pure fluids that

$$\frac{A^\lambda}{NkT} \equiv \frac{1}{NkT}\langle U^{\text{p}}\rangle_{\text{ref}} = 0, \tag{5.124}$$

and, by following the procedure of introducing the pairwise additivity approximation in terms of correlation functions, cf. Section 5.4.2,

$$\begin{aligned}
\frac{A^{\lambda\lambda}}{NkT} &= -\frac{1}{2}\frac{1}{NkT}\left(\frac{1}{kT}\right)\langle (U^{\text{p}})^2\rangle_{\text{ref}} \\
&= -\frac{1}{4}n^2\frac{1}{NkT}\left(\frac{1}{kT}\right)\iint \langle (\phi_{12}^{\text{p}})^2\rangle_{\omega_1\omega_2}g^{\text{ref}}\,\mathrm{d}\boldsymbol{r}_1\mathrm{d}\boldsymbol{r}_2 \\
&\quad -\frac{1}{2}n^3\frac{1}{NkT}\left(\frac{1}{kT}\right)\iiint \langle \phi_{12}^{\text{p}}\phi_{13}^{\text{p}}\rangle_{\omega_1\omega_2\omega_3}g_3^{\text{ref}}\,\mathrm{d}\boldsymbol{r}_1\mathrm{d}\boldsymbol{r}_2\mathrm{d}\boldsymbol{r}_3
\end{aligned} \tag{5.125}$$

and finally

$$\begin{aligned}
\frac{A^{\lambda\lambda\lambda}}{NkT} &= \frac{1}{6}\frac{1}{NkT}\left(\frac{1}{kT}\right)^2\langle (U^{\text{p}})^3\rangle_{\text{ref}} \\
&= \frac{1}{12}n^2\frac{1}{NkT}\left(\frac{1}{kT}\right)^2\iint \langle (\phi_{12}^{\text{p}})\rangle_{\omega_1\omega_2}^3 g^{\text{ref}}\,\mathrm{d}\boldsymbol{r}_1\mathrm{d}\boldsymbol{r}_2 \\
&\quad + \frac{1}{6}n^3\frac{1}{NkT}\left(\frac{1}{kT}\right)^2\iiint \langle \phi_{12}^{\text{p}}\phi_{13}^{\text{p}}\phi_{23}^{\text{p}}\rangle_{\omega_1\omega_2\omega_3}g_3^{\text{ref}}\,\mathrm{d}\boldsymbol{r}_1\mathrm{d}\boldsymbol{r}_2\mathrm{d}\boldsymbol{r}_3,
\end{aligned} \tag{5.126}$$

where g_3^{ref} is the so-called triplet correlation function of the reference system and is defined as discussed in Section 5.4.1 for g_h with $h = 3$. Extension to mixtures is done, as usual, by a double summation and a triple summation in the mole fractions for the pair correlation and the triplet correlation terms, respectively.

We note that the first-order expansion term is zero due to the definition (5.121) of the reference system, which leads to (5.123). As shown in Exercise 5.3, this is consistent with the properties of the multipolar forces. Other types of anisotropic forces, such as the induction and, in particular, repulsive interactions, have individual nonzero contributions on angle averaging. By way of (5.121) these should strictly be incorporated in the reference system. However, because a conformal potential reference system, i.e., the Lennard-Jones, is to be used here, such contributions can more practically be taken into account by evaluating the associated first-order perturbation terms. We further note that, although we used pairwise additivity, higher-order correlation functions appear as artifacts of the perturbation expansion. The integrations over the angular coordinates can be performed analytically if use is made of the representation of the pair potential as an expansion in terms of spherical harmonics; cf. Section 2.4. It can then be shown that these integrations simply reduce to numerical factors specific to the particular potential energy model considered; cf. Exercise 5.9. In principle, further terms in (5.126) appear from formulating $(U^P)^3$ in terms of pair potentials, such as those associated with $(\phi_{12}^P)^2 \phi_{13}^P$ and others. Any terms, however, containing one index only once, such as index 3 here, will disappear on angle averaging for the multipolar forces due to (A 4.8); cf. Exercise 5.9. For the other types of forces these further terms are neglected, because the multipole effects dominate the long range anisotropy of the pair potential. The other long-range anisotropies are considered to be small as compared to them. The short-range anisotropy is taken into account only in the first-order term, because this reproduces the effect better than the Padé approximation [6]. The integrations over the correlation functions of the reference system are universal for a universal reference system. They thus have to be performed numerically only once by computer simulations for the Lennard–Jones system, and the results have been fitted to simple empirical functions of density and temperature. The equation of state for the reference system can be approximated by an equation of state for argon [27]. This takes nonadditive three-body effects of the dispersion forces into account at least approximately. In the final form, the theory therefore reduces to an analytical expression for the free energy, or, by taking the adequate derivatives, to analytical expressions for the equation of state, the internal energy, etc. This version of perturbation theory can, therefore, be used like any analytical equation of state, although many additive terms may have to be included if complicated intermolecular force models are considered. Explicit equations for the various perturbation terms up to third order and many examples of applications have been summarized in the literature [6,20,42], including perturbation contributions of the multipole forces, the induction forces, the nonspherical dispersion forces, and the nonspherical repulsion forces. A compilation of the terms for the multipole interactions is presented in App. 12.

EXERCISE 5.9

Derive the second-order perturbation term in the λ-expansion for the dipole–dipole interaction of linear molecules.

Solution

The second-order perturbation term reads, according to (5.125),

$$\frac{A^{\lambda\lambda}}{NkT} = \left(\frac{A^{\lambda\lambda}}{NkT}\right)_A + \left(\frac{A^{\lambda\lambda}}{NkT}\right)_B,$$

where

$$\left(\frac{A^{\lambda\lambda}}{NkT}\right)_A = -\frac{1}{4} n^2 \frac{1}{NkT} \left(\frac{1}{kT}\right) \int \int \langle (\phi_{12}^P)^2 \rangle_{\omega_1 \omega_2} g^{\mathrm{ref}} \, d\mathbf{r}_1 \, d\mathbf{r}_2$$

and

$$\left(\frac{A^{\lambda\lambda}}{NkT}\right)_{\mathrm{B}} = -\frac{1}{2}n^3\frac{1}{NkT}\left(\frac{1}{kT}\right)\int\int\int\langle\phi_{12}^{\mathrm{p}}\phi_{13}^{\mathrm{p}}\rangle_{\omega_1\omega_2\omega_3}g_3^{\mathrm{ref}}\,\mathrm{d}\boldsymbol{r}_1\mathrm{d}\boldsymbol{r}_2\mathrm{d}\boldsymbol{r}_3.$$

The term $\left(A^{\lambda\lambda}/NkT\right)_A$ can be written as

$$\left(\frac{A^{\lambda\lambda}}{NkT}\right)_{\mathrm{A}} = -\pi n\left(\frac{1}{kT}\right)^2\int_0^\infty\langle(\phi_{12}^{\mathrm{p}})^2\rangle_{\omega_1\omega_2}g^{\mathrm{ref}}r_{12}^2\mathrm{d}r_{12}.$$

Here we need the unweighted orientational average of the squared perturbation potential. At this stage we keep the evaluation general and specialize to the dipole–dipole interaction later. Introducing the general functional form of the pair potential, i.e., the expansion in terms of spherical harmonics for ϕ_{12}^{p}, cf. Equation (2.125), we find

$$\langle(\phi_{12}^{\mathrm{p}})^2\rangle_{\omega_1\omega_2} = \Bigg\langle\Bigg[\sum_{l_1=0}^\infty\sum_{l_2=0}^\infty\sum_{\substack{m_1\\m_2}}\sum_{\substack{n_1\\n_2}}E_{12}(l_1l_2l_1+l_2;n_1n_2;r_{12})\sqrt{\frac{2_1+2l_2+1}{4\pi}}$$
$$\cdot\,C(l_1l_2l_1+l_2;m_1m_20)D_{m_1n_1}^{l_1}(\omega_1)^*D_{m_2n_2}^{l_2}(\omega_2)^*\Bigg]^2\Bigg\rangle_{\omega_1\omega_2}$$
$$=\sum_{l_1=0}^\infty\sum_{l_2=0}^\infty\sum_{\substack{m_1\\m_2}}\sum_{\substack{n_1\\n_2}}\sum_{l_1'=0}^\infty\sum_{l_2'=0}^\infty\sum_{\substack{m_1'\\m_2'}}\sum_{\substack{n_1'\\n_2'}}\frac{2l_1+2l_2+1}{4\pi}$$
$$\cdot\,E_{12}(l_1l_2l_1+l_2;n_1n_2;r_{12})E_{12}(l_1'l_2'l_1'+l_2';n_1'n_2';r_{12})$$
$$\cdot\,C(l_1l_2l_1+l_2;m_1m_20)C(l_1'l_2'l_1'+l_2';m_1'm_2'0)$$
$$\cdot\,\langle D_{m_1n_1}^{l_1}(\omega_1)^*D_{m_1'n_1'}^{l_1'}(\omega_1)^*\rangle_{\omega_1}\langle D_{m_2n_2}^{l_2}(\omega_2)^*D_{m_2'n_2'}^{l_2'}(\omega_2)^*\rangle_{\omega_2}.$$

We note from (A 4.7) that only terms with $l_1l_2 = l_1'l_2'$ and $m_1 = \underline{m_1'}, m_2 = \underline{m_2'}$ and $n_1 = \underline{n_1'}, n_2 = \underline{n_2'}$ survive. We thus have, using also (A 4.6) and (A 4.9),

$$\langle(\phi_{12}^{\mathrm{p}})^2\rangle_{\omega_1\omega_2} = \sum_\Lambda\langle(\phi_{12}^{\mathrm{p}}(\Lambda))^2\rangle_{\omega_1\omega_2},$$

where $\Lambda = l_1l_2l_1+l_2$ and

$$\langle(\phi_{12}^{\mathrm{p}}(\Lambda))^2\rangle_{\omega_1\omega_2} = \sum_{\substack{n_1\\n_2}}E(\Lambda;n_1n_2;r_{12})E(\Lambda;\underline{n_1}\underline{n_2};r_{12})$$
$$\cdot\,(-1)^{2(l_1+l_2)}(-1)^{-n_1-n_2}\frac{2l_1+2l_2+1}{(2l_1+1)(2l_2+1)}\frac{1}{4\pi}.$$

We simplify by formally introducing the conjugate complex form

$$\phi_{12}^{\mathrm{p}}(\Lambda)^* = \phi_{12}^{\mathrm{p}}(\Lambda) = \sum_{\substack{m_1\\m_2}}\sum_{\substack{n_1\\n_2}}E(\Lambda;\underline{n_1}\underline{n_2};r_{12})^*C(\Lambda;\underline{m_1}\underline{m_2}0)\sqrt{\frac{2_1+2l_2+1}{4\pi}}$$
$$\cdot\,D_{\underline{m_1}\underline{n_1}}^{l_1}(\omega_1)D_{\underline{m_2}\underline{n_2}}^{l_2}(\omega_2)$$

and note that the signs of the dummy indices are irrelevant because they run from $-l$ to $+l$. Transformation of the rotation matrices into their conjugate complex forms and comparison to the original spherical harmonic expansion of the pair potential yields

$$E(\Lambda;n_1n_2;r_{12}) = (-1)^{2(l_1+l_2)}(-1)^{n_1+n_2}E(\Lambda;\underline{n_1}\underline{n_2};r_{12})^*$$

and thence

$$\langle(\phi_{12}^{\mathrm{p}}(\Lambda))^2\rangle_{\omega_1\omega_2} = \frac{1}{4\pi}\frac{2l_1+2l_2+1}{(2l_1+1)(2l_2+1)}$$
$$\cdot\,\sum_{\substack{n_1\\n_2}}E(\Lambda;n_1n_2;r_{12})E(\Lambda;n_1n_2;r_{12})^*.$$

The complete perturbation pair potential will contain various types of interactions, such as multipolar forces, and induction forces. We have to sum over all these types of interactions. If we denote by s and s' two particular types of interaction, we arrive at

$$\langle \phi_{12}^{P}(r_{12}\omega_{1}\omega_{2})^{2}\rangle_{\omega_{1}\omega_{2}} = \frac{1}{4\pi} \sum_{\Lambda} \frac{2l_{1}+2l_{2}+1}{(2l_{1}+1)(2l_{2}+1)}$$
$$\cdot \sum_{\substack{n_{1}\\n_{2}}} \sum_{s} \sum_{s'} E^{s}(\Lambda; n_{1}n_{2}; r_{12}) E^{s'}(\Lambda; n_{1}n_{2}; r_{12})^{*}.$$

It is convenient to factor the expansion coefficient into a term depending only on Λ and n_{1}, n_{2} and a second term containing the r-dependence through

$$E^{s}(\Lambda; n_{1}n_{2}; r) = E^{s}(\Lambda; n_{1}n_{2}) \left(\frac{\sigma}{r}\right)^{n_{s}} \frac{1}{(\sigma)^{n_{s}}},$$

which yields for the contribution of the interaction type ss' to $\left(A^{\lambda\lambda}/NkT\right)_{A}$

$$\left(\frac{A^{\lambda\lambda}(\Lambda)}{NkT}\right)_{A}^{ss'} = -\frac{n^{*}}{4T^{*2}} \frac{1}{\sigma^{(n_{s}+n_{s'})}} \frac{2l_{1}+2l_{2}+1}{(2l_{1}+1)(2l_{2}+1)}$$
$$\cdot J^{(n_{s}+n_{s'})}(n^{*}, T^{*}) \frac{1}{\varepsilon^{2}} \sum_{\substack{n_{1}\\n_{2}}} E^{s}(\Lambda; n_{1}n_{2}) E^{s'}(\Lambda; n_{1}n_{2})^{*},$$

with

$$J^{(n_{s}+n_{s'})} = \int \frac{1}{r^{*(n_{s}+n_{s'})}} g^{\text{ref}} r^{*2} dr^{*}.$$

We now apply the above general equation to the dipole–dipole interaction. The expansion coefficient of the dipole–dipole interaction for linear molecules reads according to Exercise 2.7

$$E^{\text{mult}}(112; 00; r_{12}) = -2 \cdot \sqrt{\frac{6\pi}{5}} \cdot \mu^{2} \left(\frac{\sigma}{r_{12}}\right)^{3} \frac{1}{\sigma^{3}}.$$

We thus find

$$\left(\frac{A^{\lambda\lambda}(112)}{NkT}\right)_{A}^{\text{mult}-\text{mult}} = -\frac{1}{4} \frac{n^{*}}{T^{*2}} \frac{1}{\sigma^{6}} \frac{5}{3\cdot 3} J^{(6)}(n^{*}, T^{*}) 4 \frac{6\pi}{5} \mu^{4} \frac{1}{\varepsilon^{2}}$$
$$= -\frac{2}{3} \frac{\pi n^{*}}{T^{*2}} \frac{\mu^{4}}{\varepsilon^{2}\sigma^{6}} J^{(6)}(n^{*}, T^{*}).$$

To evaluate this term numerically as a function of temperature and density we need the dipole moment of the fluid under consideration along with the potential parameters ε and σ, and also the $J^{(6)}$-integral for the Lennard–Jones system, for which empirical correlations based on computer simulations are available.

The term $\left(A^{\lambda\lambda}/NkT\right)_{B}$ contains an angle average over two different interactions, i.e.,

$$\langle \phi_{12}^{P}\phi_{13}^{P}\rangle_{\omega_{1}\omega_{2}\omega_{3}} = \sum_{l_{1}=0}^{\infty} \sum_{l_{2}=0}^{\infty} \sum_{\substack{m_{1}\\m_{2}}} \sum_{\substack{n_{1}\\n_{2}}} \sum_{l_{1}'=0}^{\infty} \sum_{l_{3}'=0}^{\infty} \sum_{\substack{m_{1}'\\m_{3}'}} \sum_{\substack{n_{1}'\\n_{3}'}}$$
$$\cdot E_{12}(l_{1}l_{2}l_{1}+l_{2}; n_{1}n_{2}; r_{12}) E_{13}(l_{1}'l_{3}'l_{1}'+l_{3}'; n_{1}'n_{3}'; r_{13})$$
$$\sqrt{\frac{2l_{1}+2l_{2}+1}{4\pi}} \sqrt{\frac{2l_{1}'+2l_{2}'+1}{4\pi}}$$
$$\cdot C(l_{1}l_{2}; m_{1}m_{2}0) C(l_{1}'l_{3}'; m_{1}'m_{3}'0)$$
$$\cdot \langle D_{m_{1}n_{1}}^{l_{1}}(\omega_{1})^{*} D_{m_{1}'n_{1}'}^{l_{1}'}(\omega_{1})^{*}\rangle_{\omega_{1}} \langle D_{m_{2}n_{2}}^{l_{2}}(\omega_{2})^{*}\rangle_{\omega_{2}} \langle D_{m_{3}'n_{3}'}^{l_{3}'}(\omega_{2})^{*}\rangle_{\omega_{3}}.$$

We see here that this contribution will be zero for the dipole–dipole (and all other multipole) interactions, because (A 4.8) requires $l_{2} = 0, l_{3} = 0$, which is inconsistent with the multipole forces; cf. Section 2.4.

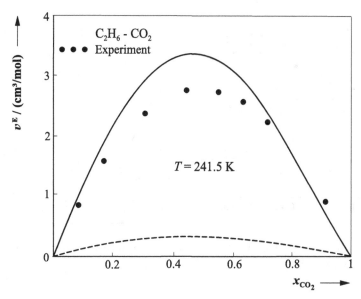

Fig. 5.19. The excess volume of C_2H_6-CO_2 from perturbation theory (—) and VDW1 conformal solution theory (- -).

As an illustration, Figures 5.19 to 5.22 show results obtained for the system C_2H_6–CO_2 [42]. Here, all types of interaction have been included. So in addition to the quadrupolar interactions we consider those due to induction and anisotropic dispersion forces and also those due to the anisotropic repulsive forces. Both molecules are assumed to be linear rigid structures with an eccentricity $r_a = r_b$. The interactions in this system are modeled with the SSR-MPA potential model, cf. Section 5.2, with the spherical pure fluid parameters ε, σ and the eccentricity r_a being fitted to pure fluid data along the saturation lines. All other molecular parameters, such as multipole moments and polarizabilities,

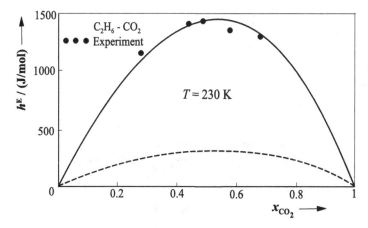

Fig. 5.20. The excess enthalpy of C_2H_6-CO_2 from perturbation theory (—) and VDW1 conformal solution theory (- -).

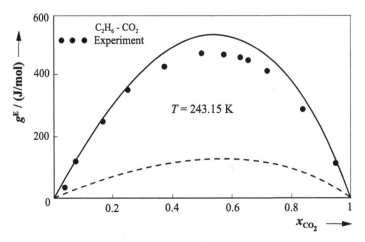

Fig. 5.21. The excess free enthalpy of C_2H_6-CO_2 from perturbation theory (—) and VDW1 conformal solution theory (- -).

are taken from independent sources [20] or calculated quantum-chemically. The conformal reference system is extended to mixtures by applying the VDW1 rules for the equation of state and similar mixing rules for the correlation functions, as outlined in App. 12. The mixing rules for the perturbation terms follow from their derivations. The combining rules for the center-to-center interactions are (5.23) and (5.24), as in the illustration of the VDW1 model.

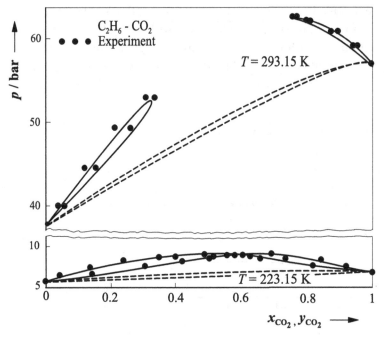

Fig. 5.22. The vapor–liquid equilibrium of C_2H_6–CO_2 from perturbation theory (—) and VDW1 conformal solution theory (- -).

The model is then fully predictive. For comparison, the conformal solution results from Figures 5.14 to 5.16 are included in the figures.

The merits of including the various anisotropic perturbation terms are evident, because, contrary to the VDW1 model, the perturbation model now predicts the thermodynamics of this system rather well. Although performing excellently in many systems with moderately anisotropic short-range interactions, it is clear that the perturbation model based on the conformal Lennard–Jones reference still contains several insufficiencies, so that failures in predictions of mixture behavior cannot be excluded. The remaining insufficiencies are for one part associated with the model for the potential energy, in particular that for the dispersion interaction; cf. Sections 2.5.5 and 2.4.2. There are, however, also obvious deficiencies in the perturbation expansion itself. In particular, the expansion is basically not adequate to account for multipole effects of strongly nonspherical molecules. Various attempts have been made to develop perturbation models considering both short-range and long-range anisotropy. As yet, no generally accepted statistical mechanical model adequate for practical application has emerged. However, computer simulation data for nonspherical molecules with point multipoles are available, and these can effectively be used to generate statistical models for treating polarity effects in molecules of anisotropic shape. We shall discuss these in the context of generalized van der Waals theory in the next section.

5.5.4 Generalized van der Waals Models

We have seen in Section 5.5.2 that splitting the potential into a hard sphere repulsive and a soft isotropic attractive term and using the λ-perturbation expansion to first order leads to the van der Waals equation of state, when some simplifying approximations for the free volume and the hard-sphere correlation function are made. These assumptions are rather unrealistic and thus, the van der Waals equation of state is not used in practical fluid phase predictions. In particular, a realistic representation of the fluid phase behavior requires the anisotropic nature of the repulsive as well as of the attractive interactions to be taken into account. We can derive a similar but more general equation of state model by making use of the potential split into hard core repulsive and soft attractive forces, both not necessarily restricted to isotropic models, in a different way. We will for this purpose avoid the formal perturbation expansion for the attractive contributions, which is difficult to evaluate when both nonspherical repulsive and attractive forces are to be taken into account. Instead, we base the derivation on more empirical, i.e., data-based ingredients, which also helps to correct inadequacies of the potential energy model. The starting point is the general equation for the canonical partition function, cf. Section 2.2, i.e.,

$$Q = Q_{\mathrm{tr}}^{\mathrm{kin}} Q_{\mathrm{r}}^{\mathrm{kin}} Q_{\mathrm{int}} Q^{\mathrm{C}} = \frac{1}{N!} \Lambda^{-3N} Q_{\mathrm{r}}^{\mathrm{kin}} Q_{\mathrm{int}} Q^{\mathrm{C}}. \tag{5.127}$$

Here

$$\Lambda = \sqrt{\frac{h^2}{2\pi m k T}} \tag{5.128}$$

is the de Broglie wavelength and takes the kinetic contribution of translation into account, Q_r^{kin} is the kinetic, i.e., the ideal gas contribution of molecular rotation. Q_{int} represents the internal degrees of freedom, which for simple molecules are again density independent and given by the ideal gas values, i.e., $Q_v^{\text{kin}} Q_{\text{ir}}^{\text{kin}}$. Finally, Q^C is the configurational partition function, defined by (2.75) as

$$Q^C = \int e^{-U(\mathbf{r}^N \omega^N)/kT} \mathrm{d}\mathbf{r}^N \mathrm{d}\omega^N \tag{5.129}$$

in the rigid molecule approximation. Here $U\left(\mathbf{r}^N \omega^N\right)$ is the intermolecular potential energy depending on the configurational coordinates of the molecular system, which here are the locations of the molecular centers and the orientational angles. The configurational partition function introduces the potential contribution of the translational and external rotational modes.

To derive the generalized van der Waals partition function, we cast the general equation (5.127) into a specific form, suitable for introducing well-defined empirical approximations. We note from Section 4.1 that we can formulate the configurational part of the partition function generally in terms of its value at $T = \infty$ as

$$\ln Q^C(T, V, N) - \ln Q^C(T = \infty, V, N) = \int\limits_{T=\infty}^{T} \frac{U^C(T, V, N)}{kT^2} \mathrm{d}T$$
$$= \left(A^C/kT\right)_{T=\infty} - \left(A^C/kT\right)_T. \tag{5.130}$$

The integral in (5.130) represents the change of free energy in bringing the system from the infinite temperature limit to the actual temperature at constant density. It thus reflects the action of the intermolecular attractive forces. For convenience we set

$$\overline{U} = -\frac{kT}{N} \int\limits_{T=\infty}^{T} \frac{U^C(T, V, N)}{kT^2} \mathrm{d}T, \tag{5.131}$$

to which we refer as the mean attractive potential energy per molecule. We note that, unlike the λ-expansion, the attractive contribution here is not formulated explicitly in terms of a potential model. We note further that at $T = \infty$ only hard core repulsive forces contribute to the configuration integral. We then find

$$Q^C(T, V, N) = Q_h^C(n)e^{-(N\overline{U})/kT}, \tag{5.132}$$

where $Q_h^C(n)$ is the configurational partition function of the hard body fluid. The mean attractive potential will in general be a function of temperature and density. Expressing the hard body partition function in terms of a free volume V_f analogous to (5.115), we finally arrive at the canonical partition function in the form [43, 44]

$$Q = \frac{1}{N!} \left(\frac{V}{\Lambda^3}\right)^N Q_v^{\text{kin}} Q_r^{\text{kin}} \left(\frac{V_f}{V}\right)_h^N e^{(-N\overline{U})/(kT)}. \tag{5.133}$$

We refer to this expression as the generalized van der Waals partition function, because it is based on the ideas that lead to the van der Waals equation of state, i.e., the split of the intermolecular potential energy into a hard repulsive and a soft attractive part. While being an exact formulation of the partition function in the rigid molecule approximation, (5.133) is also a convenient starting point for introducing well-defined empirical approximations leading to successful semiempirical equation of state models.

Generally, the molecules will have nonspherical shapes and so the free volume will be shape-dependent in addition to its dependence on density. Various models for nonspherical hard body fluids have been suggested. An extension of the hard-sphere equation of state to hard convex bodies has been given by Boublik [45]. It reads

$$Z_{\text{hcb}} = \frac{1 + y(3\alpha - 2) + y^2(1 - 3\alpha + 3\alpha^2) - y^3\alpha^2}{(1 - y)^3}, \tag{5.134}$$

where the reduced density is given by

$$y = nV_{\text{hcb}}, \tag{5.135}$$

with V_{hcb} as the volume of the hard convex body under consideration. Hard convex bodies are solid shapes for which any line intersecting two points on the surface must lie entirely within the volume of the body. They can be essentially defined in terms of three characteristic geometrical quantities, the mean radius of curvature R_{hcb}, the surface S_{hcb}, and the volume V_{hcb}. The nonsphericity parameter α in the Boublik equation is defined by a combination of these as

$$\alpha = R_{\text{hcb}}S_{\text{hcb}}/3V_{\text{hcb}}.$$

Obviously, for a hard sphere with

$$R_{\text{hcb}} = \frac{d}{2}, \; S_{\text{hcb}} = 4\pi \left(\frac{d}{2}\right)^2, \quad \text{and} \quad V_{\text{hcb}} = \frac{4}{3}\pi(d/2)^3,$$

we find $\alpha = 1$ and (5.134) reduces to (5.116). Specific values for α, reflecting the shape of the hard convex body, can be calculated for a defined hard core model of a molecule. Practically, α is adjusted to data. The hard convex body equation of state is suitable for modeling the shape of mildly nonspherical rigid molecules. An extension to mixtures is available [45].

Another extension of the hard-sphere equation of state to nonspherical hard bodies has been derived by Chapman et al. [46]. It models a hard molecule as a chain of freely jointed hard spheres, the number of which is m. The hard-chain equation of state reads for pure fluids

$$Z_{hc} = mZ_d - (m-1)\frac{2+2\eta-\eta^2}{(1-\eta)(2-\eta)}, \qquad (5.136)$$

where the reduced density is $\eta = \frac{1}{6}\pi mnd^3$ and d is the diameter of one spherical segment. It is basically suitable for small nonspherical as well as large flexible molecules. The nonsphericity of the molecule is modeled in terms of the number m of hard spheres. This parameter is adjusted to data. Z_d is the equation of state for hard spheres; cf. (5.116). The extension to mixtures is based on that for Z_d and a mole fraction average for m.

The mean attractive potential, due to dispersion forces for simple nonpolar molecules, would in a rigorous perturbation expansion have to be calculated from the attractive part of the potential energy averaged over the hard body pair correlation function; cf. Sections 5.5.2. Here, we avoid this complicated procedure and introduce a more empirical approach by expressing $\overline{U}$ by an empirical function of temperature and density, such as

$$\frac{\overline{U}}{2kT} = \sum_m \sum_n C_{n,m}\frac{(n^*)^m}{(T^*)^n}, \qquad (5.137)$$

or similar expressions. This formulation takes up the spirit of a more general high-temperature perturbation representation of the average soft part of the potential. The dimensionless temperature and density are defined, as usual, by

$$T^* = kT/\varepsilon \qquad (5.138)$$

and

$$n^* = nd^3, \qquad (5.139)$$

where ε, d are adjusted to data of the fluid under consideration. The quantities $C_{n,m}$, which essentially replace the complicated perturbation integrals, are determined universally from a data base.

Various equation of state models have been derived on the basis of generalized van der Waals theory. Using (5.134) and (5.137), we arrive at an equation of state of the form

$$Z = Z_{hcb} + \sum_m \sum_n mC_{n,m}\frac{(n^*)^m}{(T^*)^n}. \qquad (5.140)$$

This is the BACK (Boublik–Alder–Chen–Kreglewski) equation of state [47]. The quantities $C_{n,m}$ here are constants that have been obtained from fitting to data of argon, i.e., a fluid of spherical molecules. This brings in experimental data and thus replaces the formal integration and the imprecisely known dispersion interactions. On the basis of the results in Exercise 5.3, the nonspherical nature of the attractive forces is taken into account empirically by making the intermolecular energy of the spherical force field temperature-dependent, i.e., by setting

$$\varepsilon = \varepsilon^0 (1 + 0.6 \, \omega T_c / kT), \tag{5.141}$$

where ε^0 is an adjustable parameter with the physical meaning of the attractive energy parameter at $T = \infty$. Here T_c is the critical temperature and ω the Pitzer acentric factor, characterizing empirically the anisotropic nature of the interactions. Extensive tabulations exist for these quantities [48]. The hard convex body volume in Z_{hcb} is also assumed to be temperature-dependent, in the spirit of short-range perturbation theory, to allow for the soft character of the repulsive interactions. It is described by $V_{hcb} = \frac{1}{6} \pi d^3$, using an effective diameter calculated from the formula

$$d = d_0 \left[1 - 0.12 e^{-3\varepsilon^0 / kT} \right], \tag{5.142}$$

where d_0 is the hard body diameter at $T = 0 \, \text{K}$. There are again three adjustable parameters in this equation of state model, i.e., d_0, ε^0, and α.

Significant improvements over BACK within the framework of generalized van der Waals models have been obtained. In one version [49], (5.134) is used for the hard body term but with a modified temperature dependence of the hard body diameter. The attractive term is represented by a more flexible functional form than (5.137). In particular, and contrary to BACK, it is adjusted by a universal function to experimental data of some anisotropic molecules in terms of the same parameters as used for the hard body term, ε^0, d_0, and α. This can be visualized as making the quantities $C_{n,m}$ depend on α and is in line with considering the attractive term as a perturbation of the anisotropic hard body reference. This makes it better suited for nonspherical molecules. Another extension has been designed to make the BACK model applicable to molecules of stronger nonsphericity and flexibility [50].

The most promising family of equations of state today, based on the generalized van der Waals model has recently been developed by using the hard-chain fluid (5.136) as the reference and formulating the attractive term on the basis of a second-order λ-perturbation expansion as a function of the chain length m in addition to reduced temperature and density. By use of the hard-chain equation of state the model becomes suitable also for rather large and flexible molecules. The intermolecular interactions are now formulated in terms of segment–segment interactions and thus are anisotropic, as they should be. Again the perturbation integrals are not evaluated explicitly. Instead, and analogously to [49], they are adjusted in terms of expressions similar to (5.137) to pure component properties of nonspherical fluids, in particular those of the n-alkanes, as a function of density and the chain length. The three substance-specific parameters are the chain length m, the independent segment diameter d_0, to be used in (5.142), and the depth ε of the segment–segment potential, which is temperature-independent. Extension to mixtures is done by applying

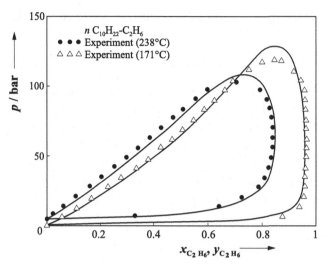

Fig. 5.23. Vapor–liquid equilibrium in the system $nC_{10}H_{22}$–C_2H_6 at two temperatures (— PC-SAFT). Data from [52].

one-fluid mixing rules for the perturbation terms, whereas theoretical mixing rules are available for the hard body reference. Further, combination rules are required for the pure fluid parameters, where we note that in applying the London formula (5.23) the polarizabilities refer to segment–segment interactions and thus are found from dividing the molecular polarizability by m. The model is referred to as the PC-SAFT (perturbed chain–statistical associating fluid theory) equation of state [51]. Figure. 5.23 shows the rather successful vapor–liquid equilibrium prediction of PC-SAFT for n-decane–ethane mixtures at two temperatures as compared to the data of [52]. Again, such good prediction performance must not be generalized to all systems without exception for this model, because basic problems such as imperfect knowledge of intermolecular forces, notably the dispersion combination rule, remain to be solved.

The above models can be extended to include multipolar long-range forces. An early approach to doing so used the results of the λ-expansion for these contributions around the conformal Lennard–Jones reference in an ad hoc manner [50]. The contributions of the multipole terms to the free energy for this reference were derived in Section 5.5.3. This can be accommodated into the formalism of generalized van der Waals theory by setting formally

$$A = A^{\text{rep, iso}} + A^{\text{att, iso}} + A^{\text{rep, aniso}} + A^{\text{att, mult}}. \tag{5.143}$$

Because $A^{\text{rep, iso}} + A^{\text{att, iso}}$ together represent the free energy of an isotropic fluid, we can consider the anisotropic part of the repulsive forces as well as the anisotropic multipole part of the attractive forces formally as perturbations of the isotropic fluid, as was done in the conformal potential perturbation expansion. If we assume further that the isotropic fluid is a Lennard–Jones (12–6) system, we can set, cf. (5.119),

$$\frac{A^{\text{att,mult}}}{NkT} = \frac{A^{\lambda\lambda}}{NkT}\left(\frac{1}{1 - A^{\lambda\lambda\lambda}/A^{\lambda\lambda}}\right), \tag{5.144}$$

using exactly the terms available for the various contributions of the anisotropic long-range forces, as summarized in App. 12. However, instead of using $A^{\text{ref}} = A^{\text{rep, iso}} + A^{\text{att, iso}}$, we replace $A^{\text{rep, iso}} + A^{\text{rep, aniso}}$ formally by A_{h}, i.e., the free energy of a nonspherical hard body system, which brings in the shape effects in a better way than by a λ-expansion with a conformal potential reference; cf. Section 5.5.3. $A^{\text{att, iso}}$ is calculated again from (5.137). We then have an equation of state model for nonspherical rigid molecules including multipolar contributions to the pair potential. It is based conceptually on perturbation theory, but has more of an empirical flavor and is simpler to use. We note in particular that a Lennard–Jones reference system is not used here explicitly, except in the various perturbation terms of the λ-expansion for the multipolar forces that are based on that system. It is assumed that the inconsistencies introduced by using a nonspherical hard body equation of state plus an empirical isotropic attraction contribution in combination with (5.144) are of minor relevance and can be absorbed sensibly into the adjustable parameters of the model, which are a shape parameter for the hard body term along with ε and σ, as usual. Written in terms of the partition function, this model is given by

$$Q = \frac{1}{N!}\left(\frac{V}{\Lambda^3}\right)^N\left[\left(\frac{V_{\text{f}}}{V}\right)_{\text{h}} e^{-N\overline{U}^{\text{iso}}/kT}e^{-N\overline{U}^{\text{mult}}/kT}\right]^N Q_{\text{r}}^{\text{kin}} Q_{\text{v}}^{\text{kin}} Q_{\text{ir}}^{\text{kin}}. \tag{5.145}$$

In [50], $(V_{\text{f}}/V)_{\text{h}}$ was represented by the Boublik hard convex body equation of state, $\overline{U}^{\text{iso}}/kT$ by (5.137), and $\overline{U}^{\text{mult}}/kT$ by the λ-expansion perturbation terms; cf. (5.144).

A refinement of this approach was developed [53, 54], in which dipolar and quadrupolar contributions were added to the free energy in a more consistent way. To the PC-SAFT equation of state, quite successful for nonpolar fluids, there was added a multipolar term, such as

$$Z_{\text{PCP-SAFT}} = Z_{\text{PC-SAFT}} + Z^{\text{mult}}, \tag{5.146}$$

which turns the model into one termed PCP-SAFT. In the sense of the λ-perturbation expansion of (5.144), the multipolar contribution was represented by a second-order and a third-order integral term, averaging the multipolar forces over the nonspherical reference system. Since these terms cannot effectively be evaluated directly, results of computer simulation studies for SS-PM models over a range of elongations and polar strengths were used to express the integrals over the correlation functions of the reference system in terms of simple analytical functions. The model has again the three pure fluid parameters of the PC-SAFT model, which have to be adjusted to data. The mixing rules for the

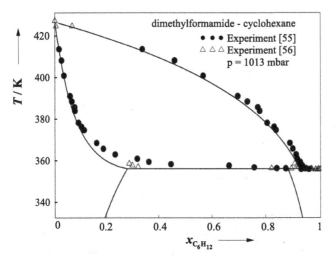

Fig. 5.24. Vapor–liquid–liquid equilibrium for C_3H_7NO–C_6H_{12} (— PCP-SAFT).

polar terms emerge from the perturbation expansion. As an illustration, Figure 5.24 shows the phase diagram of dimethyl formamide–cyclohexane at normal pressure. Here, dimethyl formamide has a strong dipole and quadrupole, whereas cyclohexane has a small quadrupole, all referred to a segment basis. The phase diagram shows a vapor–liquid coexisting with a liquid–liquid equilibrium. Data are from [55, 56]. It can be seen that PCP-SAFT is quite successful in predicting this rather complicated phase behavior. As a further illustration, Figure 5.25 shows the excess enthalpy of the system benzene–cyclohexane at two temperatures as predicted by PCP-SAFT. Data are from [57]. Although not perfect, the predicted excess enthalpies are in reasonable agreement with the data. As usually found, prediction of the excess enthalpy is more demanding than that of vapor–liquid equilibrium and provides a strong test of a model.

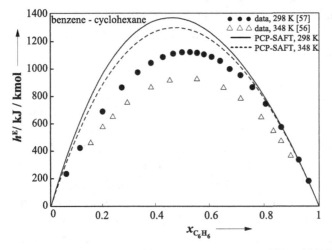

Fig. 5.25. Excess enthalpy of benzene–cyclohexane at 298 and 348 K (—, - - PCP-SAFT).

In the comparisons with data shown above, the progress achieved by incorporating sophisticated potential energy models by perturbation theory has been emphasized. It must be realized, however, that these models are not exact, neither with regard to the statistical mechanics nor to the potential energy model. So, although good predictions of mixture behavior from pure component data can be achieved in many cases from them, there are also failures, frequently due to incomplete knowledge about the potential energy in general and about the dispersion interaction between unlike molecules in particular. Still, the models based on generalized van der Waals theory have significantly better prediction performance than current empirical models. When corrected by an adjustable binary interaction parameter, as discussed in Section 5.2, this parameter tends to be smaller and less state-dependent, and the representation of the data tends to be more accurate. More research in this area, based on quantum mechanics, is clearly required.

5.6 Summary

Fluid phase behavior over a large density range is generally described in terms of an equation of state, from which, when augmented by ideal gas information, all aspects of thermodynamic behavior can be calculated. In practice, such applications are frequently encountered for systems composed of simple molecules. The interactions are then modeled on the basis of the rigid molecule approximation, which assumes that the molecules are treated as rigid geometrical structures. Some parameters, typically three, relating to the potential model have to be adjusted to pure fluid data. The dispersion interactions in mixtures are treated in terms of semiempirical combining rules, which is a severe factor of unreliability.

At low density, i.e., for real gases, the equation of state can be formulated in terms of an expansion in density. This is the virial equation of state. The virial coefficients can be calculated from integrals over the intermolecular potential energy and exact mixing rules for them may be derived. At higher densities, a reformulation of the statistical mechanical expressions for the thermodynamic functions in terms of correlation functions is useful since it can make explicit use of the pairwise additivity approximation. When pair potentials of different components follow the same functional form, the corresponding states principle, as well as conformal solution models, emerge. More specific information about the pair potential can be introduced via perturbation theory. A semiempirical version of this approach is available in the form of generalized van der Waals theory, which has become the basis for many modern molecular-based equations of state. Many satisfactory predictions of mixture behavior from pure component data have been performed on the basis of such models, much more reliable than the established empirical procedures. However, because neither the statistical mechanics nor the intermolecular potential models are strictly correct, there is a chance of false predictions, too. So, in technical applications with high demands for accuracy, equation of state models are usually applied with an empirical binary interaction parameter as a correction to the combination rule for $\varepsilon_{\alpha\beta}$. The benefit of the molecular models then manifests itself in a smaller value for this empirical parameter and a better overall representation of the data. In view of the progress in recent years it may be expected that the predictive performance of the molecular models for the equation of state will further increase in the future.

5.7 References to Chapter 5

[1] J. V. Sengers, R. F. Kayser, C. F. Peters, and J. H. White Jr. (Eds). *Equations of State for Fluids and Fluid Mixtures*. Elsevier, Amsterdam, 2000.

[2] J. A. Barker, R. A. Fisher, and R. O. Watts. *Mol. Phys.*, 21:657, 1971.

[3] W. Ameling, M. Luckas, K. P. Shukla, and K. Lucas. *Mol. Phys.*, 65:335, 1985.

[4] G. C. Maitland, M. Rigby, E. B. Smith, and W. A. Wakeham. *Intermolecular Forces*. Clarendon Press, Oxford, 1981.

[5] T. Kihara. *Intermolecular Forces*. Wiley, New York, 1972.

[6] K. Lucas. *Applied Statistical Thermodynamics*. Springer, Berlin, 1991.

[7] M. Luckas and K. Lucas. *Fluid Phase Equilibria*, 45:7, 1989.

[8] G.S. Rushbrooke. *Trans. Faraday Soc.*, 36:1055, 1940.

[9] D. Cook and J. S. Rowlinson. *Proc. Roy. Soc. A*, 219:405, 1953.

[10] T. L. Hill. *Introduction to Statistical Thermodynamics*. Addison–Wesley, Reading, MA, 1960.

[11] R. C. Tolman. *The Principles of Statistical Mechanics*. Dover, New York, 1938.

[12] J. E. Mayer and M. G. Mayer. *Statistical Mechanics*. Wiley, New York, 1940.

[13] W. Ameling, M. Luckas, K. P. Shukla, and K. Lucas. *Mol. Phys.*, 56:335, 1986.

[14] K. Bier, G. Maurer, and H. Sand. *Ber. Bunsenges. Phys. Chem.*, 84:437, 1980.

[15] J. M. H. Levelt-Sengers, M. Klein, and J. S. Gallagher. Technical Report AEDC-TR-71-39, Arnold Engineering Development Center, Tullahoma, TN, 1971.

[16] J. H. Dymond and E. B. Smith. *The Virial Coefficients of Pure Gases and Mixtures. A Critical Compilation*. Clarendon Press, Oxford, 1980.

[17] B. Volle and K. Lucas. *Forsch. Ing. Wes.*, 46:14, 1980.

[18] A. H. Stroud. *Approximate Calculation of Multiple Integrals*. Prentice–Hall, Englewood Cliffs, NJ, 1971.

[19] M. Ripke. *The Thermophysical Properties of Gases Composed of Rigid Molecules*. Ph.D. dissertation, University of Duisburg, 1994.

[20] C. G. Gray and K. E. Gubbins. *Theory of Molecular Fluids I*. Clarendon Press, Oxford, 1984.

[21] H. J. Böhm, R. Ahlrichs, P. Scharf, and H. Schiffer. *J. Chem. Phys.*, 81:1389, 1984.

[22] R. D. Mountain and G. Morrison. *Mol. Phys.*, 64:91, 1988.

[23] K. Sagarik and R. Ahlrichs. *Chem. Phys. Lett.*, 131:74, 1986.

[24] B. Schramm. Personal communication.

[25] R. Freyhof. Ph.D. thesis, Universität Karlsruhe, 1986.

[26] T. W. Leland, J. S. Rowlinson, and G. A. Sather. *Trans. Faraday Soc.*, 64:1447, 1968.

[27] C. H. Twu, L. L. Lee, and K. E. Starling. *Fluid Phase Equilibria*, 4:35, 1980.

[28] R. H. Davies, A. G. Duncan, G. Saville, and L. A. K Staveley. *Trans. Faraday Soc.*, 63:855, 1967.

[29] A. Fredenslund and J. Mollerup. *J. Chem. Soc. Faraday Trans.*, 70(1):1653, 1974.

[30] K. P. Wallis, P. Clancy, J. A. Zollweg, and W. B. Streett. *J. Chem. Thermodyn.*, 16:811, 1984.

[31] R. W. Zwanzig. *J. Chem. Phys.*, 22:1420, 1954.

[32] J. A. Barker and D. Henderson. *Mol. Phys.*, 21:187, 1971.

[33] J. Perram. *J. Mol. Phys.*, 30:1505, 1975.

[34] L. Verlet and J. J. Weis. *Phys. Rev. A*, 5:939, 1972.

[35] N. F. Carnahan and K. F. Starling. *J. Chem. Phys.*, 51:635, 1969.

[36] G. A. Mansoori, N. F. Carnahan, K. E. Starling, and T. W. Leland. *J. Chem. Phys.*, 54:1523, 1971.

[37] J. D. Weeks, D. Chandler, and H. C. Anderson. *J. Chem. Phys.*, 54:5237, 1971.

[38] K. P. Shukla, M. Luckas, W. Ameling, and K. Lucas. *Fluid Phase Equilibria*, 28:217, 1986.

[39] G. Stell, J. C. Rasaijah, and H. Narang. *Mol. Phys.*, 23:393, 1972.

[40] J. A. Pople. *Proc. Roy. Soc. A*, 221:498, 1954.

[41] K. E. Gubbins and C. G. Gray. *Mol. Phys.*, 23:187, 1972.

[42] K. P. Shukla, K. Moser, and B. Lucas. *Fluid Phase Equilibria*, 17:153, 1984.

[43] J. H. Vera and J. M. Prausnitz. *Chem. Eng. J.*, 3:1, 1972.

[44] S. I. Sandler. *Fluid Phase Equilibria*, 19:233, 1985.

[45] T. Boublik. *J. Chem. Phys.*, 63:4048, 1975.

[46] W. G. Chapman, G. Jackson, and K. E. Gubbins. *Mol. Phys.*, 65:1057, 1988.

[47] S. S. Chen and A. Kreglewski. *Ber. Bunsenges. Phys. Chem.*, 81:1048, 1977.

[48] B. E. Poling, J. M. Prausnitz, and J. P. O'Connell. *The Properties of Gases and Liquids*, 5th ed., McGraw–Hill, New York, 2001.

[49] A. Müller, J. Winkelmann, and J. Fischer. *AIChE J.*, 42:1116, 1996.

[50] M. A. Siddiqi and K. Lucas. *Fluid Phase Equilibria*, 60:1, 1990.

[51] J. Gross and G. Sadowski. *Ind. Eng. Chem. Res.*, 40:1244, 2001.

[52] H. Reamer and B. Sage. *J. Chem. Eng. Data*, 7:161, 1962.

[53] J. Gross. *AIChE J.*, 51:2556, 2005.

[54] J. Gross and J. Varabec. *AIChE J.*, 52: 1194, 2006.

[55] X. Zhang, Y. Zhen, and W. Zhao. *Nature Gas Chem. Ind. (China)*, 22: 52, 1997.

[56] B. Blanco, S. Betrán, J. I. Cabezas, and J. *Chem. Eng. Data*, 42:938, 1997.

[57] K. Elliot and C. J. Wormald. *J. Chem. Thermodyn.*, 8:881, 1976.

Fundamental Constants and Atomic Units

Fundamental Constants

Planck's constant	h	$6.626176 \cdot 10^{-34}$ Js
Speed of light in vacuum	c	$2.99792458 \cdot 10^{8}$ m/s
Avogadro's number	N_A	$6.022045 \cdot 10^{23}$ 1/mol
Boltzmann's constant	k	$1.380662 \cdot 10^{-23}$ J/K
Gas constant	$R = N_A k$	8.31441 J/mol K
Elementary charge	e	$1.602189 \cdot 10^{-19}$ C
Vacuum permittivity	ε_0	$8.854 \cdot 10^{-12}$ J^{-1} C^2 m^{-1}

Atomic Units

Length   (1 bohr) $= 0.5292$ Å
1 Å $= 10^{-8}$ cm $= 10$ nm

Energy   (1 hartree) $= 27.206$ eV
1 eV $= 1.602189 \cdot 10^{-12}$ erg
(ε / k) in K $= 1.43878$ (ε / hc) in cm^{-1}

Charge $(1e) = 1.602189 \cdot 10^{-19}$ C
1 C$^2 = 8.9876 \cdot 10^9$ N m^2

Dipole moment $(e \times 1$ bohr$) = 8.478 \cdot 10^{-30}$ C m $= 2.5418$ D

Quadrupole moment $(e \times 1$ bohr$^2) = 4.487 \cdot 10^{-40}$ C m$^2 = 1.345$ D Å

APPENDIX 2

Stirling's Formula

In 1730 J. Stirling proposed the following estimate for $n!$:

$$1 \le \frac{n!}{\sqrt{2\pi n}\, n^n e^{-n}} \le e^{\frac{1}{12n}}. \qquad (A\,2.1)$$

On this basis, we have for large n

$$n! \cong \sqrt{2\pi n}\, n^n e^{-n} \qquad (A\,2.2)$$

or

$$\ln n! = n\ln n - n + \frac{1}{2}\ln n + \frac{1}{2}\ln 2\pi \cong n\ln n - n. \qquad (A\,2.3)$$

Relative Probability of a Microstate

To derive an expression for the relative probability P_i of a microstate i we first consider a canonical ensemble, i.e., a collection of systems, each with fixed values of N, V, and T; cf. Figure A 3.1. The ensemble as a whole is isolated adiabatically so that the total energy of the canonical ensemble has a fixed value, U_c. Each of the π_c systems of the canonical ensemble finds itself in a large heat bath at temperature T provided by the other systems.

Many different microstates or quantum states are associated with each fixed macrostate of the canonical ensemble. Different quantum states of a canonical ensemble may be represented by different distributions of its systems over the many possible energy states of the single system. We denote by $\{E_i\}$ the various possible energy values of a system in the canonical ensemble. In principle, the spectrum of energy values of the system follows from its Schrödinger equation. We assume in the following that the spectrum of energy values is known. It is identical for each system of the canonical ensemble, because the values of V and $\{N_\alpha\}$ are identical. Thus $\{E_i\}$ is available as a set of values $(E_1, E_2, E_3, \ldots, E_j)$, if we assume a total of j quantum states for each system. At each moment of time each system of the canonical ensemble will be found in one of these energy states, whose number j is extremely large.

We consider a particular distribution n of the systems of the canonical ensemble over the possible energy values $\{E_i\}$; i.e., n_1 systems have energy E_1, n_2

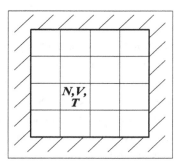

Fig. A 3.1. A canonical ensemble.

systems energy E_2, etc. The set of numbers n $(n_1, n_2, \ldots)$, referred to as the distribution n, defines a particular quantum state of the canonical ensemble. Obviously many different distributions n and thus quantum states of the canonical ensemble are possible. All of them, however, must fulfil the conditions

$$\sum_i n_i(n) = \pi_{\rm c} \tag{A 3.1}$$

$$\sum_i n_i(n) E_i = U_{\rm c}, \tag{A 3.2}$$

where the summation extends over all j quantum states.

For a given distribution n, the relative probability $(P_i)_n$ of microstate i of a system in the canonical ensemble is given by definition as

$$(P_i)_n = \frac{n_i(n)}{\pi_{\rm c}} = \frac{\text{number of systems in state } i \text{ at the distribution } n}{\text{total number of systems}}. \tag{A 3.3}$$

There exist very many distributions n that fulfil the constraints of a fixed number of systems $\pi_{\rm c}$ and a fixed total energy $U_{\rm c}$ of the canonical ensemble. The value of P_i we are looking for is the average of $(P_i)_n$ over all possible distributions n; i.e.,

$$P_i = \sum_n (P_i)_n W_n, \tag{A 3.4}$$

where W_n is the statistical weight of distribution n.

We first look at the statistical weight of a distribution n of the systems in the canonical ensemble over their energy states. In order to be able to use the second postulate, we mentally reproduce an arbitrarily large number of these canonical ensembles and put them together to form an isolated superensemble. Due to the identical energies of the constituent canonical ensembles this superensemble is a microcanonical ensemble, as shown in Fig. A 3.2. Each system of this microcanonical ensemble, i.e., each canonical ensemble, has a macrostate defined by $\pi_{\rm c}$, the number of systems, $U_{\rm c}$, the energy, and $V_{\rm c} = \pi_{\rm c} V$, the volume of the canonical ensemble. Using this mental construction of a microcanonical ensemble we find for the statistical weight of a distribution n, by definition,

$$W_n = \frac{n_{\rm mc}(n)}{\pi_{\rm mc}} = \frac{n_{\rm mc}(n)}{\sum_n n_{\rm mc}(n)}, \tag{A 3.5}$$

with $n_{\rm mc}(n)$ as the number of systems in the microcanonical ensemble, i.e., the number of canonical ensembles, that are in a state characterized by the distribution n and $\pi_{\rm mc} = \sum_n n_{\rm mc}(n)$ as the total number of all systems in the microcanonical ensemble.

To calculate the statistical weight of any distribution n we need the number of systems in the microcanonical ensemble $n_{\rm mc}(n)$ in state n. To calculate this number we make use of the fact that all these systems, which are in fact

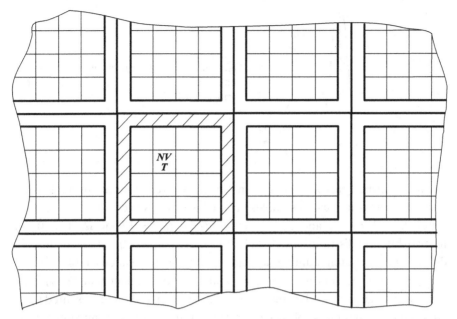

Fig. A 3.2. Microcanonical ensemble constructed from adiabatically isolated canonical ensembles.

isolated canonical ensembles, have the same energy U_c. The different distributions n of the systems of the canonical ensemble over their many possible energy values represent different quantum states of the canonical ensembles with equal energy. According to the second postulate of statistical mechanics all these quantum states have equal probability; i.e., all these distributions n have the same probability. Thus the number $n_{mc}(n)$ is directly proportional to the number of combinatorial ways $\Omega_c(n)$ that the distribution n can be arranged. Thus, if in a particular distribution n' the systems of a canonical ensemble can be distributed over the system energy values in 100 ways, whereas in a different distribution n'' they can only be distributed in 10 ways, the state represented by n' will be found 10 times as frequently as the state represented by n'', provided that both distributions are equally probable. The ratio of statistical weights is thus $W(n')/W(n'') = 10$. This is like tossing an ideal die, on which all of the many sides are equally probable, but which has 100 sides with number n' and 10 sides with number n''. Computation of a number for $\Omega_c(n)$ leads to a combinatorial problem. Mathematically we have to compute the number of possible arrangements of π_c distinguishable objects (systems) over j containers (energy values), such that each container has $n_1, n_2, \ldots n_j$ objects with $\sum n_i = \pi_c$ and $\sum n_i E_i = U_c$. The solution is

$$\Omega_c(n) = \frac{\pi_c!}{\prod_j n_j!}, \tag{A 3.6}$$

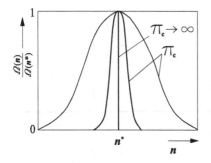

Fig. A 3.3. Number of arrangements of distribution n relative to that of the largest number of arrangements.

as can be verified easily by inspection. So, in a microcanonical ensemble composed of four canonical ensembles each of which has access to three energy states, $E_1 = 2$, $E_2 = 1$, and $E_3 = 4$, we find $\Omega_c(1, 2, 1) = 12$ and $\Omega_c(4, 0, 0) = 1$ for two different distributions $n' = (1, 2, 1)$ and $n'' = (4, 0, 0)$ satisfying the conditions of fixed system number and energy of the canonical ensemble.

To evaluate the sum in (A 3.4) we make use of the fact that $\Omega_c(n)$ is an arbitrarily large number, because $\pi_c \longrightarrow \infty$. The sum is then represented by its largest term with remarkable accuracy. This is referred to as the maximum term method and is visualized in Fig. A 3.3, which shows that the ratio of the number of arrangements of a distribution n to that of the distribution n^*, defined by having the largest number of arrangements, becomes sharply peaked at n^* for a large number of systems. The maximum term method can easily be verified for the simple example of a sum related to (A 3.4) and (A 3.6) and defined by

$$\sum_{n=0}^{n} \Omega(n) = \sum_{n_1=0}^{n} \frac{n!}{n_1! n_2!},$$

with $n = n_1 + n_2 = $ const. This sum has the exact result 2^n, which can also be written as

$$\ln \sum_{n_1=0}^{n} \Omega(n) = n \ln 2.$$

To show the equivalence with the maximum term result, i.e., for $n_1 = n_1^*$, we note that because $d \ln f = 1/f df$ we can go to logarithms and write, for $n \to \infty$, so that Stirling's approximation, cf. Appendix 2, can be used,

$$\ln \Omega(n_1) = n \ln n - n_1 \ln n_1 - n_2 \ln n_2$$

and

$$\frac{\partial \ln \Omega(n_1)}{\partial n_1} = 0,$$

giving $n_1^* = n/2$ and thus introducing

$$\ln \Omega(n_1^*) = n \ln 2,$$

in agreement with the exact result.

We thus find that the number of arrangements in distribution n^* is so much larger than that in any other distribution that all other terms in the sum of (A 3.4), except the one associated with n^*, are irrelevant. With $W(n^*) = 1$ we thus write (A 3.4) as

$$P_i = \frac{n_i(n^*)}{\pi_c}. \tag{A 3.7}$$

We finally need the distribution n^* of the systems in the canonical ensembles over their energy values for which the number of arrangements is a maximum. Mathematically this amounts to finding the maximum of the function $\Omega_c(n)$ as given by (A 3.6) under the constraints $\sum n_i(n) = \pi_c$ and $\sum n_i(n)E_i = U_c$, where the sum extends over all quantum states of a system in the canonical ensemble. We again work with logarithms. From (A 3.6) we have

$$\ln \Omega_c(n) = \ln \frac{\pi_c!}{\prod_i n_i!} = \ln \{\pi_c!\} - \sum_i \{\ln(n_i)!\}. \tag{A 3.8}$$

The logarithms of the factorials can be evaluated using Stirling's formula, and we find

$$\ln \Omega_c(n) = \pi_c \ln \pi_c - \sum_i n_i \ln n_i. \tag{A 3.9}$$

The maximum of $\ln \Omega(n)$ is found from

$$d \ln \Omega_c(n) = 0 = - \sum_i d(n_i \ln n_i) = - \sum_i dn_i(\ln n_i)$$
$$\underbrace{- \sum_i \frac{n_i}{n_i} dn_i}_{0} = - \sum_i (\ln n_i) dn_i. \tag{A 3.10}$$

Here we ignore the discrete nature of the variables n_i because for our applications they will be enormously large and thus closely spaced. This permits us to regard Ω_c as a continuous function of continuous variables n_i, and we can employ a standard mathematical calculus. In common problems of finding the extremum of a function all dn_i are independent of each other and the result would be $\ln n_i = 0$ and thus $n_1 = n_2 = \cdots = n_j = 1$. This would mean that each energy value would be occupied by one system, which is obviously wrong because the total number of systems in the canonical ensemble is arbitrary, whereas that of energy values is fixed as j. A correct evaluation of the extremum condition (A 3.10) must take into account that the dn_i are not independent of each other. Due to the constraints of fixed number of systems and fixed total energy they must satisfy the relations

$$d\pi_c = \sum_i dn_i = 0, \tag{A 3.11}$$

$$dU_c = \sum_i E_i dn_i = 0. \tag{A 3.12}$$

Thus, only $(j - 2)dn_i$ are independent. Finding an extremum under constraints is most easily accomplished by using a particular mathematical method, the method of Lagrangean multipliers. Use of the method of Lagrangean multipliers in solving (A 3.10) requires that we multiply each of the two auxiliary conditions with one undetermined parameter α and β, respectively, to give

$$\sum_i \alpha\, dn_i = \sum_i \beta E_i\, dn_i = 0.$$

Adding these equations to the extremum condition gives

$$\sum_i \ln n_i\, dn_i + \sum_i \alpha\, dn_i + \sum_i \beta E_i\, dn_i = \sum_i (\ln n_i + \alpha + \beta E_i)\, dn_i = 0.$$

The as yet undetermined multipliers are now chosen to eliminate the first two terms, i.e., those for $i = 1, 2$, which requires that

$$\ln n_1 + \alpha + \beta E_1 = 0$$

and

$$\ln n_2 + \alpha + \beta E_2 = 0.$$

Then the remaining dn_i $(i = 3, 4, \ldots, j)$ are independent and we find for the distribution with the largest number of arrangements

$$\ln n_i + \alpha + \beta E_i = 0 \qquad i = 1, 2, \ldots, j, \tag{A 3.13}$$

i.e., for all i. Clearly, the distribution with the maximum number of arrangements resulting from this calculation depends on the Lagrangean multipliers α and β, which are still undetermined. These two parameters will finally be obtained from the two constraints (A 3.11) and (A 3.12). From Eq. (A 3.13) the general solution for the distribution with the largest number of arrangements is given by

$$n_i(n^*) = e^{(-\alpha - \beta E_i)}.$$

The first Lagrangean multiplier, α, is found by introducing this into the constraint of a fixed number of systems in the canonical ensemble, which gives

$$\sum_i n_i(n^*) = \pi_c = \sum_i e^{(-\alpha - \beta E_i)}$$

and thus determines the Lagrangean multiplier α to be

$$e^{-\alpha} = \frac{\pi_c}{\sum_i e^{-\beta E_i}}. \tag{A 3.14}$$

The second Lagrangean multiplier, β, may be expressed formally by using the constraint of fixed total energy of the canonical ensemble through

$$\sum_i n_i(n^*) E_i = U_c = \frac{\sum_i e^{-\beta E_i} E_i}{\sum_i e^{-\beta E_i}} \pi_c. \tag{A 3.15}$$

This equation cannot be solved for β explicitly. Obviously β depends on the energy values of the system and on its average macroscopic energy. The physical interpretation of this parameter becomes clear when a more general canonical ensemble composed of replicas of two nonidentical systems A and B is considered. The systems A and B are, like all others, in thermal equilibrium. The concept of a canonical ensemble implies that, irrespective of the nature of a single system, the total energy of the composed canonical ensemble is constant; e.g.,

$$\sum_i [n_i(A)E_i(A) + n_i(B)E_i(B)] = U_c(AB). \qquad (A\ 3.16)$$

When working out the probabilities of finding a system of type A in a state i and a system of type B in state j, we therefore only find a single Lagrangean parameter β common to both types of systems from (A 3.16). This is precisely the property attributed to temperature as a common property of two systems in thermal equilibrium. Thus, β will be related to temperature by a dimensional factor $\beta \sim 1/$ energy. We thus write

$$\beta = \frac{1}{kT}, \qquad (A\ 3.17)$$

with k as a constant of dimension energy/temperature. This constant must be universal, because no exclusivity has been associated with the nature of the systems A or B. It can thus be determined from the properties of any system. It will become clear by appealing to the properties of ideal gases that k is Boltzmann's constant; cf. Section 3.4.

For the probability of state i we thus finally find

$$P_i = \frac{n_i(n^*)}{\pi_c} = \frac{e^{-\alpha - \beta E_i}}{e^{-\alpha} \sum_i e^{-\beta E_i}} = \frac{e^{-E_i/kT}}{Q}. \qquad (A\ 3.18)$$

The term $e^{-E_i/kT}$ is referred to as the Boltzmann factor, and Q is the so-called canonical partition function, defined by

$$Q = \sum_i e^{-E_i/kT}. \qquad (A\ 3.19)$$

In the grand canonical ensemble, in addition to temperature and volume, the chemical potential is fixed; not, however, the number of molecules. The molecules will thus fluctuate and we expect a statistical formulation containing not only a sum over energy states but also a sum over molecules. The probability of a microstate in the grand canonical ensemble, i.e., the probability of finding a system with i molecules and the energy value E_{ij}, i.e., the energy state j associated with a system of i molecules, is given for a particular distribution n by

$$P_{ij}(n) = \frac{n_{ij}(n)}{\pi_{gc}}. \qquad (A\ 3.20)$$

Here $n_{ij}(n)$ is the number of systems in the state ij for distribution n and π_{gc} is the number of systems in the grand canonical ensemble. For all possible distributions n we have as closure conditions of the grand canonical ensemble

$$\sum_i \sum_j n_{ij} = \pi_{gc} \tag{A 3.21}$$

$$\sum_i \sum_j n_{ij} E_{ij} = U_{gc} \tag{A 3.22}$$

$$\sum_i \sum_j n_{ij} i = N_{gc}. \tag{A 3.23}$$

Obviously, there are three Lagrangean parameters to be introduced now. The third parameter, introduced to account for the condition of a definite number of molecules in the grand canonical ensemble, can be identified on physical grounds by an argument analogous to that leading to (A 3.17) between β and temperature. If two nonidentical systems A and B are contained in the grand canonical ensemble, both must be in material equilibrium and will require only one Lagrangean parameter common to both. The common property of two systems in material equilibrium is the chemical potential. The third Lagrangean parameter will therefore be associated with the chemical potential of the system, which is a universal property of the grand canonical ensemble. Proceeding along the lines for the canonical ensemble we find 2.52 to 2.54 in the text.

APPENDIX 4

Spherical Harmonics, Rotation Matrices, and Clebsch–Gordan Coefficients [1]

The general formulation of the spherical harmonics reads

$$Y_l^m(\theta, \phi) = (-1)^m \sqrt{\frac{(2l+1)(l-m)!}{4\pi(l+m)!}} e^{im\phi} P_l^m(\cos\theta), \qquad (\text{A } 4.1)$$

where $P_l^m(\cos\theta)$ is the associated Legendre polynomial. Explicit expressions for $l \leq 2$ are

$$Y_0^0(\theta, \phi) = \frac{1}{2\sqrt{\pi}}$$

$$Y_1^0(\theta, \phi) = \frac{1}{2}\sqrt{\frac{3}{\pi}} \cos\theta$$

$$Y_1^{\pm 1}(\theta, \phi) = \mp\frac{1}{2}\sqrt{\frac{3}{2\pi}} \sin\theta e^{\pm i\phi}$$

$$Y_2^0(\theta, \phi) = \frac{1}{4}\sqrt{\frac{5}{\pi}}(3\cos^2\theta - 1)$$

$$Y_2^{\pm 1}(\theta, \phi) = \mp\frac{1}{2}\sqrt{\frac{15}{2\pi}} \cos\theta \sin\theta e^{\pm i\phi}$$

$$Y_2^{\pm 2}(\theta, \phi) = \frac{1}{4}\sqrt{\frac{15}{2\pi}} \sin^2\theta e^{\pm i\phi}. \qquad (\text{A } 4.2)$$

The general definition of rotation matrices is

$$D_{mn}^l(\omega) = e^{-im\phi} d_{mn}^l(\theta) e^{-in\chi}, \qquad (\text{A } 4.3)$$

with

$$d_{mn}^l(\theta) = \sum_\kappa \frac{\sqrt{(l+m)!(l-m)!(l+n)!(l-n)!}(-1)^\kappa}{(l+m-\kappa)!(l-n-\kappa)!(\kappa-m+n)!\kappa!}$$

$$\cdot \left(\cos\frac{\theta}{2}\right)^{2l+m-n-2\kappa} \left(\sin\frac{\theta}{2}\right)^{n-m+2\kappa}, \qquad (\text{A } 4.4)$$

where $(-x!) = \infty$.

The Clebsch–Gordan coefficients are calculated from

$$C(l_1 l_2 l; m_1 m_2 m)$$

$$= \delta_{m,m_1+m_2} \sqrt{\frac{(2l+1)(l_1+l_2-l)!(l_1-l_2+l)!(-l_1+l_2+l)!}{(l_1+l_2+l+1)!}}$$

$$\cdot \sqrt{(l_1+m_1)!(l_1-m_1)!(l_2+m_2)!(l_2-m_2)!(l+m)!(l-m)!}$$

$$\cdot \sum_Z (-1)^Z [Z!(l_1+l_2-l-Z)!(l_1-m_1-Z)!(l_2+m_2-Z)!$$

$$\cdot (l-l_2+m_1+Z)!(l-l_1-m_2+Z)!]^{-1}, \tag{A 4.5}$$

where δ_{ij} is the Kronecker symbol.

Some useful theorems involving rotation matrices and Clebsch–Gordan coefficients are

$$D_{mn}^l(\omega)^* = (-1)^{m+n} D_{\underline{mn}}^l(\omega) \tag{A 4.6}$$

$$\langle D_{mn}^l(\omega)^* D_{m'n'}^{l'}(\omega)\rangle_\omega = \frac{1}{2l+1}\delta_{ll'}\delta_{mm'}\delta_{nn'} \tag{A 4.7}$$

$$\langle D_{mn}^l(\omega)\rangle_\omega = \delta_{l0}\delta_{m0}\delta_{n0}/(2l+1) \tag{A 4.8}$$

$$C(l_1 l_2 l; m_1 m_2 m) = (-1)^{l_1+l_2+l} C(l_1 l_2 l; \underline{m_1}\,\underline{m_2}\,\underline{m}) \tag{A 4.9}$$

$$\sum_{m_1}\sum_{m_2} C(l_1 l_2 l; m_1 m_2 m)C(l_1 l_2 l'; m_1 m_2 m') = \delta_{ll'}\delta_{mm'}. \tag{A 4.10}$$

[1] C. G. Gray and K. E. Gubbins. *Theory of Molecular Fluids I*. Clarendon Press, Oxford, 1984.

Higher-Order Perturbation Terms for the Intermolecular Potential Energy of Simple Molecules

The pair potential of the induction forces as obtained from second-order quantum mechanical perturbation theory reads, if we place both molecular centers on the z-axis of the space-fixed coordinate system [1],

$$\phi_{\alpha_i \beta_j}^{ind}(r_{\alpha_i \beta_j} \omega_{\alpha_i} \omega_{\beta_i}) = \sum_{l_1''=0} \sum_{l_2''=0} \sum_{l''=0} \sum_{\substack{m_1'' \\ m_2''}} \sum_{\substack{n_1'' \\ n_2''}}$$

$$\cdot E_{\alpha_i \beta_j}^{ind}(l_1'' l_2'' l''; n_1'' n_2''; r_{\alpha_i \beta_j}) \sqrt{\frac{2l_1'' + 2l_2'' + 1}{4\pi}}$$

$$\cdot C(l_1'' l_2'' l''; m_1'' m_2'' 0) D_{m_1'' n_1''}^{l_1''}(\omega_{\alpha_i})^* D_{m_2'' n_2''}^{l_2''}(\omega_{\beta_j})^*, \qquad (A\,5.1)$$

with the induction expansion coefficients defined as

$$E_{\alpha_i \beta_j}^{ind}(l_1'' l_2'' l''; n_1'' n_2''; r_{\alpha_i \beta_j})$$

$$= -\sum_{l_1=0}^{\infty}\sum_{l_2=0}^{\infty}\sum_{l_1'=0}^{\infty}\sum_{l_2'=0}^{\infty} \cdot \sqrt{\frac{4\pi(2l_1+2l_2+1)!(2l_1'+2l_2'+1)!(2l_1''+1)(2l_2'+1)}{(2l_1+1)!(2l_2+1)!(2l_1'+1)!(2l_2'+1)!(2l''+1)}}$$

$$\cdot \begin{Bmatrix} l_1 & l_2 & l_1+l_2 \\ l_1' & l_2' & l_1'+l_2' \\ l_1'' & l_2'' & l'' \end{Bmatrix} C(l_1+l_2\ l_1'+l_2'\ l''; 000) \frac{16\pi^2(-1)^{l_2+l_2'}}{r_{\alpha_i \beta_j}^{l_1+l_2+l_1'+l_2'+2}}$$

$$\cdot \sum_{\substack{n_1 \\ n_1'}}\sum_{\substack{n_2 \\ n_2'}} C(l_1 l_1' l_1''; n_1 n_1' n_1'')\ C(l_2 l_2' l_2''; n_2 n_2' n_2'') \cdot {}^{\alpha}\Pi_{l_1 l_1'}^{n_1 n_1'}\ {}^{\beta}Q_{l_2}^{n_2}\ {}^{\beta}Q_{l_2'}^{n_2'}$$

$$- \sum_{l_1=0}^{\infty}\sum_{l_2=0}^{\infty}\sum_{l_1'=0}^{\infty}\sum_{l_2'=0}^{\infty} \cdot \sqrt{\frac{4\pi(2l_1+2l_2+1)!(2l_1'+2l_2'+1)!(2l_1''+1)(2l_2'+1)}{(2l_1+1)!(2l_2+1)!(2l_1'+1)!(2l_2'+1)!(2l''+1)}}$$

$$\cdot \begin{Bmatrix} l_1 & l_2 & l_1+l_2 \\ l_1' & l_2' & l_1'+l_2' \\ l_1'' & l_2'' & l'' \end{Bmatrix} C(l_1+l_2\ l_1'+l_2'\ l''; 000) \frac{16\pi^2(-1)^{l_2+l_2'}}{r_{\alpha_i \beta_j}^{l_1+l_2+l_1'+l_2'+2}}$$

353

$$\cdot \sum_{\substack{n_1 \\ n_1'}} \sum_{\substack{n_2 \\ n_2'}} C(l_1 l_1' l_1''; n_1 n_1' n_1'') \; C(l_2 l_2' l_2''; n_2 n_2' n_2'') \cdot{}^{\beta} \Pi_{l_2 l_2'}^{n_2 n_2'} \; {}^{\alpha} Q_{l_1}^{n_1} \; {}^{\alpha} Q_{l_1'}^{n_1'} \; . \quad \text{(A 5.2)}$$

For the dispersion forces pair potential we find [1]

$$\phi_{\alpha_i \beta_j}^{\text{disp}}(r_{\alpha_i \beta_j} \omega_{\alpha_i} \omega_{\beta_j})$$

$$= \sum_{l_1''=0} \sum_{l_2''=0} \sum_{l''=0} \sum_{\substack{n_1'' \\ n_2''}} E_{\alpha_i \beta_j}^{\text{disp}}(l_1'' l_2'' l''; n_1'' n_2''; r_{\alpha_i \beta_j}) \cdot \sqrt{\frac{2 l_1'' + 2 l_2'' + 1}{4\pi}}$$

$$\cdot \sum_{\substack{m_1'' \\ m_2''}} C(l_1'' l_2'' l''; m_1'' m_2'' 0) D_{m_1'' n_1''}^{l_1''}(\omega_{\alpha_i})^* D_{m_2'' n_2''}^{l_2''}(\omega_{\beta_j})^*, \quad \text{(A 5.3)}$$

with the dispersion expansion coefficients in the London approximation, cf. Section 2.5.5, defined as

$$E_{\alpha_i \beta_j}^{\text{disp}}(l_1'' l_2'' l''; n_1'' n_2''; r_{\alpha_i \beta_j})$$

$$= -\frac{I_\alpha I_\beta}{I_\alpha + I_\beta} \sum_{l_1=0}^{\infty} \sum_{l_2=0}^{\infty} \sum_{l'=0}^{\infty} \sum_{l_2'=0}^{\infty}$$

$$\cdot \sqrt{\frac{4\pi (2 l_1 + 2 l_2 + 1)! (2 l_1' + 2 l_2' + 1)! (2 l_1'' + 1)(2 l_2'' + 1)}{(2 l_1 + 1)! (2 l_2 + 1)! (2 l_1' + 1)! (2 l_2' + 1)! (2 l'' + 1)}}$$

$$\cdot \begin{Bmatrix} l_1 & l_2 & l_1 + l_2 \\ l_1' & l_2' & l_1' + l_2' \\ l_1'' & l_2'' & l'' \end{Bmatrix} C(l_1 + l_2 \; l_1' + l_2' \; l''; 000) \frac{16 \pi^2 (-1)^{l_2 + l_2'}}{r_{\alpha_i \beta_j}^{l_1 + l_2 + l_1' + l_2' + 2}}$$

$$\cdot \sum_{\substack{n_1 \\ n_1'}} \sum_{\substack{n_2 \\ n_2'}} C(l_1 l_1' l_1''; n_1 n_1' n_1'') \; C(l_2 l_2' l_2''; n_2 n_2' n_2'') \cdot{}^{\alpha} \Pi_{l_1 l_1'}^{n_1 n_1'} \; {}^{\beta} \Pi_{l_2 l_2'}^{n_2 n_2'}. \quad \text{(A 5.4)}$$

We note that the expressions are analogous to (2.125) and (2.126) for the multipolar forces. Evaluation of explicit formulae in terms of molecular properties reflecting intramolecular charges such as multipole moments and polarizabilities is straightforward, but tedious. The so called $9j$-symbol arising in the expansion coefficients is defined as [2]

$$\begin{Bmatrix} a & b & c \\ d & e & f \\ g & h & i \end{Bmatrix} = \sum_x (2x + 1) \begin{Bmatrix} a & i & x \\ h & d & g \end{Bmatrix} \begin{Bmatrix} h & d & x \\ f & b & e \end{Bmatrix} \begin{Bmatrix} f & b & x \\ a & i & c \end{Bmatrix}, \quad \text{(A 5.5)}$$

where the $6j$-symbol is calculated from [2]

$$\begin{Bmatrix} j_1 & j_2 & j_3 \\ l_1 & l_2 & l_3 \end{Bmatrix} = \Delta(j_1 j_2 j_3) \Delta(j_1 l_2 l_3) \Delta(l_1 j_2 l_3) \Delta(l_1 l_2 j_3) W \begin{Bmatrix} j_1 & j_2 & j_3 \\ l_1 & l_2 & l_3 \end{Bmatrix}, \quad \text{(A 5.6)}$$

with

$$W \begin{Bmatrix} j_1 & j_2 & j_3 \\ l_1 & l_2 & l_3 \end{Bmatrix} = \sum_{Z>0}$$

$$\cdot \frac{(-1)^Z (Z+1)!}{(Z-j_1-j_2-j_3)!(Z-j_1-l_2-l_3)!(Z-l_1-j_2-l_3)!(Z-l_1-l_2-j_3)!}$$
$$\cdot \frac{1}{(j_1+j_2+l_1+l_2-Z)!(j_2+j_3+l_2+l_3-Z)!(j_1+j_3+l_1+l_3-Z)!} \tag{A 5.7}$$

and

$$\Delta(abc) = \left[\frac{(a+b-c)!(a-b+c)!(-a+b+c)!}{(a+b+c+1)!} \right]^{1/2}. \tag{A 5.8}$$

Second-order perturbation theory also provides an expression for the non-additive three-body induction term [1]. Third-order perturbation theory finally yields the nonadditive three-body dispersion potential. The final result reads [1], again in the London approximation, cf. Section 2.5.5,

$$
\phi^{\text{disp}}_{\alpha_i \beta_j \gamma_k} = \sum_{\substack{l_1''' \\ l_2''' \\ l_3'''}} \sum_{\substack{n_1''' \\ n_2''' \\ n_3'''}} \sum_{\substack{l_1 \\ l_2}} \sum_{\substack{l_1' \\ l_2'}} \sum_{\substack{l_1'' \\ l_2''}} \quad \sum_{\substack{m_1 \\ m_2}} \sum_{\substack{m_1' \\ m_2'}} \sum_{\substack{m_1'' \\ m_2''}}
$$

$$
\cdot \, E^{\text{disp}}_{\alpha_i \beta_j \gamma_k}(l_1''' l_2'' l_3''';l_1 l_2 l_1 + l_2;l_1' l_2' l_1' + l_2';l_1'' l_2'' l_1''
$$

$$
+ \, l_2'';\, n_1''' n_2''' n_3''';r_{\alpha_i\beta_j},r_{\alpha_i\gamma_k},r_{\beta_j\gamma_k})
$$

$$
\cdot \, D^{l_1'''}_{m_1''' n_1'''}(\omega_{\alpha_i})^* \; D^{l_2'''}_{m_2''' n_2'''}(\omega_{\beta_j})^* \; D^{l_3'''}_{m_3''' n_3'''}(\omega_{\gamma_k})^*
$$

$$
\cdot \, Y^{m}_{l_1+l_2}(\omega_{\alpha_i\beta_j})^* \; Y^{m'}_{l_1'+l_2'}(\omega_{\alpha_i\gamma_k})^* \; Y^{m''}_{l_1''+l_2''}(\omega_{\beta_j\gamma_k})^*
$$

$$
\cdot \, C(l_1 l_1' l_1''';m_1 m_1' m_1''')C(l_2 l_1'' l_2''';m_2 m_1'' m_2''')
$$

$$
\cdot \, C(l_2' l_2'' l_3''';m_2' m_2'' m_3''')C(l_1 l_2 l_1 + l_2;m_1 m_2 m)
$$

$$
\cdot \, C(l_1' l_2' l_1' + l_2';m_1' m_2' m')C(l_1'' l_2'' l_1'' + l_2'';m_1'' m_2'' m''), \tag{A 5.9}
$$

with

$$
E^{\text{disp}}_{\alpha_i \beta_j \gamma_k}(l_1''' l_2''' l_3''';l_1 l_2 l_1 + l_2;l_1' l_2' l_1' + l_2';l_1'' l_2'' l_1'' + l_2'';n_1''' n_2''' n_3''';r_{\alpha_i\beta_j},r_{\alpha_i\gamma_k},r_{\beta_j\gamma_k})
$$

$$
= 4\frac{2}{3} \frac{\nu_{\alpha\beta\gamma}}{\alpha_\alpha \alpha_\beta \alpha_\gamma}
$$

$$
\cdot \, \frac{64\pi^3(-1)^{l_2+l_2'+l_2''}}{r_{\alpha_i\beta_j}^{l_1+l_2+1} r_{\alpha_i\gamma_k}^{l_1'+l_2'+1} r_{\beta_j\gamma_k}^{l_1''+l_2''+1}} \frac{1}{(2l_1+2l_2+l)(2l_1'+2l_2'+1)(2l_1''+2l_2''+1)}
$$

$$
\cdot \, \sqrt{\frac{64\pi^3(2l_1+2l_2+l)!(2l_1'+2l_2'+1)!(2l_1''+2l_2''+1)!}{(2l_1+1)!(2l_2+1)!(2l_1'+1)!(2l_2'+1)!(2l_1''+1)!(2l_2''+1)!}}
$$

$$
\cdot \, \sum_{\substack{n_1 \\ n_2 \\ n_1'' \\ n_2''}} \sum_{\substack{n_1' \\ n_2' \\ n_2''}} C(l_1 l_1' l_1''';n_1 n_1' n_1''')C(l_2 l_1'' l_2''';n_2 n_1'' n_2''')C(l_2' l_2'' l_3''';n_2' n_2'' n_3''')
$$

$$
\cdot \, {}^\alpha \Pi^{n_1 n_1'}_{l_1 l_1'} \, {}^\beta \Pi^{n_2 n_1''}_{l_2 l_1''} \, {}^\gamma \Pi^{n_2' n_2''}_{l_2' l_2''}
$$

$$
\tag{A 5.10}
$$

as the nonadditive three-body expansion coefficients. Here $v_{\alpha\beta\gamma}$ is given by (2.257). Simplifications result for particular molecular models. Here $\omega_{\alpha_i\beta_j}$ is the orientation of the line connecting the centers of molecules α_i and β_j in the space-fixed coordinate system. Analogous interpretations hold for $\omega_{\alpha_i\gamma_k}$ and $\omega_{\beta_j\gamma_k}$, the three lines forming a triangle. If consideration is restricted to the dipole-dipole polarizability, we find that $l_1 = l_1' = l_2 = l_1'' = l_2' = l_2'' = 1$. The most important contribution is the Axilrod–Teller term given explicitly in (2.256) and (2.258). More complicated terms are listed in [3].

[1] K. Lucas. *Applied Statistical Thermodynamics*. Springer, Berlin, 1991.
[2] C. G. Gray and K. E. Gubbins. *Theory of Molecular Fluids I*. Clarendon Press, Oxford, 1984.
[3] W. Ameling, K. P. Shukla, and K. Lucas. *Mol. Phys.*, 58:381, 1986.

APPENDIX 6

Rules for Integration

The particular integral

$$I = \int_0^\infty e^{-ax^2} \mathrm{d}x$$

has the solution

$$I = \frac{1}{2}\sqrt{\frac{\pi}{a}}.$$

In transforming the integration variables x, y, z into a, b, c we have

$$\mathrm{d}x\mathrm{d}y\mathrm{d}z = J\mathrm{d}a\mathrm{d}b\mathrm{d}c$$

with J, the Jacobian, defined as

$$J = \begin{vmatrix} \frac{\partial x}{\partial a} & \frac{\partial y}{\partial a} & \frac{\partial z}{\partial a} \\ \frac{\partial x}{\partial b} & \frac{\partial y}{\partial b} & \frac{\partial z}{\partial b} \\ \frac{\partial x}{\partial c} & \frac{\partial y}{\partial c} & \frac{\partial z}{\partial c} \end{vmatrix}.$$

APPENDIX 7

Internal Rotation Contributions

The solutions of the Schrödinger equation for the energy values of hindered internal rotation are most conveniently tabulated in terms of contributions to various thermodynamic properties as a function of $1/q_{ir}$ and $u_{ir,max}/RT$. Here, q_{ir} is the molecular partition function of free internal rotation and $u_{ir,max}$ is the potential barrier, with $U_{ir} = \frac{1}{2}U_{ir,max}^{(n)}[1 - \cos(n\phi)]$. Tables A7.1–A7.3 were calculated for $n = 3$ and are taken from [1].

[1] B. J. McClelland. *Statistical Thermodynamics*. Chapman and Hall, London, 1973.

Table A.7.1. Heat Capacity $\dfrac{C^{ig}_{v,ir}}{R}$

$\dfrac{u_{ir,max}}{RT}$	0.0	0.05	0.10	0.15	0.20	0.25	0.30	0.35	0.40	0.45	0.50	0.55	0.60	0.65	0.70	0.75	0.80	0.85	0.90	0.95
													$1/q_{ir}$							
0.0	0.500	0.500	0.500	0.500	0.500	0.500	0.500	0.500	0.500	0.500	0.500	0.500	0.500	0.500	0.500	0.500	0.500	0.500	0.500	0.500
0.2	0.505	0.505	0.505	0.504	0.504	0.503	0.503	0.503	0.503	0.503	0.503	0.503	0.503	0.503	0.503	0.503	0.503	0.503	0.503	0.503
0.4	0.520	0.520	0.519	0.518	0.517	0.516	0.515	0.514	0.513	0.512	0.512	0.512	0.511	0.510	0.509	0.508	0.507	0.507	0.506	0.505
0.6	0.544	0.543	0.543	0.541	0.540	0.537	0.536	0.533	0.531	0.529	0.528	0.526	0.524	0.521	0.519	0.516	0.514	0.512	0.510	0.509
0.8	0.575	0.575	0.574	0.573	0.570	0.568	0.564	0.561	0.557	0.553	0.549	0.545	0.541	0.537	0.532	0.528	0.523	0.519	0.516	0.513
1.0	0.614	0.613	0.612	0.610	0.607	0.603	0.599	0.594	0.588	0.582	0.576	0.569	0.563	0.556	0.549	0.542	0.536	0.529	0.523	0.519
1.5	0.730	0.729	0.727	0.722	0.716	0.708	0.700	0.689	0.678	0.666	0.654	0.641	0.627	0.613	0.600	0.586	0.574	0.561	0.548	0.538
2.0	0.844	0.853	0.849	0.842	0.833	0.821	0.808	0.792	0.775	0.757	0.737	0.717	0.695	0.675	0.654	0.633	0.613	0.594	0.577	0.560
2.5	0.967	0.965	0.960	0.950	0.939	0.926	0.906	0.884	0.864	0.840	0.815	0.786	0.757	0.729	0.701	0.675	0.649	0.623	0.599	0.577
3.0	1.056	1.054	1.048	1.038	1.023	1.004	0.982	0.956	0.929	0.903	0.872	0.837	0.804	0.771	0.738	0.705	0.673	0.642	0.612	0.586
3.5	1.118	1.116	1.109	1.097	1.080	1.060	1.034	1.004	0.973	0.940	0.907	0.869	0.832	0.795	0.758	0.721	0.685	0.651	0.617	0.586
4.0	1.157	1.154	1.145	1.132	1.114	1.091	1.062	1.031	0.996	0.960	0.923	0.883	0.842	0.802	0.761	0.722	0.684	0.647	0.611	0.578
4.5	1.175	1.172	1.163	1.147	1.126	1.102	1.071	1.038	1.001	0.962	0.922	0.880	0.837	0.794	0.753	0.711	0.671	0.634	0.596	0.561
5.0	1.180	1.176	1.166	1.150	1.128	1.100	1.067	1.035	0.992	0.951	0.910	0.864	0.821	0.776	0.733	0.691	0.650	0.611	0.574	0.537
6.0	1.165	1.161	1.149	1.130	1.103	1.072	1.036	0.996	0.953	0.907	0.859	0.812	0.765	0.719	0.675	0.632	0.590	0.552	0.514	0.480
7.0	1.140	1.135	1.121	1.099	1.070	1.034	0.993	0.948	0.899	0.849	0.799	0.748	0.699	0.652	0.607	0.564	0.523	0.484	0.448	0.416
8.0	1.115	1.110	1.094	1.069	1.036	0.996	0.950	0.900	0.847	0.793	0.739	0.687	0.635	0.586	0.540	0.497	0.457	0.420	0.385	0.354
9.0	1.095	1.089	1.072	1.044	1.006	0.961	0.910	0.855	0.799	0.742	0.685	0.629	0.576	0.527	0.481	0.437	0.397	0.361	0.328	0.298
10.0	1.080	1.073	1.054	1.023	0.982	0.933	0.878	0.820	0.758	0.695	0.635	0.579	0.526	0.475	0.428	0.385	0.346	0.311	0.280	0.251
12.0	1.059	1.051	1.028	0.992	0.944	0.887	0.823	0.756	0.687	0.620	0.557	0.498	0.441	0.389	0.343	0.302	0.266	0.233	0.205	0.180
14.0	1.047	1.038	1.011	0.968	0.913	0.848	0.778	0.704	0.631	0.560	0.492	0.430	0.374	0.324	0.279	0.241	0.207	0.177	0.152	0.132
16.0	1.039	1.029	0.998	0.950	0.888	0.816	0.739	0.660	0.582	0.508	0.439	0.377	0.322	0.273	0.230	0.195	0.163	0.137	0.115	0.098
18.0	1.034	1.022	0.987	0.932	0.864	0.786	0.703	0.620	0.538	0.462	0.392	0.331	0.276	0.229	0.190	0.157	0.130	0.108	0.088	0.072
20.0	1.030	1.016	0.978	0.919	0.844	0.760	0.671	0.583	0.499	0.421	0.353	0.292	0.240	0.196	0.159	0.129	0.105	0.085	0.068	0.055

Table A.7.2. Entropy $\dfrac{\Delta s_{ir}^{ig}}{R} = \dfrac{s_{ir,free}^{ig}}{R} - \dfrac{s_{ir}^{ig}}{R}$

$\dfrac{u_{ir,max}}{RT}$	$1/q_{ir}$																			
	0.0	0.05	0.10	0.15	0.20	0.25	0.30	0.35	0.40	0.45	0.50	0.55	0.60	0.65	0.70	0.75	0.80	0.85	0.90	0.95
0.0	0.000	0.000	0.000	0.000	0.000	0.000	0.000	0.000	0.000	0.000	0.000	0.000	0.000	0.000	0.000	0.000	0.000	0.000	0.000	0.000
0.2	0.003	0.003	0.003	0.003	0.002	0.002	0.002	0.002	0.002	0.002	0.002	0.002	0.002	0.002	0.002	0.001	0.001	0.001	0.001	0.001
0.4	0.010	0.010	0.009	0.009	0.009	0.009	0.008	0.008	0.007	0.006	0.006	0.006	0.006	0.006	0.006	0.005	0.005	0.005	0.005	0.005
0.6	0.022	0.022	0.022	0.022	0.020	0.020	0.020	0.019	0.017	0.017	0.016	0.015	0.014	0.013	0.013	0.012	0.011	0.011	0.010	0.010
0.8	0.039	0.039	0.039	0.038	0.036	0.035	0.034	0.034	0.033	0.031	0.028	0.029	0.026	0.025	0.024	0.021	0.019	0.016	0.016	0.014
1.0	0.060	0.059	0.059	0.058	0.056	0.056	0.054	0.053	0.051	0.048	0.046	0.044	0.041	0.038	0.037	0.033	0.030	0.027	0.025	0.022
1.5	0.127	0.127	0.126	0.125	0.122	0.119	0.116	0.114	0.108	0.103	0.100	0.093	0.088	0.083	0.076	0.069	0.064	0.058	0.055	0.048
2.0	0.210	0.201	0.209	0.206	0.202	0.198	0.192	0.187	0.179	0.171	0.163	0.155	0.146	0.137	0.127	0.117	0.107	0.098	0.090	0.080
2.5	0.302	0.301	0.299	0.294	0.290	0.286	0.277	0.268	0.257	0.246	0.234	0.223	0.211	0.198	0.185	0.172	0.157	0.144	0.131	0.119
3.0	0.395	0.394	0.391	0.386	0.381	0.373	0.362	0.352	0.340	0.326	0.310	0.305	0.278	0.262	0.244	0.227	0.209	0.192	0.175	0.159
3.5	0.486	0.485	0.482	0.475	0.467	0.458	0.446	0.434	0.421	0.402	0.383	0.364	0.344	0.324	0.303	0.282	0.261	0.239	0.218	0.198
4.0	0.571	0.570	0.567	0.559	0.550	0.539	0.525	5.09	0.493	0.473	0.451	0.430	0.407	0.384	0.360	0.334	0.310	0.286	0.262	0.238
4.5	0.650	0.649	0.644	0.637	0.626	0.614	0.597	0.581	0.562	0.538	0.515	0.489	0.464	0.439	0.412	0.383	0.356	0.329	0.300	0.274
5.0	0.722	0.720	0.715	0.706	0.694	0.681	0.663	0.645	0.622	0.598	0.573	0.546	0.517	0.487	0.457	0.427	0.397	0.367	0.336	0.307
6.0	0.844	0.842	0.836	0.827	0.813	0.796	0.776	0.752	0.727	0.699	0.670	0.637	0.604	0.571	0.536	0.502	0.467	0.433	0.398	0.364
7.0	0.945	0.943	0.939	0.936	0.924	0.909	0.866	0.840	0.810	0.779	0.745	0.708	0.673	0.635	0.597	0.560	0.521	0.483	0.445	0.408
8.0	1.029	1.027	1.018	1.005	0.987	0.965	0.940	0.910	0.877	0.843	0.806	0.766	0.726	0.686	0.644	0.604	0.562	0.521	0.481	0.441
9.0	1.100	1.097	1.088	1.074	1.054	1.029	1.001	0.968	0.932	0.894	0.854	0.811	0.768	0.726	0.681	0.637	0.593	0.550	0.507	0.465
10.0	1.162	1.159	1.149	1.133	1.111	1.085	1.052	1.016	0.978	0.937	0.893	0.848	0.803	0.756	0.709	0.663	0.617	0.572	0.528	0.484
12.0	1.266	1.262	1.250	1.231	1.205	1.173	1.138	1.095	1.050	1.003	0.954	0.903	0.852	0.802	0.749	0.699	0.650	0.602	0.555	0.509
14.0	1.351	1.347	1.333	1.312	1.282	1.245	1.204	1.156	1.104	1.051	0.998	0.942	0.887	0.832	0.778	0.724	0.673	0.620	0.751	0.523
16.0	1.423	1.419	1.403	1.379	1.346	1.304	1.256	1.204	1.148	1.090	1.031	0.972	0.912	0.853	0.797	0.741	0.685	0.633	0.581	0.532
18.0	1.487	1.481	1.464	1.437	1.399	1.354	1.301	1.243	1.183	1.121	1.057	0.995	0.931	0.870	0.810	0.751	0.695	0.641	0.588	0.538
20.0	1.543	1.537	1.518	1.487	1.445	1.396	1.338	1.277	1.212	1.146	1.078	1.012	0.946	0.882	0.820	0.760	0.702	0.646	0.593	0.542

Table A.7.3. Gibbs Free Energy $\frac{\Delta g_{fr}^{ig}}{RT} = \frac{g_{fr}^{ig}}{RT} - \frac{g_{fr,free}^{ig}}{RT}$

$\frac{u_{fr,max}}{RT}$	0.0	0.05	0.10	0.15	0.20	0.25	0.30	0.35	0.40	0.45	0.50	0.55	0.60	0.65	0.70	0.75	0.80	0.85	0.90	0.95
0.0	0.000	0.000	0.000	0.000	0.000	0.000	0.000	0.000	0.000	0.000	0.000	0.000	0.000	0.000	0.000	0.000	0.000	0.000	0.000	0.000
0.2	0.097	0.077	0.059	0.043	0.031	0.023	0.017	0.013	0.009	0.006	0.005	0.003	0.003	0.002	0.001	0.001	0.001	0.001	0.001	0.000
0.4	0.190	0.164	0.138	0.113	0.089	0.066	0.049	0.036	0.028	0.022	0.018	0.012	0.009	0.007	0.005	0.003	0.003	0.003	0.002	0.001
0.6	0.278	0.246	0.213	0.182	0.150	0.119	0.093	0.072	0.057	0.045	0.034	0.026	0.021	0.015	0.011	0.008	0.006	0.004	0.003	0.002
0.8	0.360	0.322	0.285	0.248	0.211	0.176	0.144	0.116	0.093	0.073	0.056	0.045	0.034	0.026	0.020	0.014	0.010	0.007	0.005	0.003
1.0	0.438	0.395	0.352	0.310	0.270	0.231	0.196	0.162	0.132	0.105	0.082	0.066	0.052	0.040	0.031	0.023	0.016	0.011	0.007	0.004
1.5	0.614	0.561	0.508	0.457	0.407	0.360	0.313	0.271	0.227	0.190	0.155	0.126	0.101	0.079	0.061	0.046	0.032	0.023	0.016	0.009
2.0	0.764	0.702	0.642	0.583	0.526	0.471	0.417	0.365	0.316	0.269	0.227	0.187	0.152	0.121	0.095	0.072	0.053	0.037	0.025	0.014
2.5	0.892	0.823	0.755	0.690	0.627	0.566	0.505	0.447	0.391	0.340	0.290	0.243	0.201	0.162	0.128	0.099	0.073	0.052	0.034	0.020
3.0	1.001	0.925	0.852	0.781	0.712	0.645	0.580	0.517	0.456	0.399	0.343	0.292	0.244	0.200	0.160	0.125	0.093	0.066	0.044	0.025
3.5	1.095	1.013	0.934	0.857	0.783	0.712	0.642	0.575	0.513	0.449	0.390	0.334	0.281	0.232	0.188	0.148	0.111	0.079	0.053	0.030
4.0	1.176	1.088	1.004	0.922	0.843	0.770	0.694	0.624	0.558	0.490	0.428	0.369	0.313	0.260	0.212	0.167	0.128	0.092	0.061	0.034
4.5	1.247	1.154	1.065	0.979	0.896	0.816	0.739	0.665	0.596	0.525	0.460	0.398	0.340	0.285	0.232	0.185	0.142	0.104	0.068	0.038
5.0	1.309	1.212	1.118	1.028	0.940	0.856	0.777	0.700	0.626	0.555	0.488	0.423	0.361	0.303	0.249	0.199	0.153	0.111	0.074	0.041
6.0	1.414	1.308	1.206	1.108	1.014	0.924	0.838	0.755	0.676	0.601	0.530	0.460	0.395	0.334	0.276	0.222	0.171	0.125	0.083	0.045
7.0	1.501	1.386	1.277	1.171	1.071	0.974	0.884	0.797	0.714	0.635	0.559	0.488	0.420	0.356	0.295	0.237	0.184	0.134	0.089	0.047
8.0	1.575	1.452	1.335	1.224	1.117	1.016	0.920	0.828	0.742	0.660	0.582	0.508	0.438	0.371	0.308	0.248	0.193	0.141	0.093	0.049
9.0	1.639	1.509	1.385	1.268	1.156	1.051	0.950	0.855	0.765	0.680	0.600	0.523	0.450	0.383	0.317	0.256	0.199	0.146	0.096	0.049
10.0	1.695	1.558	1.429	1.305	1.189	1.079	0.975	0.876	0.783	0.696	0.613	0.535	0.461	0.391	0.325	0.263	0.205	0.149	0.098	0.050
12.0	1.791	1.642	1.501	1.368	1.242	1.124	1.013	0.909	0.811	0.720	0.633	0.552	0.475	0.404	0.336	0.271	0.211	0.153	0.100	0.051
14.0	1.872	1.711	1.559	1.417	1.284	1.159	1.042	0.933	0.831	0.736	0.647	0.563	0.485	0.412	0.342	0.276	0.215	0.156	0.102	0.051
16.0	1.942	1.770	1.609	1.458	1.317	1.187	1.065	0.952	0.846	0.748	0.657	0.571	0.491	0.417	0.346	0.280	0.217	0.158	0.103	0.051
18.0	2.002	1.821	1.650	1.492	1.346	1.210	1.083	0.966	0.857	0.757	0.664	0.577	0.496	0.420	0.349	0.282	0.219	0.159	0.104	0.051
20.0	2.057	1.865	1.687	1.522	1.369	1.228	1.098	0.977	0.867	0.764	0.669	0.581	0.499	0.422	0.351	0.283	0.220	0.161	0.104	0.051

$1/q_{fir}$

APPENDIX 8

Quasichemical Approximation for the Degeneracy in a Lattice

From (4.45), the degeneracy in a lattice associated with a particular value of N_{12} is given by

$$g(N_1, N_2, N_{12}) = C(N_1, N_2)h(N_1, N_2, N_{12}),$$

with $h(N_1, N_2, N_{12})$ given by (4.43). We wish to evaluate $C(N_1, N_2)$ in such a way that g fulfills the ideal mixing condition (4.44). For this purpose we replace the sum in (4.46) by the maximum term, as justified in App. 3. To deal with the factorials we use the logarithm of h with Stirling's formula, cf. App. 2, and find that

$$\ln h = \frac{z}{2} N \ln \left(\frac{z}{2} N\right) - \frac{N_{11}}{2} \ln \frac{N_{11}}{2} - \frac{N_{22}}{2} \ln \frac{N_{22}}{2} - N_{12} \ln \frac{N_{12}}{2},$$

where the conservation equations (4.30) and (4.31) have been used. For a constant number of molecules, but a variable number of contacts, the maximum condition can be formulated as

$$\begin{aligned}
\frac{\partial \ln h}{\partial N_{12}} &= \frac{\partial}{\partial N_{12}} \left[-\frac{N_{11}}{2} \ln \frac{N_{11}}{2} - \frac{N_{22}}{2} \ln \frac{N_{22}}{2} - N_{12} \ln \frac{N_{12}}{2} \right] \\
&= \frac{1}{2} \ln \frac{N_{11} N_{22}}{(N_{12})^2} \overset{!}{=} 0.
\end{aligned} \tag{A 8.1}$$

The maximum condition for $\ln h$ thus gives

$$N_{12}^2 = N_{11} N_{22}.$$

By introducing the conservation equations for N_{11} and N_{22}, this can be transformed into

$$N_{12} = \frac{z N_1 N_2}{N}. \tag{A 8.2}$$

This leads to

$$
\ln h = \frac{z}{2} N \ln \left(\frac{N^2}{N} \frac{z}{2} \right) - \frac{N_{11}}{2} \ln \left(\frac{z N_1{}^2}{2N} \right) - \frac{N_{22}}{2} \ln \left(\frac{z N_2{}^2}{2N} \right) - N_{12} \ln \left(\frac{z N_1 N_2}{2N} \right)
$$

$$
= \ln \left(\frac{z}{2N} \right) \left[\frac{z}{2} N - \frac{N_{11}}{2} - \frac{N_{22}}{2} - N_{12} \right] + \frac{z}{2} N \ln N^2 - \frac{N_{11}}{2} \ln N_1^2
$$

$$
- \frac{N_{22}}{2} \ln N_2^2 - N_{12} \ln(N_1 N_2)
$$

$$
= \frac{z}{N} \left[N^2 \ln N - N_1 N \ln N_1 - N_2 N \ln N_2 \right]
$$

$$
= z \left[\ln N! - \ln(N_1)! - \ln(N_2)! \right]
$$

$$
= \ln \left[\frac{N!}{N_1! N_2!} \right]^z,
$$

and thus

$$
\sum_{N_{12}} h(N_1, N_2, N_{12}) = \left[\frac{N!}{N_1! N_2!} \right]^z, \tag{A 8.3}
$$

which gives

$$
C(N_1, N_2) = \left[\frac{N!}{N_1! N_2!} \right]^{1-z}. \tag{A 8.4}
$$

For the degeneracy we then get

$$
g(N_1, N_2, N_{12}) = \left[\frac{N!}{N_1! N_2!} \right]^{1-z} \frac{(\frac{z}{2} N)!}{N_{11}/2! \, N_{22}/2! \, [(N_{12}/2)!]^2}. \tag{A 8.5}
$$

Making use of the conservation equations in the form $N_{12} = z x_{12} N_2 = z x_{21} N_1$ and $x_{11} + x_{21} = x_{22} + x_{12} = 1$, this can be written as

$$
g(N_1, N_2, N_{12}) = \left(\frac{N!}{N_1! N_2!} \right)^{1-z} \frac{(\frac{z}{2} N)!}{(z \frac{N_1}{2} - \frac{N_{12}}{2})! (z \frac{N_2}{2} - \frac{N_{12}}{2})! \left[(\frac{N_{12}}{2})! \right]^2}
$$

$$
= \left(\frac{N!}{N_1! N_2!} \right)^{1-z} \frac{(\frac{z}{2} N)!}{(\frac{z}{2} x_{11} N_1)! (\frac{z}{2} x_{21} N_1)! (\frac{z}{2} x_{12} N_2)! (\frac{z}{2} x_{22} N_2)!}. \tag{A 8.6}
$$

Working out the factorials by Stirling's formula, we finally find

$$
\ln g = (1 - z) N (-x_1 \ln x_1 - x_2 \ln x_2)
$$

$$
+ \frac{zN}{2} \left(-x_{11} x_1 \ln x_{11} x_1 - x_{21} x_1 \ln x_{21} x_1 - x_{12} x_2 \ln x_{12} x_2 - x_{22} x_2 \ln x_{22} x_2 \right)
$$

$$
= N (-x_1 \ln x_1 - x_2 \ln x_2)
$$

$$
+ \frac{zN}{2} \left(x_{11} x_1 \ln \frac{x_1}{x_{11}} + x_{21} x_1 \ln \frac{x_1}{x_{21}} + x_{12} x_2 \ln \frac{x_2}{x_{12}} + x_{22} x_2 \ln \frac{x_2}{x_{22}} \right). \tag{A 8.7}
$$

Off-Lattice Formulation of the Quasichemical Approximation

The general relation between the chemical potential of a component i in a mixture and the canonical partition function is, cf. (2.62),

$$\mu_i = \left(\frac{\partial A}{\partial N_i}\right)_{T,V,N_i^*} = -kT\left(\frac{\partial \ln Q}{\partial N_i}\right)_{T,V,N_i^*}.$$

This can be written as

$$\mu_i = -kT\frac{\ln Q - \ln Q_{(-i)}}{N_i - (N_i - 1)}$$

$$= -kT\ln\frac{Q}{Q_{(-i)}}. \tag{A 9.1}$$

Here $Q_{(-i)}$ is the partition function of the system after one molecule of component i has been removed. We base the further derivation on the assumption that we deal with independent molecular pairs, not with single molecules. Under this assumption, which is equivalent to the quasichemical approximation, and in the limit of large N we can relate the partition function $Q_{(-i,-j)}$, where one molecule of type i and one molecule of type j have been removed from the system, to the sum of the corresponding chemical potentials by

$$\mu_i + \mu_j = -kT\ln\left(\frac{Q}{Q_{(-i)}}\frac{Q_{(-i)}}{Q_{(-i,-j)}}\right) = -kT\ln\frac{Q}{Q_{(-i,-j)}}, \tag{A 9.2}$$

without taking account of any effect that this may have on the rest of the molecules in the system.

In order to derive a formal relation between Q and $Q_{(-i,-j)}$ we consider the probability of finding a particular ij-pair, as given by (2.50) in the classical approximation,

$$P_{ij} = \frac{\int \cdots \int e^{-U/kT}d\mathbf{r}^N_{(-i,-j)}}{\int \cdots \int e^{-U/kT}d\mathbf{r}^N},$$

where $d\boldsymbol{r}^N_{(-i,-j)}$ stands for the position variables of all molecules except those belonging to the pair ij. We assume the molecules in a liquid to be arranged in a space-filling manner with polyhedral surface contacts, as shown in Fig. 4.4. We further use the pairwise additivity approximation for the contact energies ϕ_{ij} between two molecules i and j with $U = \sum_i \sum_{j<i} \phi_{ij}$. Because all pairs are independent of each other, we can take ϕ_{ij} out of the integral in the numerator to find

$$P_{ij} = \frac{e^{-\phi_{ij}/kT} Q_{(-i,-j)}}{Q}.$$

Because, further, the average probability of finding a pair with one molecule being i and the other being arbitrary is equal to the mole fraction x_i, we have

$$\sum_j P_{ij} = \frac{N_i}{N} = \frac{1}{Q} \sum_j e^{-\phi_{ij}/kT} Q_{(-i,-j)}$$

and thus

$$Q = \frac{N}{N_i} \sum_j e^{-\phi_{ij}/kT} Q_{(-i,-j)}. \tag{A 9.3}$$

If we substitute (A 9.3) into (A 9.2) we get

$$\mu_i = -kT \ln \left\{ \sum_j e^{\frac{-\phi_{ij}+\mu_j}{kT}} \right\} - kT \ln \frac{N}{N_i}. \tag{A 9.4}$$

For a pure component i this reduces to

$$\mu_{0i} = -kT \ln \left\{ e^{\frac{-\phi_{ii}+\mu_{0i}}{kT}} \right\} = \frac{1}{2}\phi_{ii}.$$

We now find the activity coefficient of component i as, cf. (2.29),

$$kT \ln \gamma_i = \mu_i - \mu_{0i} - kT \ln x_i$$

$$= -kT \ln \left\{ \sum_j e^{\frac{-\phi_{ij}+\mu_j}{kT}} \right\} - \frac{1}{2}\phi_{ii}$$

$$= -kT \ln \left\{ e^{\frac{-\phi_{ij}}{2kT}} \sum_j e^{\frac{-\phi_{ij}+\frac{\phi_{ii}+\phi_{jj}}{2}+\mu_j-\frac{\phi_{ij}}{2}}{kT}} \right\} - \frac{\phi_{ii}}{2}$$

$$= -kT \ln \left\{ \sum_j x_j \tau_{ij} \gamma_j \right\}, \tag{A 9.5}$$

which is (4.59) in the text, with

$$\tau_{ij} = e^{-\omega_{ij}/2kT}$$

and

$$\omega_{ij} = \phi_{ij} + \phi_{ji} - \phi_{ii} - \phi_{jj}.$$

The same type of reasoning also applies when we switch from molecules i, j to molecular groups α, β. Written in terms of group activity coefficients, (A 9.5) becomes

$$\ln \gamma_\alpha = -\ln \left\{ \sum_\beta x_\beta \tau_{\alpha\beta} \gamma_\beta \right\}, \qquad (A\ 9.6)$$

with

$$\tau_{\alpha\beta} = e^{-\omega_{\alpha\beta}/2kT}, \qquad (A\ 9.7)$$

$$\omega_{\alpha\beta} = \phi_{\alpha\beta} + \phi_{\beta\alpha} - \phi_{\alpha\alpha} - \phi_{\beta\beta}. \qquad (A\ 9.8)$$

For consistency with the quasichemical approximation we here have assumed the free segment approximation to be valid. For a component i consisting of various groups indexed by β we have the general relation

$$\mu_{i,\text{att}} - kT \ln x_i = -kT \sum_\beta \frac{\partial \ln Q}{\partial N_\beta} \frac{\partial N_\beta}{\partial N_i} = \sum_\beta n_{\beta,i} \mu_\beta, \qquad (A\ 9.9)$$

where N_i and N_β are the numbers of molecules and groups, respectively, μ_β is the chemical potential of group β in the mixture, and $n_{\beta,i}$ is the number of groups β in molecule i. We note that (A 9.9) gives the chemical potential minus the ideal mixing term, because it is added up from independent group chemical potentials and thus does not include the permutational factors associated with the various components in the general formulation of the partition function. This gives

$$\ln \gamma_{i,\text{att}} = \frac{\mu_{i,\text{att}} - kT \ln x_i - \mu_{0i,\text{att}}}{kT} = \sum_\beta n_{\beta,i} \frac{\mu_\beta - \mu_\beta^{0i} - \mu_{\beta\beta} + \mu_{\beta\beta}}{kT}$$

$$= \sum_\beta n_{\beta,i} (\ln \gamma_\beta - \ln \gamma_\beta^{0i}), \qquad (A\ 9.10)$$

where $\mu_\beta = \mu_{\beta\beta} + kT \ln \gamma_\beta$, $\mu_\beta^{0i} = \mu_{\beta\beta} + kT \ln \gamma_\beta^{0i}$, and μ_β^{0i} is the chemical potential of β segments in pure component i, whereas $\mu_{\beta\beta}$ is the chemical potential of a pure system of β segments. When, instead of groups, surface segments are considered, the equations remain the same. However, we have to replace the number of groups by the number of surface contacts, i.e., by $N_{\beta,i}$, in (A 9.10). Because $N_{\beta,i}$ is a measure for the number of contacts, it is given by $n_{\beta,i} z q_\beta$. (A 9.10) with (A 9.6) for the segment activity coefficients is (4.93) with (4.92) in the text.

The preceding derivation has not only produced an equation for the activity coefficient but, more generally, one for the chemical potential; cf. (A 9.9) with (A 9.4) for $i = \beta$. We can therefore calculate the Henry coefficient and other

Table A 9.1. Henry coefficients and free energy of solvation in water: Comparison of COSMO-RS to experimental data for $T = 298.15$ K (in MPa and kJ/mol, respectively)

Solute	H^* exp	$\Delta g^{s,0}$ exp	H^* COSMO-RS	$\Delta g^{s,0}$ COSMO-RS
Ammonia	0.1	−10.0	0.0086	−16.1
Methanol	0.025	−13.4	0.0163	−14.5
Sulfur dioxide	4.6	−0.5	2	−2.3
Methane	3965	16.2	3439	15.9
Oxygen	4208	16.4	9056	18.3
Butane	4749	16.1	5148	16.9
Xenon	1256	13.4	750	12.1
Benzene	32	4.3	38	4.7

thermodynamic properties in the reference state of the ideally dilute solution from this model. The Henry coefficient in the mole fraction scale at the pressure p is given from classical thermodynamics as, cf. (2.39),

$$H_i^*(T, p)/p = \exp\frac{\mu_i^*(T, p) - \mu_{0i}^{ig}(T, p)}{RT}, \qquad (A\ 9.11)$$

where $\mu_i^* = \mu_i^{ids} - RT \ln x_i$ is the chemical potential of solute i in the standard state of the ideally dilute solution ($x_i = 1$), with μ_i^{ids} the chemical potential of i at $x_i = 0$. Further, μ_{0i}^{ig} is the chemical potential in the pure ideal gas state. The Gibbs free energy of solvation is given as $\Delta g_i^s = \mu_i^* - \mu_{0i}^{ig}$ and is thus simply related to the Henry coefficient. The chemical potentials in (A 9.11) can be calculated from the COSMO-RS model. For internal consistency both, i.e., μ_i^* and μ_{0i}^{ig}, must be evaluated with respect to the same reference state, i.e., the COSMO reference state; cf. Section 4.2. As discussed in Section 2.5, we can evaluate $-\Delta g_i^s$, i.e., the solvation free energy, from the COSMO model, which corresponds to μ_{0i}^{ig} with respect to the COSMO reference state. Also, we can evaluate μ_i^* as $\mu_i^{ids} - kT \ln x_i$ on the basis of (A 9.9) and (A 9.10) from

$$\mu_{i,att}^{ids} - kT \ln x_i = kT \sum_{\beta} N_{\beta i} \ln \gamma_\beta + \sum_{\beta} N_{\beta i} \mu_{\beta\beta},$$

where $\ln \gamma_\beta$ is evaluated at $x_i = 0$ and $\mu_{\beta\beta} = \frac{1}{2}\phi_{\beta\beta}$, both with respect to the COSMO reference state. The energy due to dispersion interactions is either subtracted from μ_{0i}^{ig} or added to μ_i^*. Table A 9.1 presents some results obtained from COSMO-RS [1] in comparison to data [2].

It can be seen in comparison with Table 2.2 that COSMO-RS represents a significant improvement over COSMO for the nonpolar solutes in water, because the dispersion and hydrogen interactions are accounted for and also the electrostatic interactions are more realistically modeled by the misfit term. The polar fluids represented by COSMO-RS are also closer to experiment, with the exception of ammonia, but the improvement over COSMO is smaller, as expected. In view of the large range of values for the Henry coefficient, more than a factor of 10^4, the agreement is considered to be fair, although not yet sufficient to replace tabulated values of experimental data in the standard state of ideal dilute solution. Analogous calculations performed for the solvent *n*-hexadecane give an even closer agreement between COSMO-RS and experiment, and again emphasize the significant improvement over COSMO; cf. Section. 2.5.

[1]   COSMOTHERM. Computer package. *COSMOlogic*, 2003.
[2]   C. J. Cramer and D. G. Truhlar. *Reviews in Computational Chemistry Vol. VI.* VCH Publishers, New York, 1995.

Combinatorial Contribution to the Excess Entropy in a Lattice

According to (4.71), the combinatorial contribution to the entropy of mixing for molecules consisting of equally sized segments on a lattice is given by

$$\Delta S^{\mathrm{M}}_{\mathrm{comb}} = k \ln \frac{\Omega(N_1, N_2)}{\Omega(N_1, 0)\Omega(0, N_2)}. \tag{A 10.1}$$

To simplify the argument we assume the solvent molecules (1) to consist of just one segment, $r_1 = 1$, whereas for the structured molecules we have $r_2 > 1$ and generalize later. We note that $\Omega(N_1, 0) = 1$, because there is only one distinguishable way of arranging N_1 identical solvent molecules over N_1 lattice sites.

We study the number of different arrangements in the mixture by first placing the N_2 structured molecules and after that the N_1 solvent molecules. The number $\Omega(N_1, N_2)$ can then be replaced by the number of distinguishable arrangements of just the N_2 structured molecules over the total of Nr sites, because there is only one way of arranging the identical N_1 solvent molecules after having placed the structured molecules, and that is by filling up the remaining unoccupied sites. Here we have used $r = r_1 x_1 + r_2 x_2 = x_1 + r_2 x_2$. We label the structured molecules by $1, 2, \ldots, N_2$ and introduce them one at a time and consecutively into the lattice. We further denote by Ω_{i+1} the number of possible arrangements for the $(i + 1)$th structured molecule in the lattice with i molecules being already placed. In a random distribution all subsequent placing events are independent of each other and we find

$$\Omega(N_1, N_2) = \frac{1}{N_2!} \prod_{i=0}^{N_2-1} \Omega_{i+1}. \tag{A 10.2}$$

Here $N_2!$ corrects for the indistinguishability of the structured molecules in the usual way and symmetry effects due to reversing the order of the segments are neglected, because they cancel in the excess functions anyhow. To proceed we derive an expression for Ω_{i+1}, i.e., the numbers of ways the structured molecule $i + 1$ can be arranged when $0, 1, 2, \ldots, i$ molecules have already been placed in the lattice.

To do this we consider subsequent cases of increasing complexity. We first treat structured molecules without taking into account that the volume segments of a molecule can either have internal contacts, i.e., contacts with volume segments of the same molecule, or external contacts, in which a molecule may interact with another one. Then, with i structured molecules already placed in the lattice, the probability of finding another empty lattice site to place the first volume segment of the $(i + 1)$th structured molecule in a random trial is $(Nr - ir_2)/Nr$. The associated number of empty sites is $(Nr - ir_2)$. This unit will have z nearest neighbors, of which $z(Nr - ir_2)/Nr$ will be empty in the average of all random distributions. This then is the number of possible locations for the second volume segment. Because the second volume segment is connected to the first, one neighboring lattice site is occupied and the third unit can be placed in $(z - 1)(Nr - ir_2)/Nr$ ways, if the molecule is fully flexible. Here we have neglected the placing of the first and the first plus the second segment. These approximations are justified in view of their negligible relative effect due to the additivity of the single logarithmic contributions. The same expression holds for the number of possible arrangements for all further segments $4, 5, \ldots, r_2$. Also, we replace z by $(z - 1)$ in the factor associated with the second segment. So we find that

$$\Omega_{i+1} = (Nr - ir_2)\frac{z(Nr - ir_2)}{Nr}\left[\frac{(z - 1)(Nr - ir_2)}{Nr}\right]^{r_2-2}$$

$$= (Nr - ir_2)^{r_2}\left(\frac{z - 1}{Nr}\right)^{r_2-1} \tag{A 10.3}$$

and further

$$\ln \prod_{i=0}^{N_2-1} \Omega_{i+1} = N_2(r_2 - 1)\ln\frac{z - 1}{Nr} + r_2 \sum_{i=0}^{N_2-1} \ln(Nr - ir_2). \tag{A 10.4}$$

To find a closed expression for the sum we replace it by an integral,

$$\sum_{i=0}^{N_2-1} \ln(Nr - ir_2) \simeq \int_0^{N_2} \ln(Nr - ir_2)\mathrm{d}i = -\frac{1}{r_2}\int_{Nr}^{N_1} \ln n\, \mathrm{d}n$$

$$= -\frac{1}{r_2}[N_1 \ln N_1 - N_1 - Nr \ln(Nr) + Nr].$$

Introducing this into (A 10.2) gives

$$\ln \Omega(N_1, N_2) = -N_2 \ln N_2 + N_2 - N_1 \ln N_1 + N_1 + Nr \ln(Nr)$$

$$- Nr + N_2(r_2 - 1)\ln\frac{z - 1}{Nr}$$

$$= N \ln(Nr) + N_2(r_2 - 1)\ln(z - 1) - Nr - N_2 \ln N_2$$

$$+ N_2 - N_1 \ln N_1 + N_1 \tag{A 10.5}$$

and, by setting $N_1 = 0$,

$$\ln \Omega(0, N_2) = N_2 \ln(N_2 r_2) + N_2(r_2 - 1)\ln(z - 1) - N_2 r_2 - N_2 \ln N_2 + N_2.$$
(A 10.6)

For the combinatorial entropy of mixing this leads to

$$\left[\frac{\Delta S^M}{k}\right]_{\text{comb}} = -N_1 \ln \phi_1 - N_2 \ln \phi_2,$$
(A 10.7)

and for the excess entropy

$$\left[\frac{S^E}{k}\right]_{\text{comb}} = -N_1 \ln \frac{\phi_1}{x_1} - N_2 \ln \frac{\phi_2}{x_2},$$
(A 10.8)

where

$$\phi_1 = \frac{x_1}{x_1 + x_2 r_2}$$
(A 10.9)

and

$$\phi_2 = \frac{x_2 r_2}{x_1 + x_2 r_2}$$
(A 10.10)

are the volume fractions of the components and N_1, N_2 are again the molecule numbers.

Although these results are strictly derived under the assumption of N_1 solvent molecules consisting of just one segment, the same result is obtained by exchanging solvent and structured molecules. This indicates that the restriction to simple solvent molecules is irrelevant. Indeed, the same relations are obtained when the analogous statistical arguments are carried through with both types of molecules being structured.

Generalization to multicomponent mixtures of arbitrary components gives

$$\left[\frac{\Delta S^M}{k}\right]_{\text{comb}} = -\sum N_i \ln \phi_i$$
(A 10.11)

and

$$\left[\frac{S^E}{k}\right]_{\text{comb}} = -\sum N_i \ln \frac{\phi_i}{x_i},$$
(A 10.12)

where

$$\phi_i = \frac{x_i r_i}{\sum_j x_j r_j}$$
(A 10.13)

is the volume fraction of component i. Clearly, if we put $r_i = 1$, we recover with $\left(S^E/k\right)_{\text{comb}} = 0$ the familiar ideal mixing entropy for noninteracting, equally sized molecules. These are the Flory–Huggins equations for athermal polymer solutions [1,2]. They lead to a positive combinatorial excess entropy for $r_i \neq r_j$.

Although the Flory–Huggins combinatorial excess entropy gives a first qualitatively correct account of the combinatorial contribution to the excess free energy of a liquid mixture of structured molecules, it is clear that it is no more than a crude approximation. The probability assumed in this model of finding an empty lattice site, i.e., $(Nr - ir_2)/Nr$, does not take notice of the fact that the empty lattice sites are not distributed randomly, but rather this distribution is structured in relation to the structured molecules. This nonrandom distribution of empty lattice sites is due to the fact that the r_2 volume units of a structured molecule are connected, leading to larger coherent regions of empty lattice sites as compared to the Flory–Huggins model. Such regions will increase the number of distinguishable ways of placing the structured molecules. This effect can be taken into account by differentiating between external and internal contacts between the segments of molecules. The probability of finding an internal contact site i within a structured molecule of type 2 is generally given by $z(r_2 - q_2)/zr_2$, with reference to Fig. 4.18. For $r_2 = q_2$ there are no internal contacts. Accordingly, the probability of an external contact in one structured molecule is $1 - z(r_2 - q_2)/zr_2$. Considering i structured molecules already placed in the lattice will introduce the probability of finding any type of occupied contact as zir_2/zNr. Thus, in a lattice with i structured molecules, the probability of having noninternal contacts is $1 - [(r_2 - q_2)/r_2](ir_2/Nr)$. This is the factor by which the Flory–Huggins probability of finding an empty lattice space has to be divided in order to take molecular connectivities into account; i.e.,

$$\frac{Nr - ir_2}{Nr} \Big/ \left(1 - \frac{r_2 - q_2}{r_2}\frac{ir_2}{Nr}\right) = \frac{Nr - ir_2}{Nr - i(r_2 - q_2)}. \tag{A 10.14}$$

Equation (A 10.14) can also be found by a slightly different, but analogous argument. We are looking for the constrained probability of finding an empty site for the second segment of a chain molecule under the condition that an empty site for the first segment was found before. This is equivalent to postulating that there was no internal contact between the two sites, because an internal contact requires that two neighboring sites are occupied. If we call this constrained probability $P_B(A)$ and the probability of no internal contacts $P(B)$, we realize that the product of these two probabilities is equal to the product of $P_A(B) = 1$; i.e., the probability of a noninternal contact between the first and the second site under the condition of the second site being empty, and $P(A)$, i.e., the probability of finding an empty site for either the second or the first segment. We know from the above arguments leading to the Flory–Huggins model that $P(A) = (Nr - ir_2)/Nr$. Because $P(B) = (zNr - (r_2 - q_2)zi)/zNr$, we find again (A 10.14).

The number of distinguishable ways to arrange the $(i + 1)$th structured molecule is thus

$$\Omega_{i+1} = (Nr - ir_2)^{r_2} \left\{\frac{z - 1}{[Nr - i(r_2 - q_2)]}\right\}^{r_2 - 1}. \tag{A 10.15}$$

It is larger than (A 10.3) in the Flory–Huggins model. We then have

$$\ln \prod_{i=0}^{N_2-1} \Omega_{i+1} = r_2 \sum_{i=0}^{N_2-1} \ln(Nr - ir_2) - (r_2 - 1) \sum_{i=0}^{N_2-1} \ln[Nr - i(r_2 - q_2)]$$
$$+ N_2(r_2 - 1)\ln(z - 1).$$

We again replace the sums by integrals. The first has been considered before. For the second sum we get

$$\sum_{i=0}^{N_2-1} \ln[Nr - i(r_2 - q_2)] \simeq \int_0^{N_2} \ln[Nr - i(r_2 - q_2)]di$$

$$= -\frac{1}{r_2 - q_2} \int_{Nr}^{N\bar{q}} \ln n\, dn$$

$$= \frac{1}{r_2 - q_2}[Nr \ln(Nr) - Nr - N\bar{q}\ln(N\bar{q}) + N\bar{q}],$$

where $N\bar{q} = N_1 + N_2 q_2$. We thus find

$$\ln \Omega(N_1, N_2) = -N_2 \ln N_2 + N_2 + N_1 \ln(Nr) + r_2 N_2 \ln(Nr) - N_2 r_2 - N_1 \ln N_1$$
$$- \frac{r_2 - 1}{r_2 - q_2}\left\{N_1 \ln(Nr) + r_2 N_2 \ln(Nr) - N_2(r_2 - q_2)\right.$$
$$\left. -(N_1 + N_2 q_2) \ln(N\bar{q})\right\} + N_2(r_2 - 1)\ln(z - 1)$$
(A 10.16)

and also

$$\ln \Omega(0, N_2) = -N_2 \ln N_2 + N_2 + r_2 N_2 \ln(N_2 r_2) - N_2 r_2$$
$$- \frac{r_2 - 1}{r_2 - q_2}\left\{N_2 r_2 \ln(N_2 r_2) - N_2(r_2 - q_2) - N_2 q_2 \ln(N_2 q_2)\right\}$$
$$+ N_2(r_2 - 1)\ln(z - 1).$$
(A 10.17)

The combinatorial entropy of mixing then follows as

$$\left[\frac{\Delta S^M}{k}\right]_{\text{comb}} = \ln \Omega(N_1, N_2) - \ln \Omega(N_1, 0) - \ln \Omega(0, N_2)$$

$$= N_1 \ln(Nr) + N_2 r_2 \ln(Nr) - N_1 \ln N_1 - N_2 r_2 \ln(N_2 r_2)$$
$$- \frac{r_2 - 1}{r_2 - q_2}\left\{N_1 \ln(Nr) + N_2 r_2 \ln(Nr) - (N_1 + N_2 q_2)\ln N\bar{q}\right.$$
$$\left. - N_2 r_2 \ln(N_2 r_2) + N_2 q_2 \ln(N_2 q_2)\right\}$$

$$= -N_1 \ln \frac{N_1}{Nr} - N_2 r_2 \ln \frac{N_2 r_2}{Nr}$$
$$- \frac{r_2 - 1}{r_2 - q_2}\left\{N_1 \ln \frac{Nr}{N\bar{q}} + N_2 r_2 \ln \frac{Nr}{N_2 r_2} + N_2 q_2 \ln \frac{N_2 q_2}{N\bar{q}}\right\}.$$

We now add

$$(r_2 - 1)N_2 \ln \frac{N_2 r_2}{Nr} - (r_2 - 1)N_2 \ln \frac{N_2 r_2}{Nr} = 0$$

and find

$$
\begin{aligned}
\left[\frac{\Delta S^M}{k} \right]_{\text{comb}} &= -N_1 \ln \frac{N_1}{Nr} - N_2 \ln \frac{N_2 r_2}{Nr} - \frac{r_2 - 1}{r_2 - q_2} \left\{ N_1 \ln \frac{Nr}{Nq} \right. \\
&\quad \left. + (r_2 - q_2)N_2 \ln \frac{N_2 r_2}{Nr} + r_2 N_2 \ln \frac{Nr}{N_2 r_2} + N_2 q_2 \ln \frac{N_2 q_2}{Nq} \right\} \\
&= -N_1 \ln \frac{N_1}{Nr} - N_2 \ln \frac{N_2 r_2}{Nr} - \frac{r_2 - 1}{r_2 - q_2} \left\{ N_1 \ln \frac{Nr}{Nq} \right. \\
&\quad \left. + N_2 q_2 \ln \frac{N_2 q_2 / Nq}{N_2 r_2 / Nr} \right\}.
\end{aligned}
$$

For molecules without ring formation we can relate the parameters r and q via the coordination number z by

$$q_i z = r_i z - 2(r_i - 1). \tag{A 10.18}$$

We then finally get

$$\left[\frac{\Delta S^M}{k} \right]_{\text{comb}} = -N_1 \ln \phi_1 - N_2 \ln \phi_2 - \frac{z}{2} \left\{ N_1 \ln \frac{\theta_1}{\phi_1} + q N_2 \ln \frac{\theta_2}{\phi_2} \right\}, \tag{A 10.19}$$

where we used the no-ring condition by introducing the coordination number, and

$$\theta_2 = \frac{x_2 q_2}{x_1 + x_2 q_2} \tag{A 10.20}$$

is the surface fraction of component 2 with $\theta_1 = x_1 / (x_1 + x_2 q_2)$. As before, we generalize this equation to multicomponent mixtures of arbitrary components by inspection and find for the combinatorial part of the excess entropy

$$\left[\frac{S^E}{k} \right]_{\text{comb}} = -\sum \left[N_i \ln \frac{\phi_i}{x_i} + \frac{z}{2} N_i q_i \ln \frac{\theta_i}{\phi_i} \right], \tag{A 10.21}$$

with

$$\theta_i = \frac{x_i q_i}{\sum_k x_k q_k}. \tag{A 10.22}$$

Equation (A 10.21) has been derived by Guggenheim [3]. When no internal contacts are taken into account, we have $r_i = q_i$ and (A 10.21) reduces to the Flory–Huggins equation. The same is true when r_i / q_i is independent of i. Stavermann [4] has shown that the same formal expression is also valid for structured molecules with ring formation, i.e., without the constraint of

(A 10.18), although (A 10.16) for the number of arrangements and thus the entropy is then no longer valid. Equation (A 10.21) is therefore referred to as the Guggenheim–Stavermann equation.

[1]   P. J. Flory. *J. Chem. Phys.*, 9:660, 1941.
[2]   M. L. Huggins. *J. Chem. Phys.*, 9:440, 1941.
[3]   E. A. Guggenheim. *Mixtures*. Oxford University Press, Oxford, 1952.
[4]   A. J. Stavermann. *Recl. Trav. Chim. Pays-Bas*, 69:163, 1950.

Integration Variables
for Three-Body Interactions

We consider a closed triangle formed from the molecules 1, 2, and 3; cf. Fig. A 11.1. Integration over $d\mathbf{r}_1$, $d\mathbf{r}_2$, $d\mathbf{r}_3$ can be written as integration over $d\mathbf{r}_1$, $d\mathbf{r}_{12}$, $d\mathbf{r}_3$, because the Jacobian of this transformation is unity. The volume element $d\mathbf{r}_3$ can be considered as that of a sphere around molecule 1; i.e.,

$$d\mathbf{r}_3 = r_{13}d\alpha\,(r_{13}\sin\alpha)\,d\phi dr_{13} = -r_{13}^2 dr_{13}d(\cos\alpha)d\phi.$$

Here ϕ is the azimuthal angle extending over 2π. We thus find after integration over $d\phi$ that

$$\iiint d\mathbf{r}_1 d\mathbf{r}_2 d\mathbf{r}_3 = -\iiint d\mathbf{r}_1 d\mathbf{r}_{12} 2\pi r_{13}^2 dr_{13}d(\cos\alpha)$$
$$= -8\pi^2 V \iiint r_{12}^2 r_{13}^2 dr_{12} dr_{13}d(\cos\alpha), \qquad \text{(A 11.1)}$$

where we put $d\mathbf{r}_1 = V$ in the final equation.

When using the integration variables r_{12}, r_{13}, and $\cos\alpha$ we must calculate the distance r_{23} from

$$r_{23} = \sqrt{r_{12}^2 r_{13}^2 - 2r_{12}r_{13}\cos\alpha}. \qquad \text{(A 11.2)}$$

If we wish to replace the integration over $\cos\alpha$ by one over r_{23}, we set

$$2r_{23}dr_{23} = -2r_{12}r_{13}d(\cos\alpha)$$

and thus

$$\iiint d\mathbf{r}_1 d\mathbf{r}_2 d\mathbf{r}_3 = 8\pi^2 V \iiint r_{12}r_{13}r_{23}dr_{12}dr_{13}dr_{23}. \qquad \text{(A 11.3)}$$

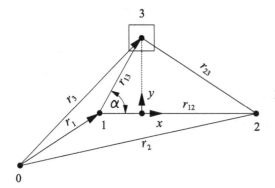

Fig. A 11.1. Integration over $r_1 r_2 r_3$.

In particular, we can derive the following relations:

$$\iint f(r_1, r_2) d\mathbf{r}_1 d\mathbf{r}_2 \iint f(r_1, r_3) d\mathbf{r}_1 d\mathbf{r}_3$$

$$= 16\pi^2 V^2 \iint f(r_{12}) f(r_{13}) r_{12}^2 r_{13}^2 dr_{12} dr_{13}$$

$$= -8\pi^2 V^2 \int_{-\infty}^{\infty} \int_{-\infty}^{+\infty} \int_{1}^{-1} f(r_{12}) f(r_{13}) r_{12}^2 r_{13}^2 dr_{12} dr_{13} d(\cos\alpha)$$

$$= V \iiint f(r_1, r_2) f(r_1, r_3) d\mathbf{r}_1 d\mathbf{r}_2 d\mathbf{r}_3. \tag{A 11.4}$$

APPENDIX 12

Multipole Perturbation Terms for the High-Temperature Expansion

Explicit expressions for the multipole terms in the high-temperature expansion around the Lennard–Jones fluid up to the quadrupole in the rigid molecule approximation for linear molecules are given as [1]

$$
\left(\frac{A^{\lambda\lambda}}{NkT} \right)^{\text{mult–mult}} = -\frac{n}{(kT)^2} \sum_\alpha \sum_\beta x_\alpha x_\beta
$$

$$
\left[\frac{2\pi}{3} \frac{\mu_\alpha^2 \mu_\beta^2}{\sigma^3} J_{\alpha\beta}^{(6)} + 2\pi \frac{\mu_\alpha^2 \theta_\beta^2}{\sigma_{\alpha\beta}^5} J_{\alpha\beta}^{(8)} + \frac{14\pi}{5} \frac{\theta_\alpha^2 \theta_\beta^2}{\sigma_{\alpha\beta}^7} J_{\alpha\beta}^{(10)} \right] \qquad (\text{A } 12.1)
$$

$$
\left(\frac{A^{\lambda\lambda\lambda}}{NkT} \right)^{\text{mult–mult–mult}}
$$

$$
= \frac{n}{(kT)^3} \sum_\alpha \sum_\beta x_\alpha x_\beta
$$

$$
\times \left[\frac{8\pi}{5} \frac{\mu_\alpha^2 \mu_\beta^2 \theta_\alpha \theta_\beta}{\sigma_{\alpha\beta}^8} J_{\alpha\beta}^{(11)} + \frac{48\pi}{35} \frac{\mu_\alpha^2 \theta_\alpha \theta_\beta^3}{\sigma_{\alpha\beta}^{10}} J_{\alpha\beta}^{(13)} + \frac{144\pi}{245} \frac{\theta_\alpha^3 \theta_\beta^3}{\sigma_{\alpha\beta}^{12}} J_{\alpha\beta}^{(15)} \right]
$$

$$
+ \frac{n^2}{(kT)^3} \sum_\alpha \sum_\beta \sum_\gamma x_\alpha x_\beta x_\gamma \left[\frac{32\pi^3}{135} \sqrt{\frac{14\pi}{5}} \frac{\mu_\alpha^2 \mu_\beta^2 \mu_\gamma^2}{\sigma_{\alpha\beta}\sigma_{\alpha\gamma}\sigma_{\beta\gamma}} K_{\alpha\beta\gamma}(222;333) \right.
$$

$$
+ \frac{192\pi^3}{315} \sqrt{3\pi} \frac{\mu_\alpha^2 \mu_\beta^2 \theta_\gamma^2}{\sigma_{\alpha\beta}\sigma_{\alpha\gamma}^2 \sigma_{\beta\gamma}^2} K_{\alpha\beta\gamma}(233;344)
$$

$$
- \frac{96\pi^3}{135} \sqrt{\frac{22\pi}{7}} \frac{\mu_\alpha^2 \theta_\beta^2 \theta_\gamma^2}{\sigma_{\alpha\beta}^2 \sigma_{\alpha\gamma}^2 \sigma_{\beta\gamma}^2} K_{\alpha\beta\gamma}(334;445)
$$

$$
\left. + \frac{32\pi^3}{2025} \sqrt{2002\pi} \frac{\theta_\alpha^2 \theta_\beta^2 \theta_\gamma^2}{\sigma_{\alpha\beta}^3 \sigma_{\alpha\gamma}^3 \sigma_{\beta\gamma}^3} K_{\alpha\beta\gamma}(444;555) \right].
$$

$$
(\text{A } 12.2)
$$

In these terms there appear various integrals over the pair and triplet correlation functions for the reference fluid, which here is the Lennard–Jones (12,6) fluid. These integrals have been evaluated based on computer simulations and fitted to correlation functions of the form [2]

$$J^{(n)} = f(T^*, n^*)$$
$$K(ll'l''; nn'n'') = f(T^*, n^*).$$

Application to mixtures can be based on VDW1 conformal solution theory; cf. Section. 5.4. Somewhat more accurate results are obtained by an extension of that model, referred to as the mean density approximation (MDA) [3]. The major difference of that approximation from VDW1 is that the temperature is reduced with $\varepsilon_{\alpha\beta}$ instead of ε_x, thus giving an individual integral $J_{\alpha\beta}$ for each interaction. We then find for the J-integrals

$$J^{(n)}_{\alpha\beta,\text{MDA}}(T, n, \{x_i\}) = J^{(n)}_0 \left(\frac{kT}{\varepsilon_{\alpha\beta}}; n\sigma_x^3 \right), \tag{A 12.3}$$

whereas the K-integrals are approximated by

$$K_{\alpha\beta\gamma,\text{MDA}}(T, n, \{x_i\}) = \left[K_0 \left(\frac{kT}{\varepsilon_{\alpha\beta}}; n\sigma_x^3 \right) K_0 \left(\frac{kT}{\varepsilon_{\alpha\gamma}}; n\sigma_x^3 \right) K_0 \left(\frac{kT}{\varepsilon_{\beta\gamma}}; n\sigma_x^3 \right) \right]^{1/3}. \tag{A 12.4}$$

The index 0 denotes the pure fluid integrals. Expressions similar to (A 12.1) and (A 12.2) are available for other types of interaction such as induction, dispersion and repulsion forces [1].

[1] K. Lucas. *Applied Statistical Thermodynamics*. Springer, Berlin, 1991.
[2] M. Luckas, K. Lucas, U. Deiters, and K. E. Gubbins. *Mol. Phys.*, 57: 241, 1986.
[3] T. W. Leland. *Adv. Cryogen. Eng.*, 21: 466, 1976.

Index

Printed in the United States
By Bookmasters